Mathematical Modelling Programs in Latin America

Milton Rosa • Francisco Cordero
Daniel Clark Orey • Pablo Carranza
Editors

Mathematical Modelling Programs in Latin America

A Collaborative Context for Social Construction of Knowledge for Educational Change

Springer

Editors
Milton Rosa
Universidade Federal de Ouro Preto
Ouro Preto, Minas Gerais, Brazil

Francisco Cordero
Centro de Investigación y de Estudios
Avanzados del IPN (CINVESTAV-IPN)
Ciudad de México, Mexico

Daniel Clark Orey
Universidade Federal de Ouro Preto
Ouro Preto, Minas Gerais, Brazil

Pablo Carranza
Universidad Nacional de Río Negro
General Roca, Río Negro, Argentina

ISBN 978-3-031-04273-7 ISBN 978-3-031-04271-3 (eBook)
https://doi.org/10.1007/978-3-031-04271-3

This Springer imprint is published by the registered company Springer Nature Switzerland AG
The registered company address is: Gewerbestrasse 11, 6330 Cham, Switzerland

Foreword[1]

It was a great pleasure and honor when my colleagues Milton Rosa, Daniel Clark Orey, Francisco Cordero, and Pablo Carranza invited me to write the foreword for this marvelous book that brings a collection of chapters dealing with mathematical modelling programs in Latin America, and which emphasizes a collaborative context for social construction of mathematical knowledge for educational change in society, which seeks for social justice and promotes total peace in its four dimensions: Inner Peace, Social Peace, Environmental Peace, and Military Peace, which are intimately related.

This book is a very important contribution to the area of mathematical modelling. The essays address issues and the consequences in mathematical practices and conceptions developed through the conduction of modelling investigations. The Summary of this book shows the breadth of this collection, and it would be very difficult to comment on each essay. The titles are illustrative of their contents, and each one has extensive bibliographical references.

A special characteristic of this book is related to its empirical evidence on how categories of mathematical modelling are developed in Latin America, which assesses the horizontal and reciprocal relations between mathematics (school/non-school contexts) and the real world. This book is very original in both the choice of theoretical contents and the methodological treatment given to all of them. It adds relevant aspects to the mathematical modelling approach that is based on three Latin American modelling programs: ethnomodelling, transversality of knowledge, and reasoned decision-making.

Undoubtedly, these three mathematical modelling programs, independently and collectively, provide educational gains, each one with its levels of specificity and loyal to its principles. Each chapter, with its respective theoretical and

[1] Professor Ubiratan D'Ambrosio graciously accepted an invitation to write the *Foreword* to open this book. We believe that this foreword is one of the very last texts he wrote before his passing on May 12th, 2021, and we are greatly honored and thankful to the D'Ambrosio family for this honor in including his thoughts here.

methodological foundations: ethnomathematics, socioepistemology, and the attribution of meaning to learning, reveals its relevance to the context of Latin America.

Thus, I will use this foreword to comment on relevant aspects of mathematical modelling, which are treated with different nuances in this entire collection. I refer to mathematical modelling as the study of scientific, mathematical and, by extension, technological phenomena in direct relation to its social, economic, political, environmental, and cultural contexts.

I claim that the great advantage of working in this border zone lies in the possibility of combining academic and dynamic treatment with historical, anthropological, and sociocultural studies to improve education. Indeed, this synthesizes the strength of this collection of chapters.

Mathematical modelling helps individuals to transcend their *pulsions of survival* that go beyond the *anatomo-physiological mechanisms*.[2] This *pulsion of transcendence*, unique to the species *homo*, is difficult to explain. It gives rise to sophisticated and creative communication, and language, and to emotions, beliefs, and preferences. Some scholars call it *will* while others call it *consciousness*.

There are other denominations, some related to myths and religions. Consciousness is an elusive concept. Many scientists say that terms such as will, mind, consciousness, and other words describing subjective mental experiences cannot be defined. It allows us to predict what will happen in the future and to socially coordinate plans for the future.

Consciousness helps to extract meaningful information from the sense organs. There are limitations, such as the recognition of will and of consciousness in anyone except oneself. Self-recognition is the ability to recognize oneself as distinct from another entity, as well as to plan, pay attention, recall memories of specific events, and take the perspective of another creature.

Environmental conditions require activation of consciousness, but also from learning and environmental influences within human's own lifetime. To provide for all similar contingencies would require an impossible roster of instructions and wasteful volume of specific directions.

In most of the contingencies, decisions must be made *ad hoc* through the use of instruments, both material and intellectual, such as counting, classifying, inferring, and modelling, which are responsible for the development of *ad hoc* solutions. Hence, we may understand the construction of knowledge as a three-step process:

1. How are ad hoc practices and solution of problems developed into methods?
2. How are methods developed into theories?
3. How are theories developed into scientific invention?

---

[2]The anatomo-physiology mechanisms are related to the functions and adaptations of the endocrine, nervous, digestive, and cardiovascular systems, as well as the interactions among these systems provide an understanding of the normal physiological responses of these systems and their adaptations to perturbations.

While methods are essentially a rational and coordinated use of techniques, theories are impregnated modes of explanation and understanding, based on myths, on spirituality, and even religions, as well as on science and mathematics and in ideology, which are all mentifacts. It is undeniable that the human species are characterized by the pulsions of survival of the individuals and of the species, like all the living species, and by the pulsion of transcendence that is unique to the human species.

Both, survival and transcendence, are in mutual interaction. The acquisition of the pulsion of transcendence is the focus of mythologies and religions, a fascinating research theme. Every individual and society, and the human species in general develop strategies to cope with the ample reality. I clarify that every time I say ample reality, I mean everything, the complex of natural and supernatural phenomena and facts, the physiological, sensorial, emotional, and psychic reactions to the environment in the broad sense, as well as in the social interactions.

Indeed, everything, the *physis* ($\approx$ nature) and the *nomos* ($\approx$ social norms) are in permanent change. *Nomos* is as indispensable ground for *physis* as *physis* is an indispensable ground for *nomos*. They are in a symbiotic relation. *Physis* provides the potentialities, *nomos* the actuality of humanity. I also consider that people select facts and phenomena of reality that inform individuals and groups.

Obviously, no one has full access, awareness, and knowledge of reality; no one is omniscient. Our natural limitations give us access to selected facts and phenomena. The reason and the form of selection are extremely complex. They go from an uneven capability of individuals and groups to receive information, in some cases related to sensorial qualities or deficiencies, and in other cases to the interest in the information received. The interest may be because of needs, or preference, or merely by chance.

Anyway, the information received is processed, in a way not yet well understood. The individuals or groups exert an action of generating artifacts and mentifacts from the selected part of reality (sociofacts). They are incorporated into reality as representations, which inform the individuals or groups and this cycle goes on. The main question is then, how individuals and groups deal with representations of selected facts and phenomena faced in their daily lives.

In a representation, reality is restricted to selected facts and phenomena, and the result is a sort of *isolated individualized reality*, and to deal with it, individuals attribute codes or parameters to the selected facts and phenomena. These parameters may be of a mathematical nature, such as mathematical forms and mathematical symbols. The isolated individualized reality, with the mathematical symbols attributed to the selected facts and phenomena, is a mathematical model of it, which is an artifact that represents a model of a complex social reality.

Through models, humans try to give explanations of myths and mysteries, and these explanations are organized as arts, techniques, theories, and strategies, which help them to explain and deal with facts and phenomena. Intellectual resources allow the individuals to deal with the models and the parameters they created, which are representations of facts and phenomena of the reality in the broad sense. The most common intellectual resources are based on observing, comparing, classifying, ordering, measuring, and quantifying, and individuals and groups deal with the representations of selected facts and phenomena. These parameters are described in terms of formal mathematics.

In this context, I call mathematical modelling as the process of dealing with models in which the parameters associated with are the objectives of coping with and explaining selected facts and phenomena of reality. They are also explained in terms of formal mathematics. The practice of mathematical modelling is an iterative method starting with reality, with which we started by selecting parameters, constructing a model, proceeding to its mathematical analysis, verifying results through control procedures, and reformulating the model by repeating the analyses and control until we reach a satisfactory perception of the selected system taken from reality.

In each step, the practitioner reformulates the choice of parameters and resumes the process, which eventually allows for a better understanding of the selected facts and phenomena of reality, which is the goal that justifies our practices as educators, mathematicians, and scientists. The ensemble of the strategies to cope with ample reality is a complex system of knowledge and behavior, generated by individuals and socialized in a group of individuals with some affinity. This is the scenario discussed in all the chapters of this book.

The basic questions are how individuals and groups of individuals develop their means for surviving in their own natural and sociocultural environments, satisfying the pulsion of survival [body] and how they go beyond survival, acquiring the pulsion of transcendence [mind]. Survival and transcendence are the quintessence of human life and are dealt with a complex system of knowledge and behavior generated and organized by both each individual (from birth to death) and by the affinity group. The way they are generated, organized, and socialized relies in language and rhetoric in a broad sense.

These questions must be faced with *transcultural* and *transdisciplinarian* strategies, which borrow methods of research from the sciences, cognition, mythology, anthropology, history, sociology (politics, economics, and education), and from sociocultural studies in general. These strategies rely on the analyses of the history of ideas, and of the evolution of behavior, and knowledge of the human species, in every natural and sociocultural environment. All these aspects are present in the chapters of this book.

Thus, mathematical modelling discusses the *corpora of knowledge* developed by humanity to survive and to transcend. It represents a transdisciplinarian and transcultural investigation knowledge area supported by new theoretical and methodological developments. It is conceptually designed as a broad investigative program of the evolution of ideas, knowledge, and practices developed by the human species in different natural and sociocultural contexts.

Essentially, mathematical modelling implies with the development of an analysis of how groups of humans generated ways, styles, arts, and techniques of *doing* and *knowing*, learning and explaining, dealing with situations, and of solving problems of their natural and sociocultural environments.

Although the chapters of this book are assembled in parts dealing with: (a) *Ethnomathematics and Ethnomodelling: Empirical Work, Theoretical-Methodological Approaches, and Research Questions*; (b) *Interdisciplinary Ecosystems: Empirical Work, Theoretical-Methodological Approaches, and Research Questions*; and (c) *Mathematics and People: Empirical Work, Theoretical-Methodological Approaches, and Research Questions*, the authors discuss specific cases that fit into the broader concept of mathematical modelling in diverse contexts.

I understand that knowledge is a cumulative succession of strategies developed by humans who live in different social, natural, and cultural environments in response to the pulsions of their survival and transcendence. The main objective of knowledge is to understand, to explain, and to cope with selected facts and phenomena of reality as a whole through the development of mathematical models.

Mathematical modelling is such a strategy that deals with facts and phenomena, and it is important to understand how knowledge is generated (cognition), how it is individually and socially organized (epistemology), and how it is expropriated by power structure, institutionalized, and given back to the people who generated it through filters (politics). These steps are treated in an integrated and holistic way that makes mathematical modelling as a strategy for building-up diverse knowledge systems in different cultural contexts. I recognize this in the collection of chapters.

In this book, well-known specialists in this area address different topics in mathematical modelling, which apply to different natural, economic, political, environmental, and sociocultural environments. Altogether, the chapters mutually complement each other, and the readers may draw conclusions that support these theoretical and methodological proposals for the development of mathematical modelling. This approach enables the creation of diverse ways of dealing with the social construction of mathematical knowledge in Latin America through the development of mathematical modelling.

The authors of the 18 chapters in this book, who represent the diversity of Latin America, are from nine countries: Argentina, Brazil, Chile, Colombia, Costa Rica, Cuba, Ecuador, Honduras, and Mexico. They were invited to share their ideas, perspectives, and discuss investigations that represent a rich sample of three Latin American perspectives on mathematical modelling. This book is a valuable and necessary contribution to educational scholarship.

I congratulate the editors and each individual author for this remarkable accomplishment that shows how historical evolution of knowledge enables the development of alternative mathematical knowledge systems that provide explanations of daily problems, and situations, and which can lead to the elaboration of models as representations of facts present in our own reality.

São Paulo, Brazil
April 2021

Ubiratan D'Ambrosio

Contents

Contributors

Eleany Barrios-Borges Centro de Investigación y de Estudios Avanzados del Instituto Politécnico Nacional, Ciudad de México, Mexico

Juan Ramón Cadena Central University of Ecuador, Quito, Ecuador

Pablo Carranza Universidad Nacional de Río Negro, General Roca, Río Negro, Argentina

Henry Chávez-Martínez Centro de Investigación y de Estudios Avanzados del Instituto Politécnico Nacional, Ciudad de México, Mexico

Francisco Cordero Centro de Investigación y de Estudios Avanzados del Instituto Politécnico Nacional, Ciudad de México, Mexico

Diego Pereira de Oliveira Cortes Secretaria de Estado de Educação de Minas Gerais, São Joaquim de Bicas, Minas Gerais, Brazil

Ubiratan D'Ambrosio Emeritus Professor - Universidade Estadual de Campinas, São Paulo City, São Paulo, Brazil

Adriana Atenea de la Cruz-Ramos Secretaria de Educación del Estado de Chiapas, Tuxtla Gutiérrez, Chiapas, Mexico

Nayane Rodrigues de Deus Colégio Estadual Odolfo Soares, Ponte Alta do Tocantins, Tocantins, Brazil

Érika Dagnoni Ruggiero Dutra Colégio América do Norte, Manhuaçu, Minas Gerais, Brazil

Alcione Marques Fernandes Universidade Federal do Tocantins, Arraias, Tocantins, Brazil

Falconery Giacoleti-Castillo Centro de Investigación y de Estudios Avanzados del Instituto Politécnico Nacional, Ciudad de México, Mexico

Jaime Huincahue Universidad Católica del Maule, Talca, Región del Maule, Chile

Mariana Letourneau Universidad Nacional de Río Negro, General Roca, Río Negro, Argentina

Ronald Patricio Chasiloa Llumiquinga Unidad Educativa Píntag General, Quito, Ecuador

Zulma Elizabete de Freitas Madruga Universidade Federal do Recôncavo da Bahia, Amargosa, Bahia, Brazil

Sindi Marcía-Rodríguez Centro de Investigación y de Estudios Avanzados del Instituto Politécnico Nacional, Ciudad de México, Mexico

Arturo Mena-Lorca Pontificia Universidad Católica de Valparaíso, Valparaíso, Chile

Jaime Mena-Lorca Pontificia Universidad Católica de Valparaíso, Valparaíso, Chile

E. Johanna Mendoza-Higuera Unidades Tecnológicas de Santander, Bucaramanga, Santander, Colombia

Ana Paula Santos de Sousa Mesquita Universidade Federal de Ouro Preto, Ribeirão das Neves, Minas Gerais, Brazil

Fabio Miguel Universidad Nacional de Río Negro, General Roca, Río Negro, Argentina

Ailén Morales Universidad Nacional de Río Negro, General Roca, Río Negro, Argentina

José Luis Morales-Reyes Universidad de Costa Rica, San José, San José, Costa Rica

Astrid Morales-Soto Pontificia Universidad Católica de Valparaíso, Valparaíso, Chile

Jaime Moreno Universidad Nacional de Río Negro, General Roca, Río Negro, Argentina

Mónica Navarro Universidad Nacional de Río Negro, General Roca, Río Negro, Argentina

Claudio Opazo-Arellano Universidad Austral de Chile, Puerto Montt, Los Lagos, Chile

Daniel Clark Orey Universidade Federal de Ouro Preto, Ouro Preto, Minas Gerais, Brazil

Irene Pérez-Oxté Universidad Abierta y a Distancia de México, Ciudad de México, Mexico

Cristiane Castro Pimentel Escola Estadual Brigadeiro Felipe, Arraias, Tocantins, Brazil

Jéssica Rodrigues Universidade Federal de Ouro Preto, Belo Horizonte, Minas Gerais, Brazil

Milton Rosa Universidade Federal de Ouro Preto, Ouro Preto, Minas Gerais, Brazil

Miguel Solís Universidad Autónoma de Chiapas, Tuxtla Gutiérrez, Chiapas, Mexico

Part I
Introduction

Based on empirical evidence, this section proposes a category of mathematical modelling that assesses the horizontal and reciprocal relations between mathematics and the real world, which provide epistemological and ontological changes in which mathematical knowledge of the *others* is valued, respected, and recognized. These changes also oblige both mathematics teachers and students to understand how mathematical knowledge producers use and apply modelling processes that enable them to build their own mathematical categories in the contexts that are governed by reciprocal relations between academic and functional knowledge, which is part of their daily life. The dimensions of these relations guide educational change in order to modify mathematical teaching and learning processes in autonomous actions compared to the emulations of typical mathematical procedures in the classrooms. This proposed approach is based on three Latin American modelling programs: *ethnomodelling*, *transversality of knowledge*, and *reasoned decision-making processes* in which these dimensions are both theoretically and respectively based on the connection between mathematics and culture, the attribution of meaning to the learning process, and elements of socioepistemology.

Chapter 1
Modelling in the Life of People: An Alternative Program for Teaching and Learning of Mathematics

Francisco Cordero, Milton Rosa, Daniel Clark Orey, and Pablo Carranza

1.1 Introduction

The *Common Core State Standards for Mathematics* (CCSSM) defines modelling as the means for using mathematics or statistics to describe a real-world situation and to deduce additional information from the given phenomena by using mathematical and/or statistical calculation and analysis (Common Core Standards Writing Team, 2013). The definition is workable for mathematical modelers and for those who believe that the teaching and learning of mathematics can be improved if teachers and students are taught to model their world mathematically.

Modelers are confident in creating relationships between school mathematics and the reality of the world. The following questions became important to our common dialogue: *How do people or citizens model mathematically? How is modelling used by a child, a young person, a university student? Within the working and professional bounds, how is modelling used at work and in other academic disciplines? And in general, how do people use modelling in their real world?*

The thesis of our approach here, consists of appreciating that disciplinary development, generically, of mathematics education, has considered and used to study and understand mathematical knowledge in school and outside of it, and, on the other

F. Cordero (✉)
Centro de Investigación y de Estudios Avanzados del Instituto Politécnico Nacional, Ciudad de México, Mexico
e-mail: fcordero@cinvestav.mx

M. Rosa · D. C. Orey
Universidade Federal de Ouro Preto, Ouro Preto, Minas Gerais, Brazil
e-mail: milton.rosa@ufop.edu.br; oreydc@ufop.edu.br

P. Carranza
Universidad Nacional de Río Negro, General Roca, Río Negro, Argentina
e-mail: pcarranza@unrn.edu.ar

M. Rosa et al. (eds.), *Mathematical Modelling Programs in Latin America*,
https://doi.org/10.1007/978-3-031-04271-3_1

hand, to integrate mathematical modelling to teaching approaches, so that mathematics is seen in the life of the learners and the diverse communities they live in.

Both orientations coincide with an important educational principle that attempts to relate mathematics to the real world, and in our case the diversity of *real worlds* found in Latin America. However, the tension between these orientations is embedded in the relational construct of mathematics. For example, one tension assumes that the school and/or academic body of facts learned in this environment is the only and true knowledge that measures the experiences in our everyday lives, while the other tension privileges traditional educational actions over mathematical modelling.

In this chapter, we propose, with empirical evidence, that there is a category of modelling of peoples' or citizens' that deals with their own mathematics, which values the horizontal and reciprocal relationships between mathematics and their own real world condition, and with their growing practices that constitutes re-significations of mathematical knowledge developed in their own contexts (Carranza, 2017, Cordero, 2016, Rosa & Orey, 2020).

With these horizontal and reciprocal relationships between mathematics and the real world, we seek to ensure that the learning process *makes sense* to both learners and educators. In this context, the term *sense* is associated with a purpose or intentionality and, in this case, related to the concepts of mathematics (Develay, 1994; Jacques, 1987).

Thus, we work with what we call *axes of the didactic proposals* that in principle enable students to attribute meanings of their own learning. One of these axes consider the functionality of learning not only for a time in the future but also for the present. In another axis, we incorporate *others* as participants in the act of learning. The term *others* is related to someone external to the classroom environment, such as: a villager, a school, a school library, and a theater, among others.

The conjunction of these two axes (future-present time horizon and the *others*) leads to addressing real problems of the students' culture and environment. In these addressed problems, the concepts of mathematics emerge as rational tools for the decision-making process that occurs by considering general areas such as cultural, classroom-reality-approaches, and other knowledge domains.

In order to appreciate the dimension of the aforementioned areas and their power to address problems of the teaching and learning of mathematics. Firstly, we present three mathematical modelling programs developed in Latin America, which includes perspectives from Argentina, Brazil, and Mexico. Secondly, we explain that each program addresses mathematical knowledge outside of school and reveals that knowledge that emerges in diverse communities, such as ethnic groups or settlers, students with commitments to social welfare, and students and professionals from other knowledge domains. Thirdly, we consider these three programs as axes that generates a three-dimensional space that forms a corpus of mathematical knowledge for educational change.

1.2 Three Latin American Mathematical Modelling Programs and Their Orientations and Conceptions

It is necessary to present three Latin American mathematical modelling programs and their orientations and conceptions that the authors base their work on and are familiar with.

1.2.1 Ethnomodelling

The ethnomodelling application provides an opportunity to examine local and global knowledge systems to get an idea of the forms of mathematics used in various contexts and cultural groups through a *symmetrical dialogical relationship and with otherness* (Rosa & Orey, 2017b).

This holistic context is created from the analysis of reality as a whole, which allows members of different cultures to participate in the modelling process, which enables these members to study and understand the aspects and components of systems taken from their own reality, as well as to comprehend their interactions that enable the approximation and also the relationship between these components with the *knowing* and *doing* process of their daily life through ethnomathematics.

According to this context, ethnomodelling allows modelers to value and respect the use of ethnomathematical practices (local) and the application of modelling tools and techniques (global) so that these members can perceive their reality in a holistic way. Thus, the main objective of ethnomodelling research is not only to solve problems, nor the creation of a simple understanding of alternative mathematical systems, but also to enable modelers to better their own understanding of the importance and role of mathematics in their own diverse societal, cultural, and school/academic contexts (Rosa & Orey, 2020).

For example, *mathematization* is one of the most important stages of mathematical modelling, and it allows the translation of diverse problems and situations for the mathematical language through its connection with ethnomathematics by means of ethnomodelling. Therefore, there is a need to highlight the processes of mathematization as they develop through the use of techniques developed by the members in distinct cultural groups and their natural encounter with ethnomathematics.

1.2.2 Mathematical Modelling in the Relationship Between the Classroom and Approach to Reality

This relationship addresses real problems as a type of didactic proposal that involves the analysis of its potentialities and difficulties. In this way, we develop proposals where concrete solutions to real problems are elaborated, which, as a result of realism, turn out to be mostly of the interdisciplinary type.

There are several lines of analysis possible in our proposal. In our case, we particularly focus on analyzing the emergence of mathematical concepts beginning with the needs of interdisciplinary projects. Thus, two logical lines arise: one turns out to be the logic of the project to be developed. One line convenes the knowledge of mathematics that brings rationality to the decision-making process. Another is formed by the disciplinarity logic that deploys concepts to give sustainability to the mathematical components that projects demand.

In this case, mathematical modelling becomes a fundamental tool that produces reasoning that would be difficult to achieve otherwise. In the addressed projects, these modellings not only describe relationships, but also to build arguments for the decisions to be made within their frameworks (Carranza, 2015, 2017). They organize and examine data and evidence in analyzing the emerging mathematical concepts in argumentative processes. However, other potential dimensions of analysis include: new roles for educators, teamwork, the role of altruism and social welfare, repositioning of students, teachers, and knowledge facing the community, among others.

1.2.3 The Category of Socioepistemological Modelling

This category includes a process that accompanies the legitimation of mathematics in use of communities of knowledge, which occur in different scenarios, as well as the crossing between these scenarios and their uses and utilities. We call the first scenario as *epistemological plurality* and the second scenario as *transversality of uses of mathematical knowledge*. Both aspects define the mathematical functionality of the communities of mathematical knowledge that occur in the different scenarios, such as schools, work, and towns and cities (Cordero, 2016).

With this category of modelling, *reality* is projected to what is usual in all these scenarios, where routine applications are expressed, such as the daily life of the disciplinary specialist, the worker, and the people. In this context, it important to standardize these scenarios in the pedagogical action of mathematics education in order to consider all educational levels, and the diversity of disciplines, as well as the work and life of people (Cordero, 2016; Mendoza & Cordero, 2018; Zaldívar et al., 2014).

Generally speaking, with this projection of reality, functional knowledge means it offers useful applicability in the worldly life situations of the people, such as work and profession (Arendt, 2005). This useful knowledge is composed of uses and meanings, which are re-signified in the transit of situations. In this regard, it can be said that functional knowledge is the result of the transversality of the use of people's knowledge in different situations, which work as re-significators.

These situations are composed of significations and re-significations with their respective procedures, which are regulated by an instrument: both are built according to the operations that the participants can perform, with the conditions that they can capture and transform, and with the concepts that are progressively built (Cordero, 2016). With that structure, it carries out multiple realizations and adjusts its own structure to produce a desirable pattern. It is a medium that supports the development of reasoning and argumentation (Suárez & Cordero, 2010).

This is the process in which mathematical knowledge transcends and is re-signified because it values the elements in the environment of the object to which they give meaning. In accordance with this point of view, the teaching and learning of mathematics would benefit if horizontal and reciprocal relationships between different mathematical knowledge are included in this process.

1.3 The Context of the Social Construction of Knowledge of the Three Latin America Mathematical Modelling Programs

It is necessary to discuss the context of the social construction of knowledge of the three Latin American mathematical modelling programs.

1.3.1 Ethnomathematics and Ethnomodelling

The ethnomathematics program offers a broader view of mathematical knowledge by covering ideas, notions, procedures, processes, methods, and practices rooted in different cultural environments. In this context, Rosa and Orey (2017a) highlight the importance of the development of critical reflection on the social, cultural, environmental, economic, and political dimensions of mathematics in a *dynamic and glocalized society.*

The proposal of the ethnomathematics program, as outlined by D'Ambrosio (1985), is to make mathematics a living and humanistic discipline that examines, values and works with real situations, in time and space, through analysis, questioning, and criticism of the phenomena present in everyday life (D'Ambrosio, 1990). The application of ethnomathematical techniques and modelling tools allow us to examine systems taken from our reality and give us an idea of the various ways of doing mathematics in a holistic way.

In this context, Rosa and Orey (2017a) highlight that ethnomathematics is related to the study of mathematical ideas and procedures that consider the cultural context in which mathematical notions and practices emerge through the *mathematization* of local mathematical practices. Mathematization is related to knowledge systems related to the daily life of the members of each cultural group and that can be *mathematized* and translated into the language of school and academic mathematics.

As mentioned earlier, the use of mathematization that is present in the daily life of members of distinct cultural groups aims at the expansion and the perfecting of mathematical knowledge since it leads to the strengthening of their cultural identity (D'Ambrosio, 1990). Hence, modelling is one of the possible strategies enabling the approximation and relationship between *knowing* and *doing* between different mathematical systems.

This context enables modelling to be perceived as a set of representations of reality that are generated, via inferences, with the use of mental representations that allow to value and respect the ethnomathematical knowledge developed in everyday situations. This approach contextualizes locally developed mathematical knowledge since it studies mathematical phenomena that occur in various (global) cultural contexts (Rosa, 2010).

In this regard, mathematical knowledge can be understood as resulting from local (emic) rather than global (etic) origins that allow the proposition of *acts of translation* between those two perspectives (Eglash et al., 2006). This approach seems to be reasonable since ethnomathematics often makes use of modelling in order to establish relationships between local conceptual frameworks and the mathematical knowledge included in global designs.

In this context, Rosa and Orey (2020) argue that mathematical ideas, procedures, and practices include geometric principles in craftsmanship, architectural concepts, and practices that are found in activities and artifacts of local and global cultures, which can be translated between different mathematical knowledge systems through ethnomodelling. Consequently, this knowledge is related to a *glocal* position, with a multicultural vision, through cultural dynamism among members of different cultures.

1.3.2 The Interdisciplinary Ecosystem

Modelling occupies a fundamental place because of its power to discover relationships and behaviors of objects and phenomena of the real problem, as well as to build arguments for decision-making. In projects carried out with students in real contexts. We have observed, on the one hand, an indispensable and strong interaction between several disciplines to address problems that context demands to solve with rationality.

Furthermore, and if we focus within the convened disciplinary fields, we have found that such convocations are not necessarily coincidental to the logic of didactic transpositions of disciplines. On the contrary, we can say that the dynamics of the convocation are characterized by a network between disciplines in which meanings in the context of concepts is the bridge that weaves them in a symbolical way.

Thus, a kind of ecosystem emerges where the concepts live and coexist in the dynamics of the project for their contribution in the resolution of posed problems. This coexistence then occurs within an ecosystem logic different from the one established a priori and manifested in conceptions of by educators, programs, and textbooks.

The notion of ecosystem used here highlights, on the one hand, the constitution of a coherent and stable ensemble and, on the other hand, its interrelation. In this perspective, we have observed that the realization of projects based on a problem of the context imply changes not only in the didactic triangle (students-teachers-knowing), but also, and very strongly in the whole framework of the school ecosystem in which this pedagogical action is inserted, which includes physical spaces, management responsibilities, execution times, and commitments and responsibilities of the whole institution.

1.3.3 Mathematics and People

In the *socioepistemological mathematical modelling program*, the interpretation of *reality* constructs the *horizontal and reciprocal* relationship between school/academic and everyday mathematical knowledge of the disciplinary specialists, workers, and people. For example, a community of bionic engineers in their everyday specialist knowledge of *control systems* constructs a category of mathematical knowledge called *reproduction of behaviors*. This category is *reciprocally related* to the *stability* of a differential equation.

The first expresses the mathematical *uses* and *meanings* of the engineering community and the second expresses the *mathematical object* of school mathematics. For those engineers, the differential equation is *an instruction that organizes behaviors* while for school mathematics it is a way *to find a solution that is not known* (Mendoza & Cordero, 2018). These are the *realities* that the modelling program addresses (Buendía & Cordero, 2005; Cordero, 2016; Cordero et al., 2015). Thus, the *modelling program* is based on theoretical-methodological constructs of the *socioepistemological theory of educational mathematics*.

For example, Cantoral (2013) has provided the foundations of the *socioepistemological theory*, which consists of four principles: (a) normative of social practice, (b) contextualized rationality, (c) epistemological relativism, and (d) progressive signification (re-signification). These principles are intended to explain the enigma of the social construction of mathematical knowledge and its institutional dissemination. A core construct of this approach is related to social practices developed in complex systems of social dimensions in which mathematical knowledge is problematized by considering wise, technical, and popular knowledge in order to synthesize them in human wisdom.

In this perspective, Cordero et al. (2016) considers that the *social practice* construct, in *socioepistemological theory* that has revealed aspects of social dimensions in school mathematics (basic, middle, and higher education) and how they may be are rooted in reality, the uses of knowledge, and in more generic terms, the people who have been forgotten and invisible throughout history. Thus, it is necessary to recover this knowledge in order to alleviate the difficulties of teaching and learning mathematics in schools (Cordero et al., 2015).

For example, in the teaching of parabolas, which is a mathematical object that appears in the mathematics courses of secondary education, between 15 and 17 years of age, it can be difficult for many teachers, in their didactic context, to develop a frame of reference to incorporate parabolas in situations of variation, approximation, and transformation in order to generate arguments of prediction, local behavior, and behavioral tendencies (Morales & Cordero, 2014; Mendoza et al., 2018).

With this approach, Cordero et al. (2016) formulates a *general socioepistemological program* called *Forgotten Subject and Transversality of Knowledge (SOLTSA*) in which the modelling program is immersed. Its main objective is to reveal the uses of mathematical knowledge and its re-significations in people's mathematical knowledge communities. The *SOLTSA Program* is developed through two simultaneous lines of work: the *Re-signification of Mathematical Knowledge* and its *Educational Impact.*

The first problematizes the categories of mathematical knowledge that happen in communities between different knowledge scenarios that necessarily come into play, such as the school mathematician, the disciplinary field, and the daily life of the community. In the second line of work, multifactors and stages are formed that contribute to the alliance of quality of the teaching of mathematics to lead to the transformation and educational change in mathematics. *Identity, socialization, and inclusion*, among others, are the multifactors for this purpose (Opazo Arellano et al., 2018; Pérez-Oxté & Cordero, 2016; Medina-Lara et al., 2018).

1.4 The Empirical Work, the Theoretical-Methodological Aspects, and Research Questions of the Three Latin America Mathematical Modelling Programs

It is necessary to discuss the empirical work, the theoretical and methodological aspects and the research question developed in the three Latin America mathematical modelling programs.

1.4.1 Ethnomodelling: Modelling from a Cultural Context

It is important to find alternative methodological and pedagogical approaches to record historical forms of local mathematical ideas and procedures that occur in different cultural contexts because Western mathematical practices are accepted globally without discussion and as unique truths. These practices are related to measurement, calculation, games, divination, navigation, astronomy, modelling,

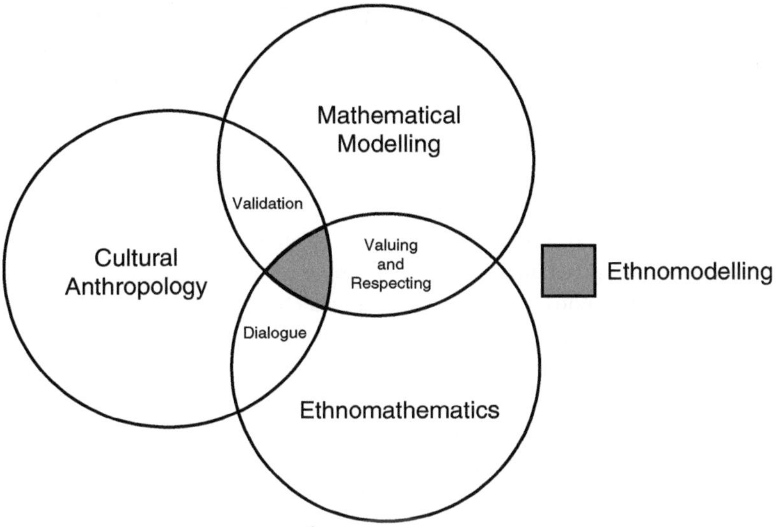

Fig. 1.1 Ethnomodelling as the intersection between three knowledge fields. (Source: Rosa and Orey (2012))

and a wide variety of mathematical procedures in cultural artifacts (Rosa & Orey, 2010).

This approach represents a process of translation and elaboration of problems and questions taken from everyday phenomena, and daily life through ethnomodelling, which includes ideas, notions, procedures, techniques, strategies, perspectives, and mathematical practices developed by members in different cultures, and which are manifested and transmitted in various ways (Rosa & Orey, 2017b).

Thus, ethnomodelling is configured as an essential element in the field of cultural anthropology, ethnomathematics, and mathematical modelling. Figure 1.1 shows ethnomodelling as the intersection between three knowledge fields.

For Rosa and Orey (2017b), translation is considered as the description of the processes of local modelling (cultural) systems that can have a mathematical representation in Western culture and vice versa and that are represented by three types of cultural visions of mathematical knowledge: local (emic), global (etic), and glocal (dialogical).

(a) *Local Mathematical Knowledge (Emic)*

Emic mathematical knowledge is related to the *knowing and doing* coming from the members of distinct cultural groups because it originates from within the culture in an inner vision according to an *intracultural*[1] posture. This knowledge is in accordance with the perceptions and interpretations considered

[1]Intraculturality promotes the recovery, strengthening, development, and cohesion within local cultures for the consolidation of a multicultural society based on equity, solidarity, complementarity, reciprocity, and social justice (Saaresranta, 2011).

appropriate by the members of such cultures (Rosa & Orey, 2012). The emic mathematical knowledge is oriented from us towards us with the perspective of the natives, which is a vision from within, inside of the culture, interior, and local.

(b) *Global Mathematical Knowledge (Etic)*

Etic mathematical knowledge is related to the *knowing and doing* coming from external observers of cultures, which originates from outside of the cultural group in an external view of its members in an *intercultural*[2] posture. Mathematical ideas and procedures are etic if they can be compared between cultures through the use of common definitions and metrics. Etic knowledge is oriented from them (researchers and educators) towards us with a perspective of external observers, which is a vision from outside, external, and global (Rosa & Orey, 2012).

(c) *Glocal Mathematical Knowledge (Dialogical)*

This knowledge presents a cultural dynamism between emic and etic mathematical knowledge, which is represented by the encounters between two or more cultures in the classrooms. The mathematical knowledge of members of distinct cultural groups is combined with the Western mathematical knowledge system, resulting in a dialogical perspective in mathematics education (Rosa & Orey, 2017b). This knowledge includes the recognition of other epistemologies and the holistic and integrated nature of mathematical knowledge developed in different cultural contexts. This approach can ensure the development of understanding of the different ways of doing mathematics through mutual dialogue and respect between global and local approaches through *glocalization*.

In this sense, Rosa and Orey (2017b) affirm that glocalization (***glo****bal* + *lo****cal***) is a dialogical approach that considers the interaction between *local* and *global* mathematical knowledge for the elaboration of ethnomodels. This approach is also related to the acceleration and intensification of interaction and integration between members of distinct cultural groups that compose society.

1.4.1.1 Ethnomodels

Ethnomodels represent examples of mathematical knowledge systems that helps members of distinct cultural groups to gain understanding and in the appropriation of reality through the use of small units of information which link the cultural heritage of these members with the evolution of mathematical procedures and practices that are developed in their own cultural context (Rosa & Orey, 2017b). These ethnomodels are emic, etic, and dialogical.

[2]Interculturality promotes the development of the interrelationship and interaction of knowledges, knowings, science, and technology specific to each culture with other cultures, which strengthens one's own identity and interaction on equal terms among all local cultures with the global cultural groups (Saaresranta, 2011).

(a) *Local Ethnomodels (Emic)*
 The emic ethnomodels are based on the contents that are important to the members of local communities and represent the mathematical thinking of the people who live in these surroundings. These ethnomodels are representations, descriptions, and analyses expressed in terms of the categories and conceptual schemes that are considered significant and appropriate for these members, as they agree with the perceptions and interpretations considered appropriate by the culture from within.
(b) *Global Ethnomodels (Etic)*
 Etic ethnomodels are based on the view of external observers about the reality that they are being modeled with the utilization of school/academic mathematics. These ethnomodels are descriptions and analyses of mathematical ideas, concepts, procedures, and practices expressed in terms of the categories that are considered significant and appropriate by the community of scientific observers. These ethnomodels must be precise, logical, complete, replicable, and independent of external observers.
(c) *Glocal Ethnomodels (Dialogical)*
 Dialogical ethnomodels present a cultural dynamism between the emic and etic perspectives. These ethnomodels include the recognition of other epistemologies and the holistic nature of mathematical knowledge developed in diverse cultures (Rosa & Orey, 2017b).

This context enabled the development of an understanding of ethnomodelling as the translation of local mathematical procedures and global mathematical practices. The Mangbetu Ivory Sculptures of Zaire are examples of the application of dialogical ethnomodels. Figure 1.2 shows the development of the Mangbetu dialogical ethnomodel.

Translation can be considered as the description of the modelling processes of local mathematical systems, which can have a representation in other alternative systems of mathematical knowledge through the elaboration of ethnomodels.

1.4.2 Modelling in Interdisciplinarity Contexts

Our interest in analyzing the potentialities and difficulties of interdisciplinary projects as a framework for the approach of disciplinary concepts is associated with proposing contexts where learning would make sense for students. Although we consider the sense as a personal construction, we also admit that there are characteristics of didactic proposals that facilitate the student's appropriation of it and thus integrate it into their experiences, expectations, and emotions.

Some authors postulate that meaning can be constructed if the proposal in question is associated by the student with existing events, thus facilitating its integration into their world. Other authors indicate that meaning is an ability that the student develops regarding to what they do in their training with what they intend

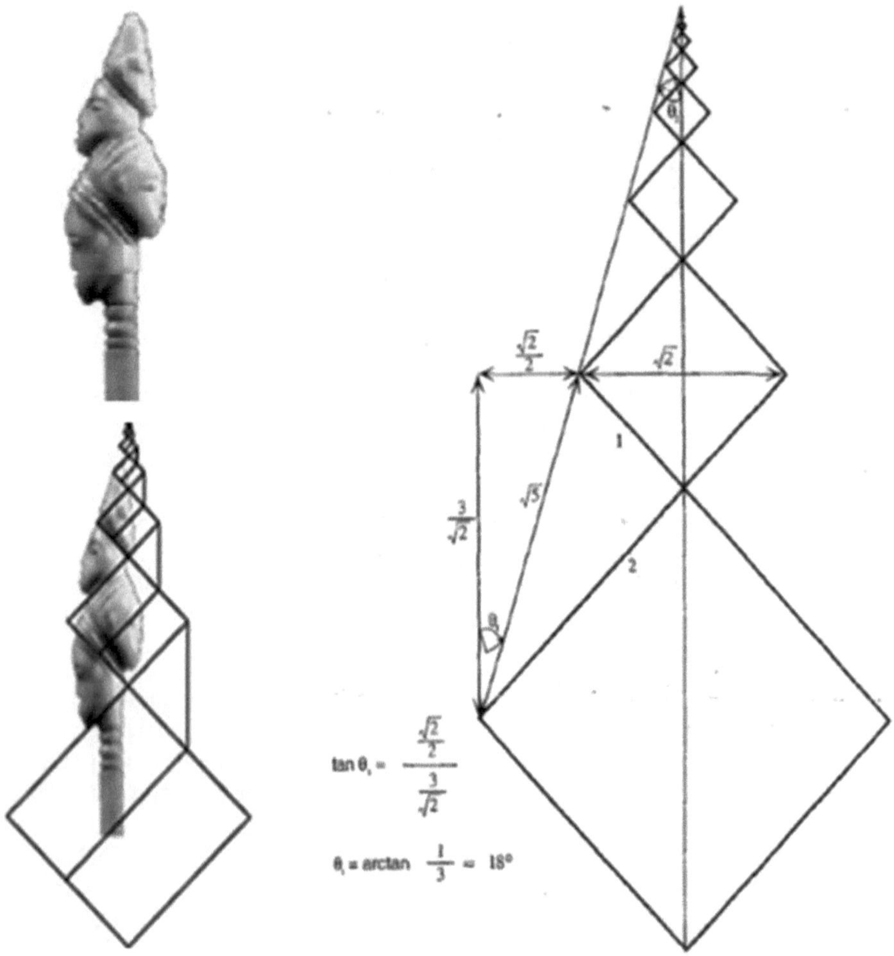

Fig. 1.2 Development of the Mangbetu dialogical ethnomodel. (Source: Adapted from Eglash (1999))

to be. Our position integrates these two perspectives by articulating the time variable in two modalities. We propose to the students, contexts with the intention that the knowings in the learning process are useful in a double temporality: for the future, but also the present.

One more dimension that is retained when designing proposals and is strongly related to the above is that referred to as the *functionality* of knowledge. We have come to understand that students must not only serve to understand the world around them, but also to intervene in it. A third dimension is also considered in the design of projects, and it refers to what we call the *transcendence* of the knowings. In the projects, we seek that the knowings be experienced as useful not only for the students but also for the community.

Some fundamental characteristics emerge from the three dimensions. One of them is that the context of the proposal is real. Thus, the interpretation here attributed to the term *real* refers to the possibility of immersion of the student in the context of the projects. Another characteristic that emerges from the previous ones is the dynamics of the relationship between the context and the knowledge *to be taught.* In general, teachers usually choose a concept and then eventually look for a context that gives meaning to it (Czocher, 2019; Stillman, 2019).

In our case, the situation is the other way around: the logic of the appearance of the concepts does not respond to a guideline of the disciplinary logic, but to the logic of the needs of each project. That is, a concept of interest is addressed when it represents a contribution to the solutions requested in that specific project. Thus, the projects modelers construct is developed in two directions: The questions to be solved in the project (the logic of the project) determine the chronology of convocation of knowledge (Blomhøj, 2019).

At the same time, the appearance of this knowledge demands approaches within the discipline (disciplinary logic). These two logics are not linked directly, but mainly by means of successive modelling; understanding modelling here, as a mental process consisting basically of relating elements of the context with disciplinary abstract entities and this is developed in order to produce new relationships at the level of abstract entities that are useful for the reference context that motivated the process (Blomhøj, 2019).

For example, one of the projects we carry out in mathematics class is the calculation, construction, and installation of Savonius type mills in rural posts in Patagonia, Argentina.[3] A fundamental aspect of the project is the one referred to ensuring the vertical of the mill against the strong winds of the region. This vertical is secured by means of tensioning cables that are fixed to concrete blocks buried in the ground. The image in Fig. 1.3 represents the elements involved in the fastening system.

A study carried out in Geogebra with the students allows for the analysis in detail of not only the relationships of efforts that occur, but also to anticipate how those efforts change if, due to the conditions of the terrain, it is necessary to place the concrete blocks closer or further away from the mill. The image in Fig. 1.4 shows the dynamic representation of the efforts made with the students.

To carry out the simulation, concepts of both mathematics and physics were indispensable. In the case of mathematics, concepts linked to vectors and transformations in the plane were called. As far as physics is concerned, the concepts referring to Newton's laws were fundamental.

[3]More information can be find ar: https://www.youtube.com/watch?v=889fvPzVK1g&t=2s

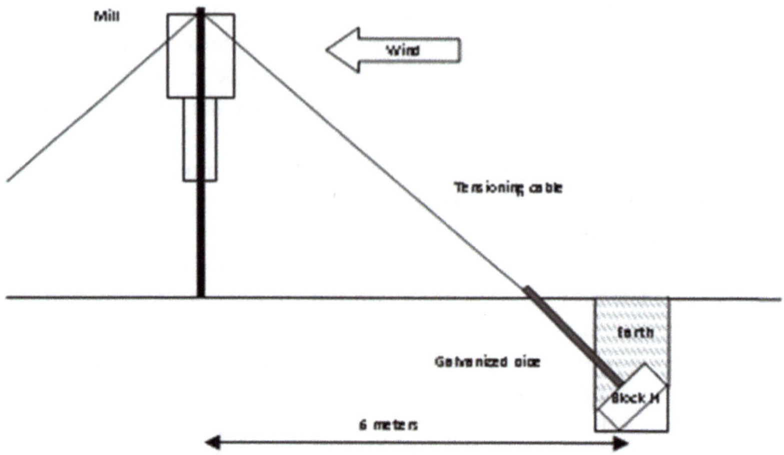

Fig. 1.3 The fastening system to secure the vertical of the mill. (Source: Elaborated by the authors)

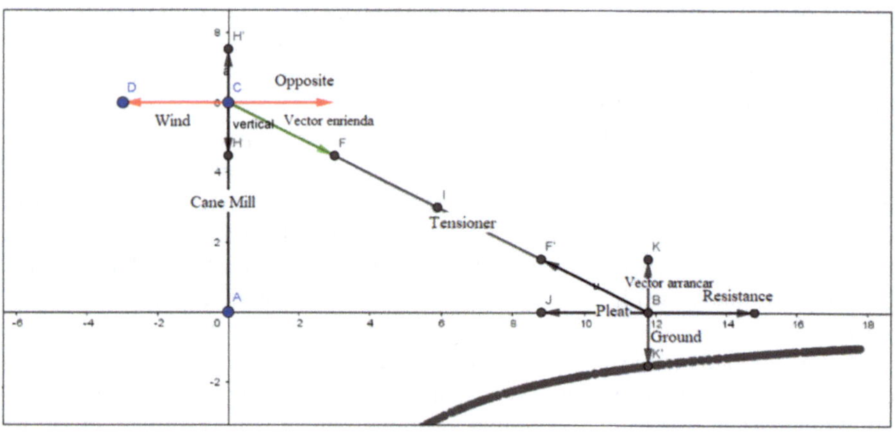

Fig. 1.4 Dynamic representation of efforts with students. (Source: Elaborated by the authors)

1.4.3 Modelling in the Transversality of Knowledge

The structure of the modelling category ζ(Mod) is composed of the uses of mathematical knowledge **U**(CM), and by the re-significations of those uses, **Res**(**U**(CM)), in specific situations (*S*). Such situations are part of that environment (horizontal and reciprocal relationships) that happen in communities of mathematical knowledge (*CCM*).

Each specific situation S_i is made up of sequential elements that construct the mathematical significance, procedure, and instrument, which derive the argumentation of the situation (*Arg*(*CM*)). In generic terms, $Arg(CM)_i$ is a **Res**(**U**(CM))$_i$ built by the CCM_i in S_i (Cordero, 2016). Indeed, it is a situational mathematical

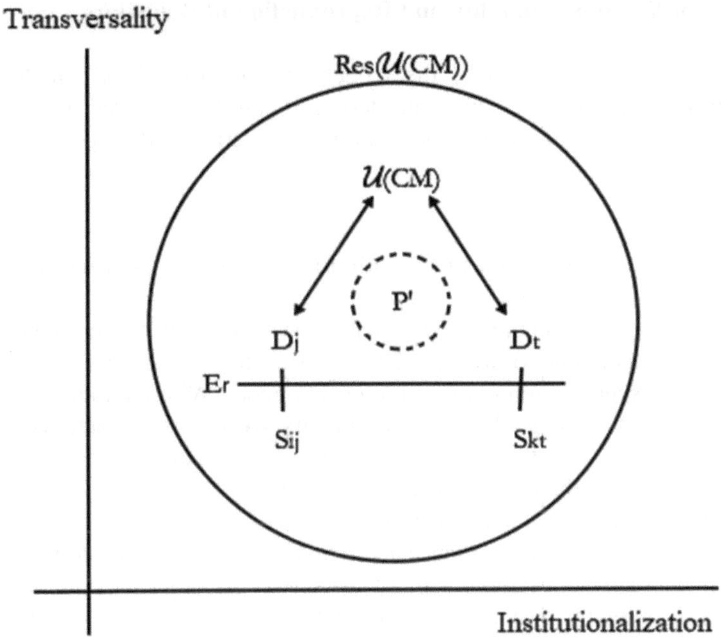

Fig. 1.5 The *Framework of Mathematical Knowledge* of ζ(Mod) (The interpretation of this framework of the mathematical knowledge of the modelling category was widely discussed considering the synthesis of three investigation experiences: the functionality of mathematical knowledge, mathematical modelling, and initial mathematics teacher education. The collaborators were F. Cordero, J. Mena-Lorca and J. Huincahue. Sabbatical stay at PUCV, 2016–2017.). (Source: Cordero (2022))

knowledge, which corresponds to the revelation of the uses and meanings of the object, typical of the community, regulated by the situation.

The **U**(CM) are re-signified in each *S*, but also when transversalities (T_i) occur between scenarios or domains of knowledge (D_i). However, in situations and transversalities, moments happen (Mo_i), and between them also the uses are re-signified. Mo_i are phases in the situational process. ζ(Mod) is composed of two axes: the institutionalization (the external knowings) and the transversality of knowledge (internal to the community), where S_{ij} situations happen, D_j domains, and alternations of scenarios: school-academic, work-profession, and city-everyday. The scheme in Fig. 1.5 shows the synthesis of the ζ(Mod) which we call the *Framework of Mathematical Knowledge* of ζ(Mod).

In addition, the principle of ζ the (Mod), P', is functional knowledge; this means that neither reality nor mathematics preexisted in the experiences of the communities. The P', in other words, is the putting into use of people's mathematics.

1.4.3.1 An Example: Stability and Reproduction of Behaviors

We outline the development of the modelling category ζ(Mod) in the episode *re-signification of stability* by considering empirical research that has been conducted with communities of engineers in the practice of their profession and in training.

1.4.3.2 Data Collection and the Definition of the Study Community

The selection of the communities consisted of the availability of their members to be video-recorded and interviewed in the scenarios of professional work. For this chapter, we present a community of bionic engineers. With ethnographic methods (Guber, 2001) and case study, their disciplinary work was characterized, through identifying routine situations where they use mathematical knowledge and the problematizations of their mathematical knowing.

The analysis of these characterizations was carried out with the constructs of the *modelling category* in the *socioepistemological theory* and considering the documentary technique and semi-structured interviews in order to problematize, the mathematical knowing was analyzed through the re-significance of these uses in the school-academic and work-profession scenarios. The problematization formed an epistemology of re-significations of the uses of mathematical knowledge that emerges in the community when considering the specificity of the scenario.

On the one hand, with the documentary technique of analysis, and on the other, with the technique of semi-structured interviews, patterns and relationships were identified between them alluding to the tendency or reproduction of a behavior in the contexts of the situations.

The patterns and relationships of behaviors were organized through an instrument (*instruction that organizes behaviors*) accompanied by their significations (*graphic and analytical patterns*) and procedures (*variation of parameters*). With the unit of analysis composed of the constructs: use, re-signification, and transversality, the *situation that expresses the re-signification of uses of stability* was formed.

The evidence of the *framework of the mathematical knowing of the* ζ(Mod), was justified in one of the laboratory practices developed by the teaching engineer. The practice has been called *control of the temperature of the light bulb* where the main problem consisted of: *once assigned a reference value, the temperature of the light bulb reaches it.* To this end, students assemble a physical model with the following elements: Arduino board, light bulb, AC solid-state relay, and temperature sensor. Figure 1.6 shows light bulb, relay, and Arduino connection scheme.

Initially, the students analyzed the behavior of the output signal in the system: the temperature of the light bulb. The teacher drew on the blackboard, a graph as a pattern of behavior of the maximum temperature that the light bulb can reach. Figure 1.7 shows the graph output signal behavior.

Fig. 1.6 Light bulb, Relay, and Arduino connection scheme. (Source: Mendoza and Cordero (2018))

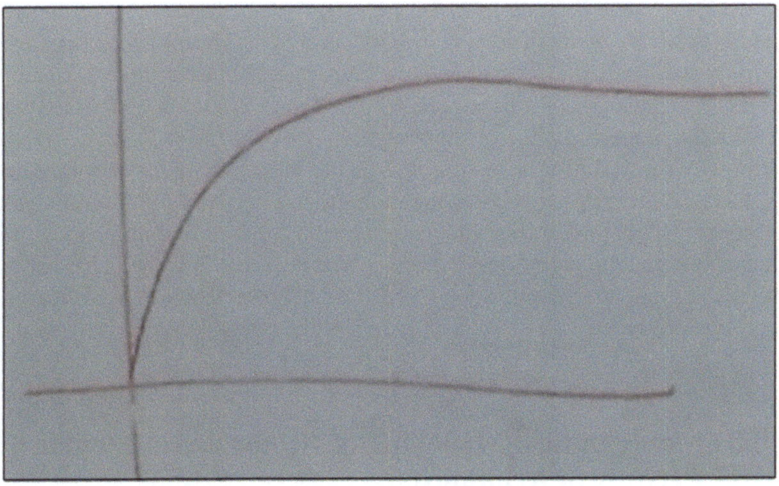

Fig. 1.7 Graph of output signal behavior. (Source: Mendoza and Cordero (2018))

The system is characterized by *system gain*, the *time constant* of the system, and the *transfer function* that relates the input and output signal. Students problematize the adjustments of the parameters of the *system gain* and the *time constant*, based on the graph provided by the Arduino software allowing it to reach its maximum temperature. Finally, they seek to control the temperature of the light bulb by making use of the on-off controller. In Fig. 1.8, it is observed how the curve, which

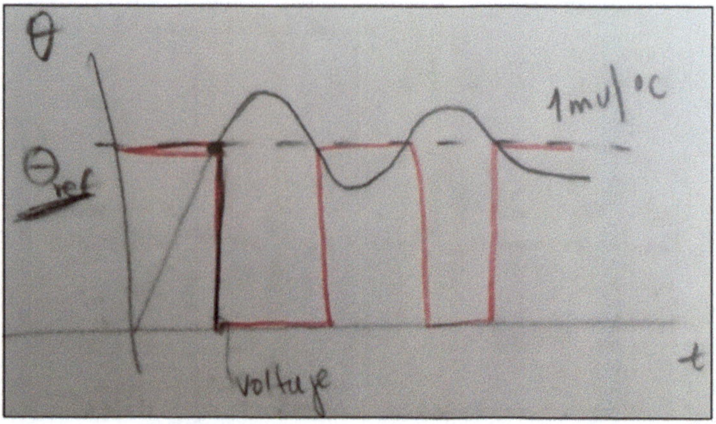

Fig. 1.8 Graph of the output signal using the on-off control. (Source: Mendoza and Cordero (2018))

represents the behavior of the output signal, at certain time intervals, surpasses the reference value and in others, it does not. This is due to the control mechanism.

In the design of the control system, three moments are observed: M_1: *System dynamics* ($\theta \rightarrow \theta ref : \theta$ *tends to behave as* θref), M_2: *Adjustment of the transfer function or behavior model*, ($a\theta' + \theta = \theta ref$), and M_3: *Control of the output signal and stability* ($a\theta' = \theta ref - \theta :$ *by control*). Each one of these moments is subject to the desired *reproduction of a behavior*, which mean that the output signal tends to behave like the reference value or input signal.

In this way the stability is *re-signified* in the behavior of the input and output signals, causing *procedures* such as the comparison between the output signal and the reference value, modifying the parameters of the differential equation that models the system, and signifying it as an *instrument* that is responsible for modelling the stability of the output signal and thus achieve that the behavior initially proposed is reproduced.

1.5 The Mathematical Modelling of People as Horizontal and Reciprocal Relationships Between School Mathematics and the Reality of the Learners: An Alternative Education

The mathematical modelling of people means the mathematics that emerges in the communities of mathematical knowledge, in different scenarios: school-academic, work-profession, and city-everyday. This mathematics is not formal/scientific mathematics, which is usually the frame of reference for school/academic mathematics. On the contrary, it is a mathematics put into use by those communities, in their

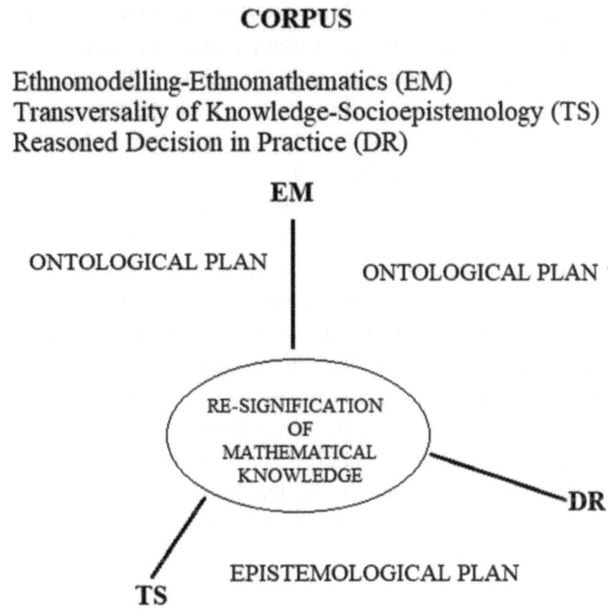

Fig. 1.9 Corpus of mathematical knowledge for educational changes. (Source: Elaborated by the authors)

scenarios and in the alternation between them. Its ontological and epistemological aspect is based on uses and situational meanings, which leads to re-significations of uses in the alternation of situations, domains, and scenarios.

These re-signification of uses of mathematical knowledge, according to the modelling programs in the context of social construction of mathematical knowledge develop a new frame of reference for an innovative school mathematics curriculum. Together the three modelling programs define a three-dimensional space that could frame mathematical knowledge used by the members of different communities.

The axes of space are the fundamental aspects of each modelling program: ethnomodelling, transversality of knowledge, and reasoned decision-making. Figure 1.9 shows the corpus of mathematical knowledge for educational changes.

The re-signification of mathematical knowledge is related to the knowledge of people in their social generality, restricted in specific situations according to the scenarios: school-academic, work-profession, and city-everyday. In that space, the mathematical knowledge of the classroom lives permanently in a reciprocal and horizontal relationship with the knowledge of the reality of the people involved.

However, this reality has to be restricted in order to standardize it to mathematics education because it considers all educational levels and the diversity of disciplines, as well as people's work and lives. In this regard, reality is interpreted in the usual way of all these scenarios, where the uses of routine mathematical knowledge are expressed; that is, the day-to-day of the disciplinarian, the worker, and the people, in cultures and in interdisciplinarity.

In this context, mathematical knowledge developed locally by people in their own sociocultural contexts needs to be valorized, valued, and respected in the classrooms because with the corpus composed of the three axes, the problems and difficulties of the teaching and learning of mathematics are addressed, where the Transversality of Knowledge (TS) and Reasoned Decision (DR) make up the epistemological plane of uses (transformation of knowledge) with the Ethnomodelling axis (EM) defines the ontological plans, respectively with local and global symmetry (transformation of people) as showed in Fig. 1.9.

This approach requires a methodology of lines of action for this purpose, which are cyclical, permanent, and reflective. On the one hand, it is necessary to problematize the mathematical knowledge developed in the communities between its different domains that necessarily come into play, such as the school mathematical discourse, the disciplinary field, and the day-to-day of the communities. Then, in these communities (CCM_k) the re-significations of uses of mathematical knowledge ($ResS_{kj}$) emerge, in different situations (S_i), domains (D_j), and scenarios (E_r).

It is important to emphasize that this knowledge emerges is related to non-school mathematics. This approach forms new frames of reference that help the generation of educational changes in the teaching and learning of mathematics. In this regard, these frames of reference need to be conducted in the school environment with moments of transversalities of mathematical knowledge (MT_{jk}), in communities of students and teachers (CCM_l (Teacher/Student)).

In this context, school and non-school mathematical knowledge are confronted and discussed, as well as the use of the mathematical knowledge of the *others* are valorized, valued, respected through educational multi-factors and stages that contribute to the alliance of quality of mathematics teaching (Cordero, 2016).

The multi-factors are elements that contribute to the educational changes of mathematics, such as disciplinary identity of teachers, inclusion of the social construction of mathematical knowledge, socialization of the use of mathematical knowledge (another epistemology), and the emancipation of the dominant mathematical knowledge. Figure 1.10 shows lines of actions for educational changes.

The educational changes of mathematics consist of valuing the horizontalities of mathematical knowledge, the construction of autonomies of knowledge, and the cultural forms of knowledge. In this regard, the episodes of students' learning in the classrooms have to be extended to the day-to-day of people in institutions and in society, as well as in educational environment.

In this approach, the role of the teachers needs to be supported and further developed in order to enable them to design and maintain educational contexts in which systems of reciprocal relations of school mathematics and its connection to reality of the learners in specific situations that enable the development of knowledge that is coordinated by two axes: permanent programs (ProPer) and to disrupt and transform the discourse of school mathematics (TrTf (dME)), which is shown in Fig. 1.10.

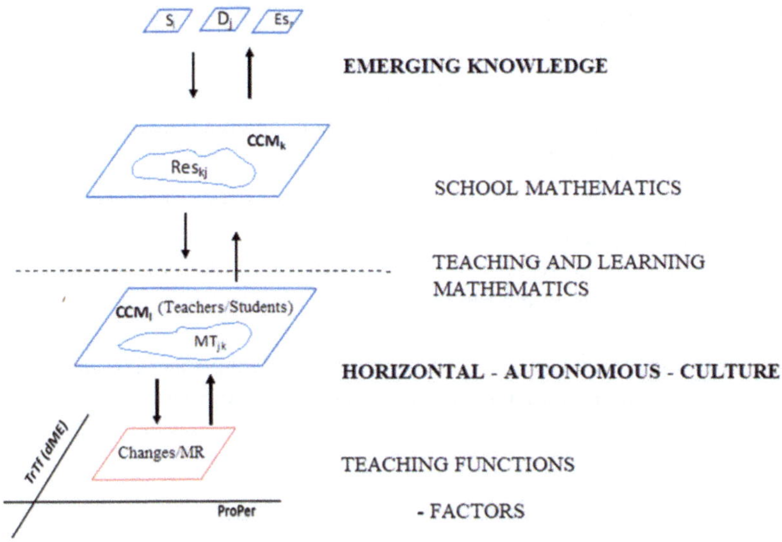

Fig. 1.10 Lines of action for educational change. (Source: Elaborated by the authors)

1.5.1 An Example: The Mathematical Modelling of People as a Mathematics Put into Use Horizontality, Autonomously, and Culturally

On the one hand, the function of the corpus of mathematical knowledge offers the development of its re-signification that entails, simultaneously ontological and epistemological transformations, and, on the other hand, the lines of action guide educational changes, such as knowledge and the individual changes.

The three Latin America mathematical modelling programs offer distinct conceptions in their epistemological, theoretical, and methodological procedures by considering specific scenarios in each community. For example,

- *Ethnomodelling* offers the manual production of a figure, such as artisanal work in the scenario of a cultural tradition of a certain community.
- The *reasoned decision* offers the lifting of a windmill according to the topography of the terrain, as a community task agreed in the classroom-reality-approach scenario of a community of technical students.
- The *transversality of knowledge* offers the design of a control system, in a school laboratory in a school-academic scenario of a community of engineers in training.

A priori, there are three different situations: (a) the craft of a figure, (b) raising of a windmill, and (c) the design of a temperature control system. Likewise, the mathematics of each modelling seem to be foreign topics, such as the mathematical iteration, the equilibrium equation, and the stability of a system or differential equation. However, the functioning of the corpus offers relationships between situations and mathematical subjects, considering the epistemological and ontological planes, simultaneously:

1.5.2 Re-significations

In the context of the three Latin America mathematical modelling programs there are three distinct types of re-significations that are complementary even though each one has its own conceptions. Thus,

- Re-signification of the mathematical iteration, which reproduces a figure with a trend (figure size): larger size to smaller size.
- Re-signification of the balance of forces, which reproduces the magnitude of a force vector according to the topography of the terrain: cable tension and anchor depth.
- Re-signification of stability, which reproduces the importance of an ideal temperature in a system: control of the temperature in a set range.

1.5.3 Emerging Knowledge: Non-school Mathematics

Mathematical iteration, balance of forces, and stability are external knowledge to the educational community, while the respective re-significations are knowledge that emerges within the local communities: it is related to the development of a non-school mathematics.

For each re-signification, a category of mathematical knowledge can be inferred, as a nucleus of the three models: *reproduction of a behavior*. This category has an epistemology and is transversal to the three situations with their respective scenarios and re-significations. Table 1.1 shows an emerging knowledge of communities in specific situations.

The reproduction of behaviors category generates a very broad spectrum of mathematical knowledge, which is not only transversal to different disciplinary domains but can also be transversal to the different educational levels: basic, middle, and higher. An iterative function, the balance of forces, and the differential equation cannot be transversal to the basic levels, but they can be the reproduction of figures in a larger or smaller grid; feel the tension of a wire in relation to a weight; or generate behaviors to a given context.

1.5.4 The Environments of Mathematical Knowledge

The models and the act of modelling itself, as presented by these three programs generate environments of mathematical knowledge that expand the mathematics in the classroom, and the minds of participants. This opens up a mathematical vision for educational systems that is deeper than the traditional emulation of standard mathematical procedures and problem solving. On the contrary, this approach integrates:

Table 1.1 Emerging knowledge of communities in specific situations

Situation	Craft of a figure	Constructing of a windmill	Design of a temperature control system
Instrument	x1 = g(x), x2 = g(x1), x3 = g(x2), ... Instructing a behavior to repeat	$\propto X + \beta Y + \delta Z = 0$ Statement that organizes behaviors	$ay' + y = f$ Statement that organizes behaviors
Signification	Geometric behavior patterns	Balancing behavior patterns	Behavior patterns of the signals to be controlled
Procedure	Geometric relationships in iteration	Parameter variation	Comparison between signals and parameter adjustment
Argumentation	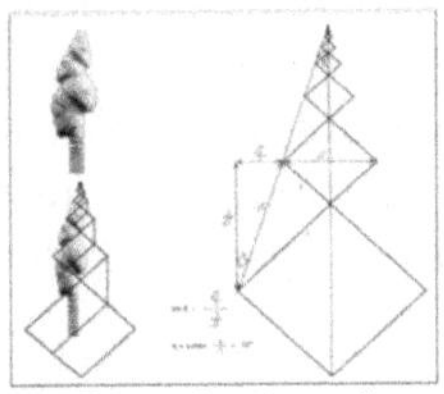 Reproduction of behaviors Reproduction of behaviors	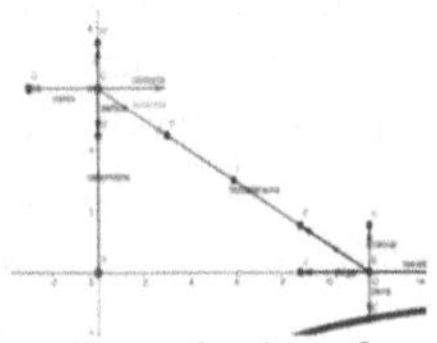 Reproduction of behaviors Reproduction of behaviors	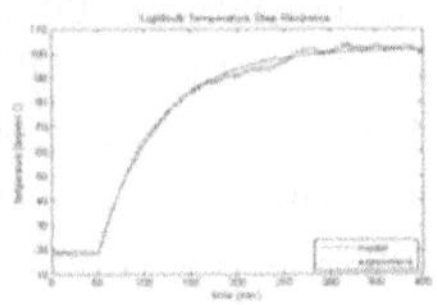 Reproduction of behaviors Reproduction of behaviors

Source: Elaborated by the authors

1. Cultural field that relates to traditional knowledge with the school/academic environments and highlights non-school mathematics in *symmetrical dialogical relationships and with otherness*, which is to say, both forms of knowledge are balanced to affect changes in individuals and communities.
2. Classroom-reality-approach in which mutual mathematical knowledge is agreed upon during dialogue in the classroom and is put into a *rational discourse* for the decision-making process among participants into the development of a pedagogical action that is tangible to reality.
3. The scope of another domain of knowledge where *it reveals the putting into use of the mathematical knowledge* of the knowing and doing process developed by community members in order to confront it with the discourse of school mathematics and consequently generate horizontal relationships between distinct knowledge.

1.5.5 *Teaching and Learning Mathematics*

In summary, the re-signification of mathematical knowledge generated by these three modelling programs that emerged in three different regions in Latin America form an epistemological and ontological basis for designing and developing school situations for the teaching and learning of mathematics, in a cyclical, continuous, and reflexive manner.

These designs consist of locating moments of transversality of knowledge to both favor and valorize the emergence and re-significations of diverse regional, cultural and curricular categories. Learning consists of creating new relationships between different forms of knowledge. Teaching must be based on, and respect for the reality and function of educators, which includes an understanding of the educational environments and diverse forms of mathematical knowledge found in Latin America, through our permanent programs of accompaniment and alliance.

It is important to highlight here that the traditional discourse of school mathematics can be disrupted and transformed in order to create a redesign of school mathematical rhetorical communication whose fundamental, epistemological, and ontological basis is related to the horizontality-autonomous-culture triad in which the principles of the mathematical modelling programs can promote in order to provoke change in the educational system.

These principles permeate the school mathematics, at all educational levels, by considering the use of diverse mathematical knowledge, the reasoned decision-making process, and the dialogical and symmetrical relations with the *others*.

1.6 Final Considerations

Undoubtedly the three Latin American mathematical modelling programs, independently, provide educational gains, each with its own levels of specificity and loyal to its principles. We both hope and imagine that over time others will emerge as this dialogue takes hold, expands, and is experimented upon across our enormous, diverse and rich region. However, the exercise of putting these first three together, organized by axes, defines a corpus of mathematical knowledge that envisions educational changes. On the one hand, an epistemological and ontological change, where mathematical knowledge of the *others* is recognized, in a horizontal plane.

There is a necessity to state here that new empirical relationships between mathematical knowledge and reality are happening across our vast region. The re-signification of mathematical knowledge must be dimensioned and valued in our classrooms. The inclusion of the environments plays a fundamental role, as it includes mathematical knowledge that emerges in the student body and the teaching staff of our schools and institutions.

On the other hand, this corpus of knowledge obliges us to respect and understand mathematics educators as a community of professionals with rich experience, creativity, and with equally diverse experience that constructs their own mathematical categories in their own environment as regulated by the reciprocal relations between the knowledge of the school and the needs of their communities.

Thus, frame of reference of the corpus outlined here guides the necessary articulations in autonomous actions in mathematical teaching, hence the importance of generating research on the role of the teachers (Opazo Arellano et al., 2018). This entails the permanence of the environment of reciprocal relationships that occurs in mathematical functionality, and the educational change of mathematics. The spectrum of the corpus is large; it must be taken advantage of, and in the case of Latin American mathematics educators is a rich resource.

References

Arendt, H. (2005). *La condición humana*. Paidós.

Blomhøj, M. (2019). Towards integration of modelling in secondary mathematics teaching. In G. A. Stillman & J. P. Brown (Eds.), *Lines of inquiry in mathematical modelling research in education* (ICME-13 monographs). Springer.

Buendía, G., & Cordero, F. (2005). Prediction and the periodical aspect as generators of knowledge in a social practice framework. A socioepistemological study. *Educational Studies in Mathematics, 58*(3), 299–333.

Cantoral, R. (2013). *Teoría Socioepistemológica de la matemática Educativa. Estudios sobre la construcción social del conocimiento*. Gedisa.

Carranza, P. (2015). Molino Savonius. Proyecto de extensión y marco didáctico en clases de matemática. Extensionismo, innovación y transferencia tecnológica. *Claves para el Desarrollo, 2*, 55–56.

Carranza, P. (2017). Proyectos interdisciplinarios con la comunidad. Posibilidades y dificultades. *Kimun, III*(5), 5–32.

Common Core Standards Writing Team. (2013). *Progressions for the common core standards in mathematics high school, modelling*. Institute for Mathematics and Education, University of Arizona.

Cordero, F. (2016). Modelación, funcionalidad y multidisciplinaridad: El eslabón de la matemática y el cotidiano. In J. Arrieta & L. Díaz (Eds.), *Investigaciones latinoamericanas de modelación de la matemática educativa* (pp. 59–88). Gedisa.

Cordero, F. (2022). *La matemática, sus usos y significados. Un programa socioepistemológico de la matemática educativa [Mathematics, its uses and meanings. A socioepistemological program of educational mathematics]*. Manuscrito en preparación.

Cordero, F., Gómez, K., Silva-Crocci, H., & Soto, D. (2015). *El Discurso Matemático Escolar: La Adherencia, la Exclusión y la Opacidad*. Gedisa.

Cordero, F., Solís, M., Buendía, G., Mendoza, J., & Zaldívar, D. (2016). *El comportamiento con tendencia, lo estable y las Ecuaciones Diferenciales Lineales. Una Argumentación Gráfica*. Gedisa.

Czocher, J. (2019). Precision, priority, and proxies in mathematical modelling. In G. A. Stillman & J. P. Brown (Eds.), *Lines of inquiry in mathematical modelling research in education* (ICME-13 monographs). Springer.

D'Ambrosio, U. (1985). Ethnomathematics and its place in the history and pedagogy of mathematics. *For the Learning of Mathematics, 5*(1), 44–48.

D'Ambrosio, U. (1990). *Etnomatemática*. Editora Ática.

Develay, M. (1994). Le sens dans les apprentissages: Du désir au passage a l'acte. *Pedagogie collégiale, 7*(4), 23–26.

Eglash, R. (1999). *African fractals: Modern computing and indigenous design*. Rutgers University Press.

Eglash, R., et al. (2006). Culturally situated designed tools: Ethnocomputing from field site to classroom. *American Anthropologist, 108*(2), 347–336.

Guber, R. (2001). La etnografía. Método, Campo y Reflexividad. Norma: Bogotá.

Jacques, F. (1987). De la Signifiance. *Revue de Métaphysique et de Morale, 92*(2), 179–218.

Medina-Lara, D., Cordero, F., & Soto D. (2018). La resignificación de la media aritmética, bajo el análisis gráfico de la representación de los datos. *Acta Latinoamericana de Matemática Educativa, 31*(1), 655–661. México: Comité Latinoamericano de Matemática Educativa

Mendoza, J., & Cordero, F. (2018). La modelación en las comunidades de conocimiento matemático. El uso de las matemáticas en ingenieros biónicos. El caso de la estabilidad. *Revista Latinoamericana de Etnomatemática, 11*(1), 36–61.

Mendoza, J., Cordero, F., Solís, M., & Gómez, K. (2018). El Uso del Conocimiento Matemático en las Comunidades de Ingenieros. Del Objeto a la Funcionalidad Matemática. *Boletim de Educação Matemática-BOLEMA, 32*(62), 1219–1243.

Morales, A., & Cordero, F. (2014). La Graficación-Modelación y la Serie de Taylor. Una Socioepistemología del Cálculo. *Revista Latinoamericana de Investigación en Matemática Educativa, 17*(3), 319–345.

Opazo Arellano, C., Cordero Osorio, F., & Silva-Crocci, H. (2018). ¿Por qué estudiar la identidad disciplinar en la formación inicial del docente de matemáticas? *Premisa, 20*(77), 5–20.

Pérez-Oxté, I. & Cordero, F. (2016). Una epistemología basada en la transversalidad de los Usos de la gráfica de una comunidad de ingenieros Químicos industriales. En F. Rodríguez, R. Rodríguez y L. Sosa (Eds.). *Investigación e Innovación en Matemática Educativa 1*(1), (pp. 24–30). Oaxaca, México: Red de Centros de Investigación en Matemática Educativa A. C.

Rosa M. (2010). *A mixed-methods study to understand the perceptions of high school leader about English language learners (ELL): The case of mathematics*. Doctorate dissertation, College of Education, California State University, Sacramento.

Rosa, M., & Orey, D. C. (2010). Ethnomodelling as a pedagogical tool for the ethnomathematics program. *Revista Latinoamericana de Etnomatemática, 3*(2), 14–23.

Rosa, M., & Orey, D. C. (2012). O campo de pesquisa em etnomodelagem: As abordagens êmica, ética e dialética. *Educação e Pesquisa, 38*(4), 865–879.

Rosa, M., & Orey, D. C. (2017a). *Influências etnomatemáticas em salas de aula: Caminhando para a ação pedagógica*. Editora Appris.

Rosa, M., & Orey, D. C. (2017b). *Etnomodelagem: A arte de traduzir práticas matemáticas locais*. Editora Livraria da Física.

Rosa, M., & Orey, D. C. (2020). Principles of culturally relevant education in an ethnomathematical perspective. *Revista de Educação Matemática, 17*(1), 1–23.

Saaresranta, T. (2011). *Educación indígena originaria campesina: Perspectivas de la educación intracultural*. Fundación PIEB.

Stillman, G. (2019). State of the art on modelling in mathematics education—Lines of inquiry. In G. Stillman & J. Brown (Eds.), *ICME 13. Lines of inquiry in mathematical modelling research in education*. Springer.

Suárez, L., & Cordero, F. (2010). Modelación-Graficación, una categoría para la matemática escolar. Resultados de un estudio socioepistemológico. *Revista Latinoamericana de Investigación en Matemática Educativa, 13*(4), 319–334.

Zaldívar, D., Cen, C., Briceño, E., Méndez, M., & Cordero, F. (2014). El Espacio de Trabajo Matemático y la situación específica de la matemática funcional: Un ejercicio de diálogo. *Revista Latinoamericana de Investigación en Matemática Educativa, 17*(4-II), 417–436.

Part II
Ethnomathematics and Ethnomodelling: Empirical Work, Theoretical-Methodological Approaches, and Research Questions

Ethnomodelling is a process of elaboration of problems and questions that emerge from real situations that form an image or sense of an idealized version of the *mathema*. The focus of this perspective essentially forms a critical analysis for the generation and production of mathematical knowledge (creativity) and forms an intellectual process for its production, the social mechanisms of institutionalization of knowledge (school/academics), and its diffusion (education). This process is modelling. In this perspective, by analyzing the role in reality as a whole, this holistic context allows those engaged in the modelling process to study systems of reality in which there is an equal effort made to create an understanding of all the components of the system as well as the interrelationships among them. In this section, the authors discuss the use of modelling as pedagogical action for an ethnomathematics program that values the tacit knowledge of the members of community by developing students' capacity to assess the process of elaborating ethnomodels in its different applications and contexts by having started with the sociocultural context, reality, and interests of the students and not by enforcing a set of external values and curriculum without context or meaning for the learners.

Chapter 2
Conceptualizing Positive Deviance in Ethnomodelling Research: Creatively Insubordinating and Responsibly Subverting Mathematics Education

Milton Rosa and Daniel Clark Orey

2.1 Initial Considerations

The recognition of the relationship between culture and mathematics can be interpreted as one of many reactions to cultural imperialism, which, in this case, imposed its version of mathematical knowledge on colonized communities around the world with the expansion of the great navigations that took place from the fifteenth century onwards (D'Ambrosio, 1990), to which members of other cultures were forced to adapt to these paradigms or perish. Consequently, mathematics can be considered as a field of knowledge that often perpetuates imperialist goals by being perceived as a secret weapon that maintains the imposition and domination of Western capitalist values in local cultures (Bishop, 1990).

In this context, school/academic mathematics is often criticized because it collaborates to reinforce an Eurocentric approach that prevails in school curricula worldwide and helps the process of globalization of particular types of mathematical ideologies and technologies (D'Ambrosio & D'Ambrosio, 2013) that support the maintenance of this cultural imperialism.

However, the development of non-prescriptive strategies to solve problems and situations faced in different social domains and in distinct cultural contexts is an important alternative method useful in identifying innovative problem solving techniques, as well as to value the diversity of ideas, procedures, and practices mathematics present in investigations in ethnomodelling (Rosa & Orey, 2017a).

Thus, this reaction to regulatory and normative impositions, as well as the opposition to cultural imperialism may be related to the development of concepts related to:

M. Rosa (✉) · D. C. Orey
Universidade Federal de Ouro Preto, Ouro Preto, Minas Gerais, Brazil
e-mail: milton.rosa@ufop.edu.br; oreydc@ufop.edu.br

M. Rosa et al. (eds.), *Mathematical Modelling Programs in Latin America*,
https://doi.org/10.1007/978-3-031-04271-3_2

1. *Creative Insubordination* (Crowson & Morris, 1982) that relates to issues of social justice.
2. *Responsible Subversion* (Hutchinson, 1990) that relates to political issues.
3. *Positive Deviance*[1] (Zeitlin et al., 1990) that relates to sociocultural issues. In this context, *globalized*—one size fits all—mathematical standards may not be realistic for implementation of curricular activities at a local level. Therefore, teachers may be forced to deviate, react creatively, responsible, subversively in meeting the educational needs of their students (Rosa & Orey, 2017b). Thus, we propose that teachers use positive deviance to develop actions in order to deal with such situations because it "involves an intentional act of breaking the rules in order to serve the greater good" (Gary, 2013, p. 26).

Despite being equivalent concepts, as they relate to the flexibility of rules and the rupture with previously established regulations, so that it is possible to achieve the well-being of members of distinct cultural groups, it is important to emphasize that these concepts do not have a uniform definition, as there are subtle differences that are related to the diverse contexts in which (mathematical) practices are developed (Rosa & Orey, 2015a).

In this context, it can be argued that these concepts encompass innovative solutions in the teaching practice of teachers and educators by helping them to confront the belief that still persists in society that mathematics is a culturally independent knowledge and that education is politically composed of exempted acts and actions (Rosa & Orey, 2019).

It is important to emphasize that these concepts also combat a single educational model of *one size fits all*, which makes it impossible to discuss the *decolonizing* processes that can provoke a continuous position of transgressing and insurgency. This decolonial concept implies a continuous struggle against the maintenance of the *status quo* of superiority, prejudiced, dominating, and oppressive societies (Rosa & Orey, 2019).

Therefore, the breadth of the concepts of creative insubordination, responsible subversion, and positive deviance encompass innovative solutions in the pedagogical action of ethnomodelling as it aims to confront the belief that still persists in contemporary society that mathematics is a culturally neutral knowledge.

From this perspective, we emphasize that the historical relationships between culture and mathematics illustrate that this field of knowledge is related to its sociocultural aspects. In this regard, D'Ambrosio (1990) states that the culturally specific nature of mathematics must be recognized, so that we can describe the ideas and mathematical procedures practiced by members of distinct cultural groups.

---

[1] In sociology, deviance describes an action or behavior that violates previously established social norms, as well as portraying its informal violations (Macionis & Gerber, 2010). In this sense, deviance is a behavioral disposition that is in disagreement with an institutionalized configuration and with the current codes of conduct. In this chapter we use the sociological concept of deviance, which is related to a set of actions or behaviors that transgress formal and informal societal norms that impose its own regulations to its own members (Akarowhe, 2018).

It is important that alternative pedagogical and methodological approaches can be used to direct local mathematical practices record the historicity of ideas and mathematical procedures developed in diverse cultural contexts with the application of innovative mathematical solutions to the challenges faced in the daily life.

In this direction, an alternative methodological approach that can contribute to the development of this discussion is related to the use of the pedagogical action of ethnomodelling in the school curriculum, can be considered as the application of the theoretical basis of ethnomathematics that adds cultural perspectives to the modelling process (Rosa & Orey, 2012).

Thus, as a process of positive deviance, ethnomodelling seeks to change, according to Marzano et al. (2005), existing external paradigms and conflicts with the values and norms prevalent in the mathematical curriculum, as it represents the development of ideas, mathematical procedures, and practices that are rooted in diverse cultural contexts.

In this regard, Rosa and Orey (2017a) state that ethnomodelling links contemporary views of ethnomathematics and, simultaneously, recognizes the need to develop a holistic view of mathematical modelling processes that must be culturally based. The insubordinate perspective of ethnomodelling shows the need for the modelling process to be culturally linked so that we can encourage investigations related to local communities to bring the cultural aspects of their practices to the teaching and learning process in mathematics.

Similarly, Rosa and Orey (2015b) argue that ethnomodelling is useful insubordinate and creative educational approach that ignores the linearity of the teaching and learning process in mathematics predominant in schools by interrupting the existing order in the development of the modelling process.

According to Hutchinson (1990), this approach can also be seen as an act of responsible subversion, as it examines how members of distinct cultural groups solve problems in their daily lives by using their own techniques and cultural artifacts.

This context reveals the presence of responsible subversion aspects in the ethnomodelling process, as it uses local mathematical knowledge and material resources developed by members of distinct cultural groups to solve everyday problems, situations, and phenomena. Thus, Rosa and Orey (2017b) affirm that this approach can also be understood as an important characteristic of positive deviance.

Consequently, deviant and positive behaviors are related to the development of local mathematical ideas, procedures, and practices, which confer legitimacy to members of distinct cultural groups who practice them in their own way in accordance with their own beliefs, values, and traditions. These behaviors are therefore accessible, acceptable, and sustainable.

For example, Lyman et al. (2005) argue that these aspects are identified as continuous movement that aims to challenge the *status quo* of academic mathematical knowledge, as it seeks to subversively modify imposed regulations by the educational system, but with responsibility, to better serve the needs of the members of the school community. This process involves the analysis of perspectives external to the pedagogical models in force in traditional educational systems.

Therefore, there is a certain sense of disruption in the bureaucratic norms and rules of school/academic mathematics by seeking to value and respect diverse problem-solving techniques, as well as to appreciate diverse modes of production of mathematical knowledge in different cultures (Rosa & Orey, 2015a). For example, Dehler and Welsh (1998) argue that this approach is related to a form of positive deviance, as it involves thoughts and/or actions that differ from the norms and regulations imposed by educational systems.

Consequently, in this chapter we seek to conceptualize creative insubordination, responsible subversion, and positive deviance in ethnomodelling investigations and pedagogical actions as both purposeful and honorable behaviors that differ from the imposition of norms and/or rules because they contain elements of innovation, creativity, and adaptability.

Additionally, we discuss the concept of positive deviance for the development of mathematics education through the exploration of its essence in the context of educational practices related to pedagogical actions of ethnomodelling.

2.2 Creative Insubordination as Social Justice Issues

The concept of creative insubordination emerged in the 1970s, when a group of health professionals made changes in public policies in the area of social well-being by minimizing the repercussions of decision-making by higher entities in relation to regulations. The practices developed locally by the professionals were also used in interventions conducted in public health.

However, although most research conducted in during that time was linked to public policies and related to health, especially in regard to nursing practices. Later concepts of creative insubordination was also used in investigations to analyze management activities in school systems (Rosa & Orey, 2015b).

In the 1980s, the concept of creative insubordination was used by Crowson and Morris (1982), to describe how school administrators (principals and vice-principals) circumvented norms or made institutional rules flexible by aiming to offer better educational services. Flexibility in regard to regulations sought to meet the needs of students, teachers, parents and/or guardians, and members of the school community by resisting to the impositions of public policies and institutional bureaucratic guidelines.

For example, through negotiation of norms or by dodging interactions with central administration, managers gained ample autonomy in relation to their superiors by providing improved school management and administrative flexibility. This enabled the development of local decision-making processes most beneficial for teachers, students, and members of the school community (Crowson & Morris, 1985).

The counter-bureaucratic aspect of creative insubordination regarding decision-making processes made the norms more flexible and refuted the rules with the objective of subverting the authorities in power relations to legitimize the search for actions that aimed at the common good of the members who participate in the development of the school (Crowson & Morris, 1985).

The use of insubordinate, alternative, and creative ways was used to achieve good results for the common good of the members of the school community through the adoption of anti-bureaucratic behaviors (Rosa & Orey, 2015a). Indeed, it is important that schools are educational spaces through which school managers and leaders can make autonomous decisions, even if their actions are inconsistent with formal institutionalized public policies (Chubb & Moe, 1990). This is an important towards the development of social justice in society and school systems.

Also Gutiérrez (2015) affirms that creative insubordination focuses on social justice, by showing its usefulness for the development of students' citizenship, as it makes it possible to question the imposition of undemocratic regulations. This approach aims to interrupt institutional practices imposed by educational institutions, as it advocates for the isonomy of students who are historically underrepresented in the school systems by promoting their access to the teaching and learning process in mathematics in an egalitarian way.

In this context, Rosa and Orey (2015a) affirm that there is a need to transform mathematics, from a tool of systemic oppression to an instrument of liberation that seeks to involve teachers and students in experimenting with the various forms of mathematical knowledge present in everyday life through a fair and humanized way.

According to Gutiérrez (2016), instead of rigidly following district mandates or implicitly imposing policies, mathematics educators can maintain a high ethical standard to make mathematical classes motivating and humane for students, teachers, and members of the school community.

This insubordinate process is creative because it is related to teaching practices that promote social justice, as it assumes that members of a community have equal rights and duties in all aspects of their social life. Similarly, it is necessary for educators to practice creative insubordination when negotiating norms and making institutional rules more flexible so that they can resolve phenomena, problems, and situations they may face in their daily school life (Gutiérrez, 2016).

However, it is important that they also look for loopholes in educational public policies, interpreting rules and norms in a way that allows them to defend the rights of their students (Gutiérrez, 2015). In this regard, educators' awareness of their own experiences related to social justice issues makes it possible to negotiate authoritarian regulations. This negotiation seeks to reflect on social inequalities, so that we can understand the specific difficulties of members of distinct cultural groups that make up school institutions.

This approach aims to implement actions that can creatively combat these regulations (Rosa & Orey, 2015a, b). Professionals also become insubordinate and creative when their values are congruent with actions taken in the development of their daily administrative and teaching functions, as they focus to meet the demands evoked by the members of the school community. It also the intent to help mathematics educators to develop pedagogical practices that seek to include students in the use of mathematical knowledge from their daily lives (Gutiérrez, 2016).

For Rosa and Orey (2017a), ethnomodelling is an alternative pedagogical action that enables us to achieve this goal. It is important to state here that mathematics educators can become insubordinate and creative by protecting students from educational public policies unrelated to the interests of the school community through the creation of networks with professionals who share the same emancipatory visions of education.

In this way, Orey (2015) argues about the need for these professionals to question the *status quo* that mathematics has acquired in contemporary society, as they seek to counterbalance power relations that operate in diverse school systems. Regarding to ethnomodelling, Rosa and Orey (2017b) state that creative insubordination questions the nature of mathematics by highlighting its humanity and uncertainty and by placing students at the center of the educational process, which enables their active participation in the construction of mathematical knowledge and challenge *deficit narratives*[2] with respect to students from minority groups.

This perspective aims to reduce prejudice, harm, and inequality arising from the disconnects between mathematical knowledge practiced in school and academic contexts and its practical use in everyday life, as it aims to pursue social justice (D'Ambrosio, 2007). Through ethnomodelling, creative insubordination allows us to question the typical rules and norms presented as absolute truths in the teaching and learning process in mathematics by adding humanistic features in the development of mathematical ideas, procedures, and practices.

2.3 Responsible Subversion as Political Issues

The results of the study conducted by Hutchinson (1990) show that responsible subversion emerged as a socio-psychological process that enabled a group of nurses to bend the rules for the benefit of their patients. The behavior of these nurses was considered responsible, because they used their best judgment to decide when and how to make these rules more flexible.

For example, the group of nurses was held responsible, even though their actions were considered subversive, as they violated medical orders and hospital policies. Thus, these professionals acted responsibly, when consciously planning the best

[2]Deficit narrative is related to the myth of genetic determinism promoted by the dominant class that provides a justification for the performance of students from minority cultural groups by promoting the reduction of expectations that parents, teachers, and school leaders have in relation to this parcel of the school population (Rosa, 2010). It argues that inequalities are caused by deficiencies in disenfranchised members of minority groups and communities. In educational systems is the belief that the education gap is caused by students' shortcomings and not by systematic injustices (Valencia, 2010).

decision-making for their patients, but their behavior was described as subversive, as they frequently inflected state laws related to nursing practices (Hutchinson, 1990).

The political characteristic of this approach is related to the fact that this group of nurses was frustrated with public health policies that affected their nursing practices, especially, when they were developed by individuals with limited knowledge and/or limited experience in health (Kunaviktikul, 2014). Thus, these professionals flexed public policies that they considered oppressive in relation to health issues as well as marginalized their patients in relation to their common well-being (Giddings, 2005).

However, this group of nurses also faced a dilemma when they found that, according to their own practices, some medical orders were not considered the best course of action for treating their patients. This tension was related to the adjustment of established rules was formally and informally moralized among this group of professionals (Hutchinson, 1990). Indeed, these nurses developed these subversions in moral and responsible terms, as there was a flexion of norms to better serve the interests of their patients.

For Hutchinson (1990), the moralization of norms and regulations exhibited by this group of clinical nurses enabled the confrontation of their experiences with a conflicting set of institutional expectations, which presented a dilemma between following the doctors' instructions or caring for the patients to better meet their medical needs. According to Beauboeuf-Lafontant (1999), the refusal of this *status quo* is an important political decision to maintain an ethical and moral sense, and it is necessary to raise awareness about the motivation that directs these professionals towards conscious decisions-making.

This fact explained the efforts these professionals made to ensure that the treatment they considered appropriate was officially sanctioned by medical deferences (Hutchinson, 1990). In this context, responsible subversion was associated with the challenges of a dominant paradigm, as well as the act of transgressing social and organizational mores traditionally accepted in society. However, politically, this established view can be challenged through a responsible subversion that is embedded in the current social order (Bloom & White, 2016).

In the field of mathematics education, D'Ambrosio and Lopes (2015) demonstrated how subversion refers to the practices of educators who, responsibly and with discernment, oppose non-pedagogical prescriptions, educational bureaucracy, and the demands of public policies that do not benefit teachers, students, and the school community. Therefore, the concept of responsible subversion also refers to the oppositional actions taken in relation to institutional norms and rules that are not committed to the educational needs of the school population.

Responsible subversion, then, seeks to combat privilege and authority that has been accorded to the traditional academic mathematical discourse. The recognition of this challenge enables the understanding of how these elements of domination influence the distribution of power in modern society (Fitzmons, 2003). This subversive action enables students to become active participants of the educational process, through a political direction that can enable full access to their rights, as well as their full participation in the citizenry with responsibility (Rosa, 2010).

This contesting action is a challenge to the established authority, as it opposes to the inclusion of members of distinct cultural groups in the decision-making process by applying discriminatory public policies. Regarding to this political aspect of responsible subversion, it is emphasized that, for D'Ambrosio (2007), mathematics education is a tool for the development of individual, national, and global well-being. In this sense, the teaching and learning process in mathematics is a pedagogical action with sociocultural and political implications.

Therefore, it is necessary to highlight that the advances of this political proposition show that the quality of mathematics education must be accessible to all students, and not just to a reduced privileged group of students (D'Ambrosio, 2007). Thus, in this act of responsible subversion, mathematics educators aim to break institutional rules that seek to minimize the *sonority of the stigmas* of discrimination frequently experienced in the school system by students from minority groups.

As proposed by Rosa and Orey (2017b), mathematics educators can be considered as subversive and responsible professionals, when they develop actions to design creative pedagogical alternatives, such as ethnomodelling, that enable the active participation of students in the teaching and learning process in mathematics. In this regard, responsible subversion is a political tool that values the mathematical practices developed by members of distinct cultural groups that compose the school community.

With respect to ethnomodelling, responsible subversion can be considered as a political instrument that seeks to combat the dehumanizing effects of the curricular bureaucratic authority. Thus, Haynes and Licata (1995) state that this subversion aims to responsibly guarantee that curricular bureaucracies do not represent a disservice to students because public policies and institutional procedures are often disconnected from the school community.

This context enables the use of responsible subversion to conduct research in ethnomodelling that aims to include the cultural aspects of mathematics in the modelling process (Rosa & Orey, 2015b). This political decision-making perspective is one of the main components of responsible subversion that fights against the inhuman effects of bureaucratic authority (Haynes & Licata, 1995), which can be triggered in these investigations.

In this regard, Rosa and Orey (2017a) state that responsible subversion in ethnomodelling can be perceived in two ways:

1. The assertion that western mathematics does not consist of a superior body of knowledge, and it is necessary to show that this kind of knowledge is not more advanced than diverse locally developed mathematical ideas, procedures, and practices that are rooted in a system of distinct values.
2. The refusal to label the perception of western mathematics as a set of neutral knowledge and, therefore, unique and hegemonic that promotes the ideology of certainty.

In this context, Svačinová (2014) affirms that these two ways might be used to identify the political level of ethnomodelling, as they aim to combat the *primitivism*[3] that is linked to the development of local mathematical practices, as well as to reject the hegemony of western mathematics. Hence, Rosa and Orey (2017a), emphasize that an important characteristic of the responsible subversion of ethnomodelling is related to the fight against colonialism that takes a political position against the dominance of western school/academic mathematics.

Consequently, ethnomodelling values *cultural dynamism*[4] between different conceptions of mathematical knowledge by recognizing the importance of cultural relativism, which aims to prevent the hegemony of this knowledge over those developed in other cultural contexts (Rosa & Orey, 2017a).

According to the assumptions of responsible subversion, it is necessary that mathematics educators become aware of when, how, and why they should act against established procedures and unfair guidelines that do a disservice to the members of the school community (D'Ambrosio & Lopes, 2015).

Therefore, it is important that educators responsibly subvert the rules available in the educational system in order to promote the development of students as critical and reflective citizens by enabling their access to civil, social, political, and social rights (Rosa & Orey, 2015a).

In this regard, to responsibly subvert education means to expand and improve the educational system, to protect educators, teachers, students, and members of the school community from the imposed norms and regulations, as well as to understand the biases that prevent these individuals to access their rights, which are routinely denied to them. It also creates a place where new policies can immerge in relation to social justice and democracy.

2.4 Positive Deviance as a Sociocultural Issue

Initially, the concept of positive deviance emerged in the literature conducting nutrition research conducted in the 1960s. In these investigations, researchers confirmed the importance of using information collected from families so that they could plan alternative nutritional programs that were in line with their local practices (Wishik & Van Der Vynckt, 1976). This concept was then refined by the

---

[3]There is a need to combat cultural primitivism that seeks to devalue locally developed mathematical ideas and procedures, which seem to lack technological potential and advanced scientific and mathematical knowledge. Hence, it is important to show that members of distinct cultural groups also develop sophisticated mathematical practices so that they can solve problem and situations they face in their daily lives (Eglash, 2000; Rosa & Orey, 2005).

[4]In the process of cultural dynamism, members of distinct cultural groups develop active interactional processes that are in an ongoing negotiation between the local and the global mathematical, scientific, and technological knowledges in a dialogical manner through the development of the dynamic of the encounters of diverse cultures (D'Ambrosio, 2007).

investigators Zeitlin et al. (1990) who observed that, despite poverty in certain communities, some poor families had well-nourished and healthy children although living in communities with a prominent level of malnutrition.

These professionals advocated the use of this concept to address issues of child malnutrition in relation specific communities through the identification of local practices that enabled the nutrition of children with the aim of amplifying them to other contexts. Subsequently, Sternin et al. (1998) used the concept of positive deviance to combat Vietnamese child malnutrition and to identify healthy children among malnourished peers, their parents and relatives, considered as *positive deviants*,[5] managed to deviate from the normal course of action to save them from malnutrition.

This approach encouraged these researchers to identify individuals who solved malnutrition problems through locally developed practices by replicating them to create contextualized solutions in their own communities. Thus, health service users considered these professionals as defenders of social inclusion, because they sought to improve the social conditions of the citizens and, also, as active participants in society through the use of locally developed practices. These positively deviant actions were frequently undertaken by the flexibilization of norms and by breaking the rules of clinical practices (Gary, 2012).

In this direction, Spreitzer and Sonenshein (2004) argue that positive deviance aims to provide a conceptual framework that helps us to understand, identify, and explain the behaviors that oppose standardization and regulations. Therefore, Gary (2012) states that positive deviance is used in different fields of knowledge to describe actions that deviate from pre-established norms and standards in a positive direction, as decisions are made to improve the solution of problems based on local practices.

According to this context, Vardi and Wiener (1996) state that a deviant and positive behavior focuses on voluntary flexibilization of social norms and standardized behaviors that threaten the well-being of the members of distinct cultural groups.

It is necessary to state that, Dodge (1985) affirms that this conceptualization is used in business, administration, management, sociology, nursing, criminology, health, healthcare, and organizational behavior. Positive deviance is also used to combat chronic societal problems, such as child malnutrition, female genital mutilation, sex trafficking, and problems related to child health and hospital infections, as well as being considered a technique to solve organizational and institutional problems and issues (Lloyd, 2011).

However, we emphasize that no consistent definition of positive deviance for contexts related to education or mathematics education. In this sense, the viability of

[5]Positive deviants are individuals who are focused, persistent, and optimistic in their pursuit of better ways to help members of their own communities. They seek for social change based on their local observations, as well as they are successful in applying strategies that enable them to create and find solutions to solve problems in their own contexts (Bloch, 2001). Similarly, teachers and educators can be considered as positive deviants

the conceptual notion of positive deviance in the arena of educational practices need to be investigated. Consequently, Rosa and Orey (2017b) state that understanding of the concept of positive deviance can contribute to the development of innovative mathematical ideas, procedures, and practices that are linked to the sociocultural context of students.

Therefore, the concept of positive deviance may be related to the ethnomodelling process, with regard to the use of local procedures and techniques to solve problems and situations faced by members of distinct cultural groups in their daily lives. It is important to point out that, in different sociocultural contexts, the characteristics associated with the concept of positive deviance (Tarantino, 2005; Gary, 2012) can be adapted to educational contexts by showing the flexibility of rules and norms through four important characteristics that are detailed below.

2.4.1 Positive Deviance Is Intentional and Honorable

Positive deviance focuses on achieving the common good, which replaces subservience to the norms and rules imposed by institutions (Bloch, 2001). As a behavior to be achieved, positive deviance has honorable intentions, regardless of the results obtained from its actions (Spreitzer & Sonenshein, 2004).

These intentionally deviant behaviors include honorable behaviors that accommodate the norms of the members of the reference group, as they are socially beneficial (Warren, 2003) for members from diverse cultures.

For example, Bloch (2001) argues that when positive deviants, such as nutritionists, nurses, school managers, teachers, and educators, realize that a given procedure may fail, these professionals are motivated to seek an innovative, intentional way in order to solve it, because they aim for a differentiated and successful and honorable service for the portion of the population they serve in their daily lives.

2.4.2 Positive Deviance Differs from Established Regulations

A central theme of positive deviance is related to its proposal to work with actions that differ from current regulations and norms (Dehler & Welsh, 1998). Most importantly, this concept is opposed to formal authoritarian systems, as it describes behaviors that deviate from commonly accepted and established norms. According to Spreitzer and Sonenshein (2004), the normative formulation of positive deviance is a behavior that significantly differs from the rules of the members of a group considered as a reference (Spreitzer & Sonenshein, 2004).

Thus, Warren (2003) highlights that there is a need for positive deviants to resist social pressures that encourage conformance to norms and regulations in order to obey the imposed rules by the system. In this regard, Bloch (2001) states that these deviants are not afraid to deviate from their path to follow a different direction, with the objective of finding a fairer way for them to play their transformative role in society more broadly.

2.4.3 Positive Deviance Contains Innovative, Creative and Adaptable Elements

Positive deviance is a source of adaptive capacity in organizational transformation (Dehler & Welsh, 1998) of members of diverse communities. This *pro-social behavior*[6] uses creativity to diverge from current norms and rules. Thus, educators considered positive deviants look for creative solutions to solve the educational problems encountered in their daily teaching practices by making the regulations imposed by the school system more flexible.

Innovation requires a departure from the accepted rules by society as innovative thinking that involves the creation and development of innovative ideas that aim at the well-being of the population (Appelbaum et al., 2007). In this direction, Lindberg and Clancy (2010) state that this deviant behavior is creative and adaptable, although it is often perceived as a resource that aims to use alternative solutions that deviate from established norms for solving the problems faced in the activities performed daily by the members of distinct cultural groups.

2.4.4 Positive Deviance Involves Risks

Positive deviance involves risks for deviants when these individuals positively deviate from the norms and regulations imposed by society (Appelbaum et al., 2007). For example, Stewart et al. (2004) describe how rules and implicit expectations are at play when nurses decide whether to extend the limits of the scope of their practices. Thus, most of these professionals practice their deviant actions with caution, as they also consider the maintenance and preservation of their work permits.

According to Kramer and Schmalenberg (2008), positive deviants are aware that they can be held responsible if there are negative outcomes for the individuals they serve, such as customers, patients, and students. However, these nurses accept these risks so that they can improve their living conditions. Consequently, these professionals also identify the risks associated with this process, such as, for example, the loss of influence, reputation, and sociopolitical position.

---

[6]Prosocial behavior or the intention to benefit *others* (Helliwell & Putnam, 2004) is a social behavior that benefits other people or society as a whole· such as helping, sharing, donating, co-operating, and volunteering (Brief & Motowidlo, 1986).

This decision-making process may contain multiple conditions related to uncertainty that provides threats to positive deviants. Even though these professionals realize that when they go beyond the limits of hospital rules and protocols, they are also taking risks, but they are not overly concerned about these risks (Kramer & Schmalenberg, 2008).

On the other hand, as the positive deviance may acquire a high emotional charge, as it evokes interruptions in the existing systems, this behavior is capable of causing administrative disapproval. So, it is important to emphasize that actions against these deviants can be decided by external professionals who are in charge of applying rules and regulations as punishment. This context shows that, for Dehler and Welsh (1998), the label of deviant may inappropriately imply that this behavior is harmful and needs to be fought.

In line with this theoretical discussion, positive deviance can be considered a problem-solving process through the use of diverse and differentiated solving problems practices (Lindberg & Clancy, 2010) developed in specific contexts. For Rosa and Orey (2017b), positive deviants discover solutions to problems and solving them by using locally developed mathematical strategies, procedures, and techniques by legitimizing their actions to solve everyday problems.

For example, Lindberg and Clancy (2010) state that positive deviance recognizes the experience acquired daily, considering it relevant to the process of solving problems faced in daily life. In this context, Rosa and Orey (2015a) emphasize that these both deviant and positive actions are also related to the assumptions of ethnomodelling.

2.5 Conceptualizing Positive Deviance in Ethnomodelling

Because it started an epistemological disturbance that may have caused a review of the academic mathematical knowledge system, deviance triggered by ethnomodelling is positive when it contributes to facing taboos that suggest the mathematics being studied as a universal field of study, without cultural roots and with a lack of traditions (Rosa & Orey, 2015b).

In this context, ethnomodelling can cause an interruption in the existing order that prevails in mathematics education, as it encourages the study of ideas and procedures, as well as the use of local mathematical practices developed in distinct cultural contexts, which are in accordance with the *tacit knowledge*[7] and perceptions of its members (Rosa & Orey, 2017a).

In accordance with this context, the emergence of ethnomodelling can also be interpreted as a reaction to forms and aspects of cultural imperialism that spread

[7]This kind of knowledge is related to the ways in which members of distinct cultural groups appropriate mathematical knowledge by relating them to their own experiences, beliefs, and cultural values (Rosa & Orey, 2017a).

internationally from European colonization. This may be linked to concepts of positive deviance, as it is related to concepts of emancipation and local mathematical knowledge by legitimizing it in its own sociocultural roots (Rosa & Orey, 2017b).

Consequently, ethnomodelling promotes a sociocultural debate that deviates from the supposedly superior *status quo* of school/academic mathematics in relation to locally developed mathematical procedures and practices, as it documents alternative forms of dissemination of mathematical thinking and reasoning (Rosa & Orey, 2017b). This approach combats ethnocentric perspectives that perpetuate the inaccessibility of members of cultural minorities to civic, political, social and cultural rights (Turner, 2016).

According to Svačinová (2014), investigating the nature of mathematical knowledge and objects supports a diversity ways to understand them in their own sociocultural contexts. In this sense, Rosa and Orey (2017b) state that the emergence of new opportunities for discussing the nature of mathematical knowledge shows that the *deviation*[8] triggered by ethnomodelling is positive because it seeks to combat the existing hegemony in mathematical modelling processes developed by school environment and academic context, while promoting the inclusion of cultural elements in this process.

In relation to this deviant behavior, the positive development of educational actions aim at valuing diverse forms of mathematical knowledge as it promotes cultural dynamism. Members of distinct cultural groups develop a local interpretation of their own culture (*emic approach*[9]), as opposed to the global analysis of external observers (*etic approach*[10]) about their beliefs, behaviors, and traditions (Rosa & Orey, 2017a).

In this perspective, positive deviance is a set of non-prescriptive practices or strategies (Fielding et al., 2006; Pascale et al., 2010) that are contextualized in the environment in which they are developed, as they aim at encouraging a search for solutions to problems faced by members of distinct cultural groups in their daily lives. In this process, mathematics is considered a social construct culturally rooted in its own traditions.

Therefore, positive deviance is an approach based on the premise that it is in the communities themselves, through collective participation, that solutions to daily

[8]Deviation is the act of deviating or a wandering from the common way and from an established rule. It is a departure, as from the right course or the path of duty, which is a noticeable or marked departure from accepted norms and rules (Rosa & Orey, 2015a).

[9]The emic approach is associated with the point of view of the internal members (insiders) of the cultural groups, as they are observers from within their own culture. Emic knowledge is obtained through observation, diffusion, and dissemination of locally developed ideas, procedures, and mathematical practices (Rosa & Orey, 2017a).

[10]The etic approach is related to the viewpoint of researchers, teachers, and educators (outsiders) in relation to customs, beliefs, and the development of mathematical and scientific knowledge of members of distinct cultural groups. They are the outsiders or the outside observers who develop concepts, theories and hypotheses about local knowledge that is considered important and meaningful to insiders (Rosa & Orey, 2017a).

problems are found. Thus, this approach aims that, through a self-discovery process, the identification and optimization of resources and existing solutions available in communities are used to solve their own problems (Sternin et al., 1998).

Similarly, Damazio (2004) suggests that it is in the community itself that schools can find the didactic elements necessary for the development of its curricula. When referring to the mathematics curriculum, Chieus (2004) emphasizes that the pedagogical work directed to diverse cultural perspectives enables a comprehensive analysis of the school context, as "pedagogical practices transcend the physical space and start to welcome the knowledge and actions present in every sociocultural context of the students" (p. 186).

Therefore, active participation in this process allows members of distinct cultural groups to identify and articulate these solutions by applying them in their daily lives with the objective of seeking to improve the quality of life of these members (Masterson & Swanson, 2000). This pedagogical action aims to make mathematics a living knowledge that works through real situations, using a critical and reflective analysis of everyday phenomena. In our view, this objective can also be achieved by applying the assumptions of ethnomodelling in the daily work of the members of these groups (D'Ambrosio, 1990).

Positive deviance enables local solutions to be developed, applied, and used to solve common problems by encompassing innovative techniques locally developed by members of distinct cultural groups. These techniques are also related to the flexibility of norms, rules, and regulations imposed by educational institutions, as they make it possible to change the behavior of these members (Rosa & Orey, 2015a).

For example, almost five decades ago, the results of the study conducted by Alinsky (1972) showed that changes in beliefs and behaviors require an unfreezing of perceptions held in relation to locally developed practices. Positive deviance involves members of diverse communities in discovering (unfreezing) successful alternatives (practices) with respect to local wisdom that, according to Rosa and Orey (2017a), may be related to the development of ideas, procedures, and mathematical practices linked to the sociocultural context of these members.

Similarly, Gerdes (2012) states that it is important to acknowledge that there is mathematical knowledge frozen in everyday practices and artifacts. For example, craftsmen who discovered a certain technique developed their own tacit knowledge, as they used hidden mathematical thinking to solve the problems they face in their daily lives. Thus, the unfreezing of this knowledge enables the (re)discovery of local mathematical procedures and techniques, which reveals the development of the mathematical potential of these members.

According to Gerdes (2012), the unfreezing of this mathematical knowledge serves as a starting point for valuing local mathematical practices in classrooms, while, at the same time, it raises awareness among researchers, teachers, and educators in reflecting on the relationship between mathematical thinking/reasoning and the artifacts produced by using this knowledge, and also between the *knowing* and *doing* mathematics regarding to the use of technological resources.

This context allows members of distinct cultural groups to deviate from society's expectations in order to explore successful alternative procedures that are related to the development of local cultural norms, rules, regulations, beliefs, and perceptions of these members (Masterson & Swanson, 2000). Therefore, researchers, teachers, and educators can be considered as positive deviants when they design alternatives methods that can achieve satisfactory results for the common good of the members of the school communities (Rosa & Orey, 2017b).

This action is in opposition and, often, poses a challenge to the established authority that opposes well-being of students through discriminatory public policies (Rosa & Orey, 2017b). Therefore, these teachers and educators can be forced to deviate from the normative rules so that they can meet the educational needs of their students. This approach is attainable through the development of positive actions that contrast to a passive acceptance of bureaucratic regulations imposed by the educational system. In this regard, Gary (2012) states that this process involves an intentional act of bending the rules to serve the common good of members of the school community and society.

Educators are positive deviants when they push institutional bureaucratic boundaries to achieve the intended results in their pedagogical practices in the classrooms. This deviant action shows the importance of these professionals leaving their own *epistemological cages*, which is a metaphor developed by D'Ambrosio (2011).

In cages, birds breed and reproduce, yet they only see what the bars allows them to observe, they only fly in the space delimited internally, they only feed on the products they find in this environment and they only communicate in a known language. The become accustomed to the world in which they live, rarely questioning it. The birds may not even know the color in which the cage is painted on its outside (D'Ambrosio, 2011).

Abandoning these cages is a challenging task, as they offer several benefits, such as recognition by peers, as well as the guarantee and maintenance of employment and promotions. Yet, the price of these benefits is high, as the bars prevent professionals from contacting and getting to know the sociocultural reality of members of other cultures, as well as making it impossible to inspire innovative paradigms for the development of their creativity (Rosa, 2019).

For D'Ambrosio (2011), frequently, researchers, teachers, and educators are imprisoned in their own cages, which make it difficult for them to become aware of the existence of other cultures, epistemologies, worldviews, perspectives, and cosmologies. In this context, remaining caged can be convenient and comfortable because it avoids contact between these professionals and members of *other* cultures.

However, there are professionals who are fearless (deviant) in relation to questioning their cages, which exist as a result of the organization of their mathematical backgrounds, training, thinking and reasoning (D'Ambrosio, 2011). It is important that these professionals discuss current norms and rules in order to make them more flexible in the teaching and learning process in mathematics.

Thus, the development of the modelling process is a pedagogical action that proposes discussion about repression by encouraging researchers, teachers, and educators to help in the recovery of student cultural dignity (Rosa, 2019). In this context, the concept of positive deviance for ethnomodelling is related to intentional and honorable behaviors that deviate from the established norms and differ from imposed rules.

This field of study provides the development of elements of innovation, creativity, and adaptability that may, however, involve risks for its deviants. Thus, the assumptions of ethnomodelling seek to bend the bureaucratic expectations of academic mathematical modelling process that aims of valuing the diverse ways in which the production of this knowledge is triggered in other cultural contexts by using the connection between ethnomathematics and modelling.

2.6 Final Considerations

The breadth of the concepts of creative insubordination, responsible subversion and positive deviance encompasses innovative solutions in relation to pedagogical action of mathematics education, including ethnomodelling, as it aims to confront the belief that still persists in society that mathematics is knowledge detached from culture. The insubordination unleashed by mathematics educators is creative because it evokes a disturbance that favors a sense of positive deviance from standardized practices and a review of regulations that, for these professionals, can be considered as responsibly subversive.

The main objective of this chapter was to discuss the concept of positive deviance in ethnomodelling by exploring and identifying the essence of this concept in the educational context, which is related to the development of teaching and learning strategies that encourage the exploration of local mathematical ideas, procedures, and practices. Thus, positive deviance can be considered as a sociocultural change, based on the observation that in distinct cultural groups there are members who develop successful procedures, techniques, and strategies that create local solutions to solve the problems faced in their daily lives.

We emphasize that positive deviance was used in nutrition, as it aimed to understand how children grow and develop in poor families and communities, in which malnutrition is a constant certainty. These families developed culturally appropriate practices that were successful so that they could nurture and care for their children, despite of their poverty and high-risk environment in which they live. Similarly, ethnomodelling involves the study of ideas, procedures, and mathematical practices that are developed in distinct cultural contexts that can be used in the pedagogical action of modelling through its connection with ethnomathematics. This context allowed researchers, teachers, and educators to challenge the traditional mathematical thinking and reasoning that is still prevalent in educational systems.

Historically, mathematical knowledge has taken different forms in diverse cultures through the development of techniques and procedures that, many times, were in opposition to the predominant mathematical system or to the rules commonly legitimized by the school context, the academic paradigms about the notions of ethnomodelling show that the development of this process is culturally rooted. Hence, a systematic study of ethnomodelling includes the development of skills that help the observation of phenomena based on distinct cultural contexts by determining innovative points of view on the teaching and learning process in mathematics, which aim to improve the *cultural sensitivity*[11] of the members of distinct cultures in this process.

The positive deviance in ethnomodelling inform how the teaching and learning process in mathematics can be transformed by valuing locally developed mathematical procedures and practices, as well as by the active participation of these members in a *glocalized society*.[12] So, there is a need for researchers, teachers, and educators to flex the Eurocentric perspective of mathematical knowledge to meet the educational needs of their students. We, therefore, propose that mathematics educators become positive deviants, so that they can guide their pedagogical actions in this direction (Rosa & Orey, 2017b). For example, the ongoing, seemingly never-ending effect of prejudice enhances school problems and perpetuates the social exclusion of students from institutional and community services.

Acts of positive deviance by educators can reduce stigma and prejudice, which can be caused by the rules imposed by inequities inside the educational system, which prevents social justice from being fully achieved. Social norms differ across communities, societies, and cultures. Thus, a certain positive act or behavior can be perceived as deviant and receive sanctions or punishments in a certain context and yet it might be understood as a traditional behavior in another context. Furthermore, as the understanding of social norms and rules by members of a society changes over time, so does the collective perception of positive deviance.

Finally, we conclude that the concept of positive deviance is useful because it offers mathematics educators a basis for empowerment and a decision-making process when actions and behaviors that are considered normal and expected by the *status quo* collide with what may be of benefit provided to students.

This concept is necessary to the development for the inclusion of local mathematical procedures and practices based on ethnomodelling, in order to assist educators in meeting cognitive, cultural, social, and pedagogical needs of participants. Certainly, a sense of positive deviance in this field of study shows the need to use a cultural perspective in the mathematics curriculum for the twenty-first century.

---

[11] Cultural sensitivity is related to the ability of members of distinct cultural groups to become aware of the differences and similarities between cultures, without attributing values or imposing rules and norms, such as positive or negative, better or worse, and right or wrong to their members (Rosa, 2010).

[12] In a glocalized society, there is a predominance of the acceleration and intensification of the interactional process between local and global knowledge developed by members of distinct cultural groups through cultural dynamism that is triggered in this dialogic process (Rosa & Orey, 2017b).

References

Akarowhe, K. (2018). Effects and remedies to cultural shock on the adolescent students. *Sociology International Journal, 2*(4), 306–309.

Alinsky, S. D. (1972). *Rules for radicals*. Random House.

Appelbaum, S. H., Iaconi, G. D., & Matousek, A. (2007). Positive and negative deviant workplace behaviors: Causes, impacts, and solutions. *Corporate Governance, 7*(5), 586–598.

Beauboeuf-Lafontant, T. (1999). A movement against and beyond boundaries: Politically relevant teaching among African American teachers. *Teachers College Record, 100*(4), 702–718.

Bishop, A. J. (1990). Western mathematics: The secret weapon of cultural imperialism. *Race & Class, 32*(2), 51–65.

Bloch, S. (2001). Positive deviants and their power on transformational leadership. *Journal of Change Management, 1*(3), 273–279.

Bloom, P. N., & White, P. J. (2016). The moral work of subversion. *Human Relations, 69*(1), 5–31.

Brief, A. P., & Motowidlo, S. J. (1986). Prosocial organizational behaviors. *The Academy of Management Review, 1*(4), 710–725.

Chieus, G. J. (2004). Etnomatemática: Reflexões sobre a prática docente. In J. P. M. Ribeiro, M. C. S. Domite, & R. Ferreira (Eds.), *Etnomatemática: Papel, valor e significado* (pp. 185–202). ZOUK.

Chubb, J. E., & Moe, T. M. (1990). *Politics, markets, and America's schools*. Brookings Institute.

Crowson, R. L., & Morris, V. C. (1982). The principal's role in organizational goal-attainment: Discretionary management at the school site level. In *Proceedings of the Annual Meeting of the American Educational Research Association* (pp. 19–23). AERA.

Crowson, R. L., & Morris, V. C. (1985). Administrative control in large-city school systems: An investigation in Chicago. *Educational Administration Quarterly, 21*, 51–70.

D'Ambrosio, U. (1990). *Etnomatemática*. Editora Ática.

D'Ambrosio, U. (2007). Peace, social justice and ethnomathematics. *The Montana Enthusiast Monograph, 1*, 25–34.

D'Ambrosio, U. (2011). A transdisciplinaridade como uma resposta à sustentabilidade. *Revista Terceiro Incluído, 1*(1), 1–13.

D'Ambrosio, U., & D'Ambrosio, B. S. (2013). The role of ethnomathematics in curricular leadership in mathematics education. *Journal of Mathematics Education at Teachers College, 4*, 19–25.

D'Ambrosio, B. S., & Lopes, C. E. (2015). *Creative insubordination in Brazilian mathematics education research*. Lulu Press.

Damazio, A. (2004). *Especifidades conceituais de matemática da atividade extrativa do carvão* (Coleção introdução à etnomatemática) (Vol. 1). Universidade Federal do Rio Grande do Norte.

Dehler, G. E., & Welsh, R. W. (1998). *Problematizing deviance in contemporary organizations: A critical perspective*. JAI Press.

Dodge, D. (1985). The over-negativized conceptualization of deviance: Programmatic exploration. *Behavior, 6*(1), 17–37.

Eglash, R. (2000). Anthropological perspectives on ethnomathematics. In H. Selin (Ed.), *Mathematics across cultures: The history of non-western mathematics* (pp. 12–22). Kluwer Academic Publishers.

Fielding, K. S., Hogg, M. A., & Annandale, N. (2006). Reactions to positive deviance: Social identity and attributional dimensions. *Group Processes & Intergroup Relations, 9*, 199–218.

Fitzmons, G. A. (2003). What counts as mathematics? Technologies of power in adult and vocational education. *ZDM, 35*(6), 323–324.

Gary, J. C. (2012). *The use of positive deviance to deliver patient-centered care*. Nurses Theses and Dissertations, Paper 28, The Graduate School, School of Nursing, The University of Texas, Tyler, TX.

Gary, J. C. (2013). Exploring the concept and use of positive deviance in nursing. *American Journal of Nursing, 113*(8), 26–34.

Gerdes, P. (2012). *Etnomatemática: Cultura, matemática, educação* (Coletânea de Textos (1979–1991)). Lulu Press.
Giddings, L. S. (2005). Health disparities, social injustice and the culture of nursing. *Nursing Research, 54*(5), 304–312.
Gutiérrez, R. (2015). Risky business: Mathematics teachers using creative insubordination. In *Proceedings of the Annual Conference of Psychology of Mathematics Education North America* (pp. 679–686). PME.
Gutiérrez, R. (2016). Strategies for creative insubordination in mathematics teaching. *Teaching for Excellence and Equity in Mathematics, 7*(1), 52–60.
Haynes, E. A., & Licata, J. W. (1995). Creative insubordination of school principals and the legitimacy of the justifiable. *Journal of Educational Administration, 33*(4), 21–35.
Helliwell, J. F., & Putnam, R. D. (2004). The social context of well-being. *Philosophical Transactions of the Royal Society London B Biological Science, 359*(1449), 1435–1446.
Hutchinson, S. A. (1990). Responsible subversion: A study of rule-bending among nurses. *Scholarly Inquiry for Nursing Practice, 4*(1), 3–17.
Kramer, M., & Schmalenberg, C. (2008). Confirmation of a healthy work environment. *Critical Care Nurse, 28*(800), 56–63.
Kunaviktikul, W. (2014). Moving towards the greater involvement of nurses in policy development. *International Nursing Review, 61*(1), 1–2.
Lindberg, C. C., & Clancy, T. R. (2010). Positive deviance: An elegant solution to a complex problem. *Journal of Nursing Administration, 40*(4), 150–153.
Lloyd, J. C. (2011). For clues to HAI prevention, seek out positive deviance. *Healthcare Purchasing News, 35*(1), 46–47.
Lyman, L. L., Ashby, D. E., & Tripses, J. S. (2005). *Leaders who dare: Pushing the boundaries.* Rowman & Littlefield Education.
Macionis, J., & Gerber, L. (2010). *Sociology* (7th ed.). Pearson.
Marzano, R. J., Waters, T., & McNulty, B. A. (2005). *School leadership that works: From research to results*. Association for Supervision and Curriculum Development.
Masterson, J. M., & Swanson, J. H. (2000). *Female genital cutting: Breaking the silence, enabling change*. International Center for Research on Women and The Centre for Development and Population Activities.
Orey, D. C. (2015). Insubordinações criativas relacionadas com a ação pedagógica do programa etnomatemática. In B. S. D'Ambrosio & C. E. Lopes (Eds.), *Ousadia criativa nas práticas de educadores matemáticos* (pp. 247–268). Mercado de Letras.
Pascale, R., Sternin, J., & Sternin, M. (2010). *The power of positive deviance: How unlikely innovators solve the world's toughest problems*. Harvard Business Press.
Rosa, M. (2010). *A mixed-methods study to understand the perceptions of high school leader about English language learners (ELL): The case of mathematics*. Dissertation (Doctorate), College of Education, California State University, Sacramento—CSUS, Sacramento, CA.
Rosa, M. (2019). Globalização e etnomatemática: Sobre o dinamismo dos encontros entre culturas. In *Anais do XIII Encontro Nacional de Educação Matemática—XIII ENEM* (pp. 1–13). SBEM.
Rosa, M., & Orey, D. C. (2005). Las raíces históricas del programa etnomatemáticas. *Relime, 8*(3), 363–377.
Rosa, M., & Orey, D. C. (2012). O campo de pesquisa em etnomodelagem: As abordagens êmica, ética e dialética. *Educação e Pesquisa, 38*(4), 865–879.
Rosa, M., & Orey, D. C. (2015a). Evidence of creative insubordination in the research of pedagogical action of ethnomathematics program. In B. S. D'Ambrosio & C. E. Lopes (Eds.), *Creative insubordination in Brazilian mathematics education research* (pp. 131–146). Lulu Press.
Rosa, M., & Orey, D. C. (2015b). Three approaches in the research field of ethnomodelling: Emic (local), etic (global), and dialogical (glocal). *Revista Latinoamericana de Etnomatemática, 8*(2), 364–380.

Rosa, M., & Orey, D. C. (2017a). *Etnomodelagem: A arte de traduzir práticas matemáticas locais*. Editora Livraria da Física.
Rosa, M., & Orey, D. C. (2017b). Concepts of positive deviance in pedagogical action in an ethnomathematics program. *REnCiMa, 8*(4), 62–79.
Rosa, M., & Orey, D. C. (2019). Aspectos de insubordinação criativa em etnomodelagem. *Educação Matemática em Revista, 24*(61), 6–25.
Spreitzer, G. M., & Sonenshein, S. (2004). Toward a construct definition of positive deviance. *American Behavioral Scientist, 47*(6), 828–847.
Sternin, M., Sternin, J., & Marsh, D. (1998). *Designing a community-based nutrition program using the hearth model and the positive deviance approach: A field guide*. Save the Children.
Stewart, J., Stansfield, K., & Tapp, D. (2004). Clinical nurses' understanding of autonomy. *Journal of Nursing Administration, 34*(10), 443–450.
Svačinová, I. (2014). Ethnomathematics: A political challenge to the philosophy of Mathematics. In L. Zámečník (Ed.), *The future of philosophy* (pp. 89–119). Univerzita Palackého v Olomouci.
Tarantino, D. P. (2005). Positive deviance as a tool for organizational change. *Physician Executive, 31*(5), 62–63.
Turner, J. (2016). Governing the domestic space of the traveller in the UK: 'Family', 'home' and the struggle over Dale Farm. *Citizenship Studies, 20*(2), 208–227.
Valencia, R. R. (2010). The construct of deficit thinking. In R. R. Valencia (Ed.), *Dismantling contemporary deficit thinking: Educational thought and practice* (pp. 1–18). Routledge.
Vardi, Y., & Wiener, Y. (1996). Misbehavior in organization: A motivational framework. *Organization Science, 7*(2), 151–165.
Warren, D. E. (2003). Constructive and destructive deviance in organizations. *Academy of Management Review, 28*(4), 622–632.
Wishik, S. M., & Van Der Vynckt, S. (1976). The use of nutritional 'positive deviants' to identify approaches for modification of dietary practices. *American Journal of Public Health, 66*, 38–42.
Zeitlin, M., Ghassemi, H., & Mansur, M. (1990). *Positive deviance in child nutrition: With emphasis on psychological and behavioural aspects and implications for development*. The United Nations University.

Chapter 3
Ethnomodelling as a Methodological Alternative to Basic Education: Perceptions of Members of a Research Group

Zulma Elizabete de Freitas Madruga

3.1 Introduction

The insertion of future teachers in research has to be done during their undergraduate coursework. The articulation between teaching, research, and extension tripod is a challenge assumed in the *Pedagogical Project of the Licentiate Degree in Mathematics* (*Projeto Pedagógico do Curso de Licenciatura em Matemática*) at the *Universidade Federal do Recôncavo da Bahia* (UFRB), in the municipality of Amargosa, state of Bahia, Brazil. With regard to research, actions are implemented to provide a space for training, which is essential for reflection and construction of the identity of the future teachers.

The development of actions aimed at research at UFRB have contributed to improving scientific thinking and the ability to generate new knowledge that helps to train students in the various aspects related to their personal and professional training. In this perspective, the Institutional Scientific Initiation Scholarship Program (PIBIC) stands out and it has helped to develop research activities in addition to contributing to the student's permanence at the University.

Considering these aspects, this chapter intends to share the actions carried out within a research group, formed primarily by students of the Licentiate Degree in Mathematics at UFRB in order to understand the perceptions of participants on Ethnomodelling, as a potential methodological alternative for Basic Education.

Z. E. d. F. Madruga (✉)
Universidade Federal do Recôncavo da Bahia, Amargosa, Bahia, Brazil
e-mail: betemadruga@ufrb.edu.br

M. Rosa et al. (eds.), *Mathematical Modelling Programs in Latin America*,
https://doi.org/10.1007/978-3-031-04271-3_3

3.2 The GEPTEMaC[1] (*Grupo de Estudos e Pesquisas Sobre Tendências da Educação Matemática e Cultura*): From the Beginning to Future Actions

The project entitled "Ethnomodelling and Problem Solving: possibilities for teaching and learning Mathematics in Basic Education", aims to plan, to develop, to experiment and reflect on the use of Ethnomodelling (Ethnomathematics combined with Mathematical Modelling) and *Problem Solving in Basic Education Mathematics* classes.

The execution of this project foresees 3 years, having started in April 2020. During the first year studies were conducted on the following trends in Mathematics Education: Ethnomathematics, Mathematical Modelling (Ethnomodelling) and Problem Solving.

The project team is made up of the coordinating professor; a collaborating professor from another Higher Education Institution (HEI); ten undergraduate mathematics students, two of whom are PIBIC scholarship holders (in their second year of work); two Master's degree students in Science and Mathematics Education; and seven teachers working in Basic Education (three masters, three specialists, and one graduate), from different regions of Recôncavo and Southern Bahia, Brazil.

Group meetings provide studies and research and are mainly moments for sharing knowledge. Meetings occur on *GoogleMeet*, synchronously; and on WhatsApp, in an asynchronous way, which enables the participation of members from different municipalities.

After a year of study and theoretical research on the trends listed in the project, this team decided to create the *Study and Research Group on Trends in Mathematics Education and Culture* (Grupo de Estudos e Pesquisas sobre Tendências da Educação Matemática e Cultura)—GEPTEMaC, registering it in the Directory of Groups in Brazil.

The various publications of theoretical research boosted the creation of the group. In a year and a half of study and research, 19 articles were produced and submitted to journals (7 published, 7 accepted for publication, and 5 submitted and waiting for the evaluation process); 6 full articles published in event proceedings, and 5 expanded abstracts and 9 abstracts published in event proceedings. In addition, 28 presentations by the group members were made at regional, national, and international scientific events.

Any production developed by group members is done through a process of collaboration, construction, and reconstruction, based on discussion with peers. It is believed that this fact is fundamental for the group's collective growth. Nine pedagogical proposals are currently being prepared on the themes of interest to the

[1] *Study and Research Group on Trends in Mathematics Education and Culture* that is registered in the *Directory of Research Groups in Brazil*. Available at: http://dgp.cnpq.br/dgp/espelhogrupo/720033. E-mail: geptemac@cfp.ufrb.edu.br

group: Mathematical Modelling, Ethnomodelling and Problem Solving, which will be published in an e-book that is under production.

The idea is that these proposals will be developed by members of the group and collaborators, with students of Basic Education, and will be later analyzed, and their results will be disseminated in the form of scientific publications.

The group's reflections, discussions, and productions are directed towards the study and search for teaching and learning strategies for Mathematics in Basic Education, supported by authors such as D'Ambrosio (2013) related to Ethnomathematics; Biembengut (2016) related to Mathematical Modelling in the perspective of Mathematics Education; Onuchic and Allevato (2014) related to Problem Solving; and Rosa and Orey (2017) related to Ethnomodelling.

The GEPTEMaC is understood as a space for initial and continuing education, which takes place through the sharing of ideas and collective and collaborative actions. On these aspects, Fiorentini (2013) highlights collective work as a starting point for overcoming and understanding numerous problems in practice, and as a means for professional training. The reflections carried out collectively favor the understanding of practices, the recognition of some lack of theoretical knowledge, among other possibilities that are presented when reflecting on the pedagogical actions and listening to the different points of view of other teachers.

It is worth highlighting the participation in the GEPTEMaC of students (undergraduate and graduate) and teachers of Basic Education and Higher Education, in the quest to break with the "culture of suspicion and misunderstanding between mathematics educators residing in the academic world and mathematics educators with action in the classrooms of the school world" (D'Ambrosio & D'Ambrosio, 2006, p. 79). According to this context, D'Ambrosio and D'Ambrosio (2006) argue that:

> Some research groups try to change this culture, inviting professors to join, as researchers, in research projects in the classroom. A collaborative relationship is established, and the Math teacher finds a voice and agency in the research group (...). Some researchers consider this practice a work of socialization of the mathematics teacher in the world of research. These teachers begin to build a role for research in their pedagogical practice, making research an essential element in their professional life (p. 79).

One of the GEPTEMaC's premises is to give equal voice to all members, there is no hierarchy, and group decisions are taken together, respecting the will of the majority. The encouragement of research and dissemination is recurrent in the group, as well as collaboration, as it is understood that in this way, the theoretical, professional, and academic growth of all involved takes place.

In addition, the results of the group's actions are developed through extension projects and partnerships with schools in the municipality and region, seeking to contribute to the qualification of both its members and other Basic Education teachers. It is understood that the creation of a Study and Research Group is important for the development of skills in students, based on collaboration and collectiveness, not only in the academic sphere, but also in the professional field.

In addition to discussions conducted among its members, the group carries out actions in the form of conversation circles and dialogue with guests who have experience in the field of Mathematics Education, striving for diversity and the exchange of knowledge and experiences among those involved.

The actions of GEPTEMaC deal with trends and methodological approaches in Mathematics Education: Ethnomodelling, Mathematical Modelling, and Problem Solving. It is noteworthy that some group participants discuss these topics in their research, such as Course Conclusion Papers (TCC) and Master's Degree Dissertations. This chapter specifically addresses the group's ideas, studies, and perceptions about Ethnomodelling.

3.3 About Ethnomodelling

Ethnomodelling focuses on the relations between Ethnomathematics and Mathematical Modelling. Ethnomathematics is understood as the art or technique of knowing, explaining and understanding different cultural contexts (D'Ambrosio, 2013). It is a natural, social, cultural, and imaginary (*ethno*) environments of explaining, learning, knowing, and dealing (*mathema*) with modes, styles, arts, and techniques (*tics*).

It is a program that aims to explain the processes of knowledge generation, organization and transmission in different cultural systems. It studies the relationships and connections between mathematical notions and other cultural elements, knowledge and mathematical know-how acquired in the development of a professional activity (D'Ambrosio, 2013).

It is understood that Mathematical Modelling (MM) enables the connection between representations and the world (Bassanezi, 2010), defined it as a dynamic process used to obtain and validate (mathematical) models. For Bassanezi (2010), modelling is a form of abstraction and generalization in order to predict trends. "Modelling essentially consists of the art of transforming reality situations into mathematical problems whose solutions must be interpreted in the usual language" (Bassanezi, 2010, p. 24).

Some researchers have published research that corroborates these relationships, such as: Madruga (2012), Albanese and Perales (2014), Biembengut (2016), Madruga and Biembengut (2016), Pradhan (2020), among others. However, the authors Rosa and Orey (2014, 2017, 2018) are the main references on Ethnomodelling.

Ethnomodelling seeks to value and understand *local* mathematical knowledge, translating it into school/academic (*global*) language, expanding the scope of this knowledge to people from other cultures or geographic space (*glocal*) (Rosa & Orey, 2017). For these authors, Ethnomodelling can be understood as the study of mathematical practices that members of the most diverse cultural groups develop through Mathematical Modelling.

Thus, the Ethnomodelling procedures involve mathematical practices used and developed in different situations and problems faced in the daily life of the members of this group. For example, Rosa and Orey (2017) state that is necessary to understand mathematical knowledge arising from social practices that are rooted in cultural relations. In this sense, Ethnomodelling studies this mathematical knowledge through an "interaction process that influences the local (emic) and global (etic) aspects of a given culture" (p. 18).

The emic approach seeks to understand the behavior of individuals from a given culture and their customs, and also to understand how these people mobilize knowledge to carry out their daily tasks; while the etic aspect seeks to analyze this behavior in an attempt to universalize it through a standardized procedure. In this context, Rosa and Orey (2017) state that:

> 1) Etic Approach: is related to the point of view of researchers, researchers and educators in relation to beliefs, customs and mathematical and scientific knowledge developed by members of a certain cultural group. 2) Emic Approach: is related to the point of view of members of different cultural groups in relation to their own customs and beliefs and also to the development of their own scientific and mathematical knowledge (p. 20).

According to Rosa and Orey (2017), the etic view is that of external observers of a given culture and they have a point of view considered culturally universal; and the emic view is of individuals who are immersed in a cultural group and have a culturally specific point of view. For these authors, from the understanding of emic and etic, individuals from a certain (local) group will enable and join together through dialogue with different cultural groups, through transculturality. D'Ambrosio (2020) states that:

> The approach to discussing integrated knowledge must be transdisciplinary. And it must, of course, contemplate the human species in all times and spaces, throughout history and the geographical occupation of the planet. It must therefore be cross-cultural. We are the same species, evolving over time and occupying different spaces (D'Ambrosio 2020, p. 153).

In this regard, "transculturality can ensure the translation of knowledge acquired by distinct cultural members to members of other cultural groups through Ethnomodelling" (Rosa & Orey, 2017, p. 18). Thus, Ethnomodelling can be considered an alternative methodological approach, which seeks to systematize mathematical knowledge from different cultural groups, allowing it to overcome global cultural and ideological barriers, contributing to the dialogue with members of other cultures.

For example, Rosa and Orey (2017, p. 19) state that: "Distinct cultural members share their own interpretation of their culture (emic approach) in contrast with the interpretation provided by researchers (...) and educators who are outsiders (etic approach) to these manifestations".

In this regard, Rosa and Orey (2018) state that it is essential that there is a dialogue between the emic and etic approaches, called the dialogic approach (glocal), through which one can understand the cultural influences in the elaboration of the ethnomodels, showing the interdependence and complementarity between the emic and the etic, through cultural dynamism.

3.4 The Ways to Understand the Participants' Perceptions

This chapter presents qualitative research, according to Bogdan and Biklen (2010), using as a data production instrument the testimony of 15 people, all members of the GEPTEMaC, including: 8 students from the Licentiate Degree in Mathematics; 2 Master's degree students from the Postgraduate Course in Science and Mathematics Education; 5 Basic Education teachers, 1 graduate student from the Mathematics Education Program, 2 specialists, and 2 masters.

The group participants answered a questionnaire with open questions, in which they discussed issues related to their participation and perceptions over the 2 years of research in the group. The questions were addressed in two blocks: (1) the importance of the group for their academic and professional training; (2) on the concept of Ethnomodelling and its potential for Basic Education; this last block being presented in this chapter.

For data processing, and in order to reach the objective of understanding the participants' perceptions about Ethnomodelling, as a potential methodological alternative for Basic Education, as well as Discursive Textual Analysis (ATD) was used. According to Moraes and Galiazzi (2013), the analysis was carried out in three stages: (a) deconstruction and unitarization (units of meaning); (b) categorization (relationships between what was unitarized); (c) construction of metatexts, based on the researcher's interpretations. Below is a summary of what each of these steps consists of, pointing out its fundamental aspects.

(a) *Deconstruction and unitarization*: This initial stage, after the constitution of the corpus of analysis (selection and organization of the material to be submitted for investigation, based on the research objectives), fragmented the text into 88 meaning units, with a view to achieve the research objective. The unitarization process requires the researcher's fidelity to what is contained in the research corpus, "a phenomenological attitude of letting the phenomenon manifested" (Moraes & Galiazzi, 2013, p. 53).
(b) *Categorization*: This step results from the process of organizing and grouping the units of meaning, which may arise from two situations: objective and deductive form—called *a priori* category; and inductive and subjective form—called emergent categories (Moraes & Galiazzi, 2013). This process requires the investigator's creative, attentive, and organized potential. The units of meaning were organized into 12 preliminary categories, and then grouped into 3 final categories, explained below. We opted for the use of emerging categories.
(c) *Metatexts*: The writing of metatexts expresses the researcher's understanding of the phenomenon of investigation, based on the categories chosen in the previous stage. At this stage, the writing process plots the description of the phenomenon, the interpretation performed by the researcher and, thus, there is the emergence of the new (Moraes & Galiazzi, 2013).

The three categories that emerged from the responses of the group participants are presented below, demonstrating their perceptions about Ethnomodelling, and its potential for teaching and learning Mathematics in Basic Education: (a) The concept of Ethnomodelling; (b) Potential of Ethnomodelling; (c) Valuing knowledge and cultures.

3.5 Ethnomodelling in the Perception of GEPTEMaC Members

In order to understand how the GEPTEMaC participants perceive Ethnomodelling as a potential methodological alternative for Basic Education, 15 testimonies of the participants were analyzed, which consisted of answers to an open questionnaire. These participants will be called research collaborators, and their answers will be coded as *C1*, *C2*, ... and *C15*. According to the procedures indicated by Moraes and Galiazzi (2013) for ATD, three categories emerged, and are summarized below.

3.5.1 The Concept of Ethnomodelling

According to the testimonies, it was clear that GEPTEMaC members perceive Ethnomodelling as the connection between Ethnomathematics and Mathematical Modelling. In this context, Rosa and Orey (2017) consider Ethnomodelling as the intersection between Cultural Anthropology, Ethnomathematics and Mathematical Modelling.

These statements corroborate to the concept that Ethnomodelling is the study of mathematical practices developed by members of different cultural groups, through modelling. In this sense, its procedures incorporate the mathematical practices developed and used in the various problems and situations faced in the daily lives of members of these groups.

However, contributors diverge on the definition of what is Ethnomodelling beyond the connection mentioned above, presenting as answers: trend (2); program (1); method (1); pedagogical alternative (1); pedagogical proposal (2); pedagogical action (1); pedagogical strategy (1); methodological approach (1); methodological alternative (1); methodological tool (2); and methodological proposal (2).

Two definitions of Ethnomodelling are in line with the idea of trend. One of the collaborators states that it is "a trend that seeks, through knowledge of a certain culture, to develop models that will help in the learning process" (*C10*), and another defines it as "a trend in the large area of research in Mathematics Education" (*C14*).

It should be noted that Mathematics Education can be considered an area of studies and research that has solid foundations in Education and Mathematics. In this regard, Mathematics Education opened space for research and discussions involving questions about the teaching of Mathematics.

From research, trends have emerged in the area of Mathematics Education that incorporate different approaches considered relevant to the teaching and learning process. Researchers in Mathematics Education have different points of view, that is, conceptions, regarding the evolution of trends in this area.

Among the trends in Mathematics Education can be highlighted: Problem Solving, History of Mathematics, Ethnomathematics, Mathematical Modelling (MM), among others. Perhaps because MM and Ethnomathematics are considered trends, collaborators believe that Ethnomodelling can also be considered as a trend in mathematics education.

Ethnomathematics, in addition to be a trend in Mathematics Education, is defined by D'Ambrosio (2013) as a research program that aims to explain the processes of knowledge generation, organization, and transmission in different cultural systems. It studies the relationships and connections between mathematical notions and other cultural elements, knowledge and mathematical know-how acquired in the development of a professional activity (D'Ambrosio, 2013).

Considering this definition, one of the collaborators stated that Ethnomodelling "is a program that enables the teaching of Mathematics, where the student's experience, reality and social knowledge is prioritized, it allows learning to be built together, between teacher and students" (*C1*).

The collaborator who considers Ethnomodelling as "a method used to bring to the classroom, the academic environment, the mathematical knowledge present in a given culture" (*C12*), must be relating to the concept of MM defended by Biembengut (2016) who defines it as a research method applied to Education that consists in the elaboration of models.

Some of the collaborators consider Ethnomodelling as a methodological alternative, proposal, strategy, or pedagogical action. The assertions that this is "a pedagogical action that proposes to (re)know the mathematical knowledge and practices historically constructed by different cultural groups and take them to the classroom through modelling" (*C8*), are in line with the ideas of what:

> (...) ethnomodelling can be considered as a set of strategies that enable the resolution of problems present in knowledge systems developed in different cultural contexts. These strategies can be considered as ways of communication, behavior, individual and collective knowledge, which through interaction can result in a pedagogical action for the teaching and learning process in mathematics (Rosa & Orey, 2018, p. 116).

Participants who consider Ethnomodelling as an approach, alternative, tool or methodological proposal, claim, for example, that "Ethnomodelling appears in the area of Mathematics Education as a methodological alternative that aims to enhance different mathematical knowledge in different social and cultural contexts" (*C6*), or that is considered as "a methodological approach that enables students to establish a relationship between their daily practice and mathematical knowledge present in cultures and cultural groups that practice mathematics far from academic mathematics" (*C15*).

It is important to highlight that Rosa and Orey (2017) also present this idea when they state that Ethnomodelling can be considered "an alternative methodological approach, which aims to record the ideas, procedures and mathematical practices that are developed in different cultural contexts" (p. 23).

In agreement with Madruga (2021), it is believed that Ethnomodelling is a proposal for the teaching of Mathematics, in order to provide a space for interaction and reflection, in the elaboration and deepening of knowledge from the most diverse cultures, in a permanent movement that turns to educational practices.

In this sense, Ethnomodelling can be a methodological proposal that uses the concepts of diversity and culture (*ethno*) in line with mathematical modelling (*tics*) in order to enhance learning (*mathema*) at different levels of education, aiming to suggest a path to teaching and learning Mathematics.

3.5.2 Potentialities of Ethnomodelling

In the Brazilian scenario, according to a survey carried out by Madruga (2021), until the first half of 2021 there were 12 scientific research published, all dissertations, which dealt with the possibilities of Ethnomodelling in the teaching of Mathematics. These, in general, highlight the potential for developing Ethnomodelling in the classroom, with positive results in terms of teaching and learning (Sonego, 2009; Reges, 2013; Altenburg, 2017; Cortes, 2017; Pimentel, 2019; Dutra, 2020; Eça, 2020; Martins, 2020; Mesquita, 2020; Santos, 2020; Barreto, 2021; Rodrigues, 2021).

Using themes from the reality and culture of students from different Brazilian regions, such as: rice planting (Sonego, 2009); candy factory (Reges, 2013); Pomeranian culture related to German immigration (Altenburg, 2017); marketers selling at a Free Fair (Cortes, 2017); coffee culture (Dutra, 2020); rural community (Martins, 2020); peripheral community (Mesquita, 2020); cocoa crop (Santos, 2020), among others. Research has shown that teaching through Ethnomodelling can provide the learning of mathematical content based on respect and cultural appreciation.

The contributors and members of GEPTEMaC agree with the authors of the aforementioned research by stating that "Ethnomodelling has very significant potential in the teaching and learning of Mathematics, especially for Basic Education students" (*C1*), and that "it seeks an improvement in the teaching-learning process of the discipline, with the incorporation in the mathematical curriculum of knowledge arising from the student's life and human values, such as, for example, cooperation, solidarity and respect" (*C4*).

The use of Ethnomodelling in teaching practice seeks to integrate the knowledge institutionalized by the academy, with the knowledge constituted and practiced by members from different sociocultural contexts through a dialogic approach (Rosa & Orey, 2017), discharging in this way, the supremacy kind of knowledge over another.

To support pedagogical actions through Ethnomodelling is to build educational scenarios that aim at the critical training of teachers and students from a sociocultural perspective. In the perception of the collaborators of this research "Ethnomodelling becomes powerful and possible because through it we will be working with the real context of our students of Basic Education" (*C5*).

And yet, "the teacher has the opportunity to be working on cultural, social and moral values that permeate the daily lives of Basic Education students. Furthermore, it seeks to contextualize the teaching of Mathematics, which is a very important factor for the university to actually reach schools" (*C6*). According to Eça and Madruga (2021),

> Ethnomodelling, with these positions demarcated by anthropological characteristics, promotes in the educational sphere the decentralization of knowledge about the figure of the teacher and shares this responsibility with everyone involved in the teaching and learning process in a participatory and active manner, thus promoting the emancipation of the students in the process rooted in dialogical principles. A fact that contributes for them to assume the role of protagonist of his own learning (Eça & Madruga, 2021, p. 9).

In line with this statement, the collaborators mentioned that Ethnomodelling allows students to "become researchers and get involved in the teaching process" (*C1*), and also "to make students more thinking and active learners to perform and solve everyday problem and situations, by conceiving and establishing possible relationships between cultural and academic mathematical knowledge" (*C7*).

Research collaborators state that, based on Ethnomodelling, "learning occurs with meaning" (*C11*); and that students can awaken "greater motivation to know and learn mathematics" (*C2*). This motivation is mentioned in the research by Santos (2020), when he states that the students felt motivated and enthusiastic about the activities carried out outside the classroom, in the context of a chocolate factory.

While Dutra (2020) states that teachers should strive "to understand the cultural aspects that are present in everyday life, providing a motivating, contextualized, and meaningful learning" (p. 49).

Several published studies dealing with Ethnomodelling as an methodological alternative for teaching Mathematics, including those by Sonego (2009), Reges (2013), Altenburg (2017), Cortes (2017), Pimentel (2019), Dutra (2020), Eça (2020), Martins (2020), Mesquita (2020), Santos (2020), Barreto (2021), and Rodrigues (2021) show different possibilities and potential for the classroom, specifically in Basic Education.

The members of GEPTEMaC also perceive these potentialities when they state that "the results obtained during and at the end of the process are perceived in the construction of the mathematical content present in the school curriculum" (*C10*); and that Ethnomodelling enables students to "explore different mechanisms of learning mathematics and perceive their presence in different areas" (*C4*). In this perspective, Santos and Madruga (2021) state that the:

> (...) contextualized use of the mathematical object with the cultural aspect contributed to the students' involvement in the teaching and learning process; in the construction of autonomy, overcoming difficulties in interacting with different types of people; for them to evaluate points of view, asking questions and contributing with colleagues and professor-researcher during the dialogues promoted in class (pp. 19–20).

In this regard, GEPTEMaC members consider Ethnomodelling as a potential methodological alternative for the teaching of Mathematics, as it values the cultural aspects of Mathematics, considering the differences between students, understanding that they learn in different ways, and that they are part of distinct cultural contexts, social, and economic contexts, and mainly that these students bring knowledge and actions that must be considered in the teaching process.

In this way, it is also highlighted that Ethnomodelling can contribute to learning process that is able to provide students with knowledge that supports their experience in society and the improvement of a critical and active view by enabling them to intervene in society and seeking to transform them positively.

3.5.3 Valuing Knowledge and Cultures

Ethnomodelling aims to provide a teaching of Mathematics that dialogues with the cultural context in which the student is inserted, contributing to the appreciation of the environment and social, cultural and economic aspects. In this context, D'Ambrosio (2020) affirms that the:

> (...) proposal of the Ethnomathematics Program is to recover the humanistic, social, and cultural character of mathematics and in all areas of knowledge. In particular, I speak in a broad sense of Mathematics as the human being's own abilities to observe, to classify and order, to evaluate, to measure and to quantify and infer. The ultimate goal of activating these capabilities is to deal with all everyday problems and situations while at the same time understanding and explaining facts and phenomena of reality in the broadest sense (p. 153).

Understanding the ways (*tics*) in which people explain and solve their daily problems (*mathema*), in the most different cultures (*ethno*), is the premise of ethnomathematics, as well as the appreciation of the most varied cultures, diversity and the search for education for peace. Ethnomodelling shares these premises, as it brings with it the assumptions of ethnomathematics in the conception of D'Ambrosio (2020) who suggests that "*tics*" of "*mathema*" can be developed in the classrooms (Madruga, 2021).

Ethnomodelling, according to Rosa and Orey (2017), seeks to value and understand local mathematical knowledge, translating it into school/academic (global) languages, as well as enabling the expansion of the reach of this knowledge to people from other cultures. or geographic spaces (glocal).

Therefore, Ethnomodelling procedures involve mathematical procedures and practices used and developed in various problem situations faced in the daily lives of these groups (Rosa & Orey, 2018), providing "an appreciation of the mathematical knowledge produced by different peoples" (*C2*). Furthermore, "it allows for the appreciation of different cultural mathematical knowledge, which is sometimes used by students implicitly" (*C7*).

In this sense, Ethnomodelling "enables students of Basic Education, as well as teachers to perceive in different contexts the local knowledge and practices to be applied in the teaching of mathematics" (*C7*), valuing "knowledge (tacit) to be somehow as a "facilitator" for learning academic knowledge" (*C2*).

Tacit (emic) knowledge comes from the experience that each person has during their lifetime. Thus, it is subjective, as it stems from the values and experience of each individual. This type of knowledge is difficult to transfer to formal and written language. It can be considered as know how because it is contextualized and analogous (Nonaka & Takeuchi, 1997).

In the development of Ethnomodelling, there is "the appreciation of the knowledge that students bring with them" (*C2*), as well as "the consideration of all the knowledge that the person already has" (*C13*).

Corroborating with the ideas of Rosa and Orey (2012), it is believed that Ethnomodelling, as well as Mathematical Modelling, can also facilitate a pedagogical structure that promotes the identification and dissemination of tacit (emic) and explicit (etic) knowledge. Explicit (etic) knowledge is understood as that which has already been transformed into formal language.

Therefore, it was passed into the form of manuals, standards, texts, and mathematical equations. In this way "it is possible to value the experiences of different peoples, including our own students in classroom practices" (*C8*), in order to "contribute so that new generations know and recognize a much more cultural mathematics, linked to everyday life of different groups" (*C4*).

It can be inferred, through the collaborators' testimonies, that they perceive that Ethnomodelling has an important role in the construction of knowledge, not just mathematical ones; and it can facilitate communication between teachers and students, enabling the conversion between tacit (emic) and explicit (etic) mathematical knowledge through the development of dialogical ethnomodels.

Consequently, this context generates an environment "where the student's experience, reality and social knowledge are prioritized, enabling learning to be built together, between professor and students" (*C1*).

Previously, Nonaka and Takeuchi (1997) suggested that the process of conversion between explicit and tacit knowledge is spiral, as it is a continuous and dynamic process that evolves through social interactions. Tacit and explicit knowledge are complementary, and the interaction between them will result in more knowledge. In this sense, the conversion of knowledge can be broken down into four ways of creating knowledge: socialization, combination, externalization, and internalization.

(a) *Socialization* comprises the conversion of tacit knowledge into another tacit one. This process occurs from the interaction between people (Nonaka & Takeuchi, 1997). It is understood that this tacit knowledge can be considered as local (emic) knowledge highlighted by Rosa and Orey (2017, 2018). In the development of Ethnomodelling, this socialization can occur through visits to certain places (Pimentel, 2019; Santos, 2020; Dutra, 2020), or conversations with people experienced in certain subjects, it is learning through experience, "valuing the culture and the knowledge that comes from the students" (*C8*). Where "the teacher and students share tacit knowledge through experiences, ideas, mental models and technical skills through the development of interactive, cooperative and contextualized activities" (Rosa & Orey, 2012, p. 277).

(b) *Externalization* is a process of articulating tacit knowledge into explicit concepts (Nonaka & Takeuchi, 1997). It can also be defined as a process of beginning the creation of global (etic) knowledge, as tacit knowledge becomes explicit, expressed in the form of metaphors, analogies, concepts, hypotheses, models or ethnomodels. Ethnomodelling can facilitate this externalization process, as it "provides the appreciation of the knowledge that students bring with them, allowing this knowledge to be somehow a *facilitator* for the learning of academic knowledge" (*C2*).

(c) *Combination* occurs when explicit knowledge is transformed into other explicit knowledge (Nonaka & Takeuchi, 1997). Thus, when changing the context, there is a recategorization or increase of explicit knowledge, in a way, transforming this knowledge.

> The different types of explicit knowledge that students possess are combined and converted into new explicit knowledge, which contains a higher level of complexity. This process allows explicit knowledge, which is combined, to be reorganized, restructured, systematized, and refined (Rosa & Orey, 2012, p. 278).

It is understood that, in Ethnomodelling, this combination occurs when students are able to articulate what has been learned, when they share different explicit knowledge, based on previous learning, integrating them into new explicit knowledge. This could be related to etic (global) knowledge itself.

(d) *Internalization* is the process of transforming explicit knowledge into tacit knowledge (Nonaka & Takeuchi, 1997). In a way, it identifies with the common concept of learning by doing. Ethnomodelling can contribute to this internalization process while "it helps students to see the application of mathematics in different contexts and everyday situations, in order to understand that this is not just an explanation and reproduction of calculations and formulas, but beyond these theories, they can be seen and used to solve problem situations" (*C7*).

In this context, Rosa and Orey (2012) affirm that "internal reflection and the exchange of information between teacher and students and between students and students favor the internalization of knowledge, facilitating the development of critical awareness through social relationships" (p. 277). It is believed that this internalization process generates learning, that is, dialogic knowledge (glocal). Figure 3.1 shows the relation of the conversion of knowledge and its relation to ethnomodelling.

Based on the investigations conducted by Nonaka and Takeuchi (1997), the conversion of knowledge can be considered as a spiral movement. According to this perspective, Rosa and Orey (2017) highlight that Ethnomodelling is a spiral process, which develops a dialogic movement that relates the students' emic knowledge (tacit) and etic knowledge (explicit).

It is important to emphasize that this approach can be effective with the support of the elaboration of ethnomodels and, consequently, generating learning, in the quest to "aggregate the cultural knowledge of a people with academic (school) knowledge,

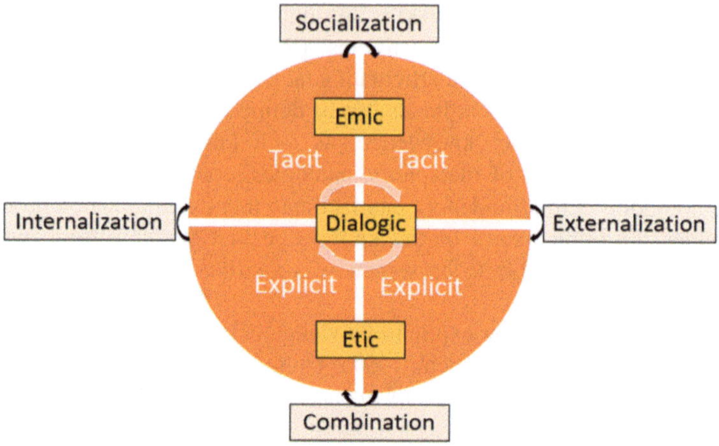

Fig. 3.1 Conversion of knowledge and its relation to Ethnomodelling. (Source: Personal file)

as well as the search for (ethno)mathematical models that are related to a study problem arising from questions cultural" (*C5*).

When models are developed as a result of the understanding of mathematical situations practiced by the members of a cultural group, these are called ethnomodels. For Rosa and Orey (2017), ethnomodels are cultural artifacts, which can be considered as pedagogical tools, and can be used to facilitate the understanding of practices and comprehension of systems taken from the reality of different cultural groups. An ethnomodel can be considered as a clear and objective way of explaining mathematical knowledge from a cultural group (externalization).

3.6 Final Thoughts

The intention of this chapter was to share the actions carried out within a research group, formed primarily by students of the Licentiate Degree in Mathematics at the Federal University of Recôncavo da Bahia, Brazil, presenting a research that aimed to understand the perceptions of the participants on Ethnomodelling, as a potential methodological alternative for Basic Education.

The results showed that the project has contributed to the initial or continuing education of the participants, both as professors and researchers, while addressing an methodological alternative considered new by the group. Research in teacher education (initial and continuing), as well as teacher participation in research groups, play a fundamental role in their professional development, supporting training that contributes to the constitution of research teachers.

In the critical and joint reflections between undergraduates, graduate students, active teachers in Basic Education and University researchers, the problematization of individual and joint actions is encouraged, as well as the practices and elaboration of research projects, followed by intervention.

Regarding Ethnomodelling, the results showed that research collaborators, members of GEPTEMaC perceive it as a connection between Ethnomathematics and Mathematical Modelling, which seeks to value and understand local mathematical knowledge, translating it into school/academic language (global) by expanding the scope of this knowledge to people from other cultures or geographic space (glocal), as highlighted by Rosa and Orey (2017, 2018).

For Rosa and Orey (2017), it is necessary to understand the mathematical knowledge arising from social practices that are rooted in cultural relations. In this sense, Ethnomodelling studies this mathematical knowledge through an "interaction process that influences the local (emic) and global (etic) aspects of a given culture" (Rosa & Orey, 2017, p. 18).

The research collaborators highlight the importance of developing Ethnomodelling in the classroom in Basic Education by considering it as a pedagogical action to enhance learning, which is able to bring the sociocultural knowledge of students closer to the school/academic Mathematics knowledge by valuing their knowledge and practices.

The data led to the understanding that Ethnomodelling can assist in the knowledge conversion process, aiding in learning by helping students to activate their tacit knowledge and transform it into explicit, passing through spiral stages (socialization, externalization, combination and internalization), through the social interactions that occur during the development of this process.

References

Albanese, V., & Perales, F. J. (2014). Pensar matematicamente: Una visión etnomatemática de la práctica artesanal soguera. *Revista Latinoamericana de Investigación en Matemática Educativa, 17*(3), 261–288.

Altenburg, G. S. (2017). *Contextualizando Cultura e Tecnologias: Um estudo etnomatemático articulado ao ensino de geometria*. Dissertação (Mestrado Profissional em Ensino de Ciências e Matemática), Faculdade de Educação, Universidade Federal de Pelotas, UFPel, Pelotas.

Barreto, F. M. (2021). *Um estudo qualitativo para entender a ação pedagógica da etnomodelagem com alunos de comunidades rurais e urbanas*. Dissertação (Mestrado Profissional em Educação Matemática), Instituto de Ciências Exatas e Biológicas, Universidade Federal de Ouro Preto, Ouro Preto. 293 f.

Bassanezi, R. C. (2010). *Ensino-aprendizagem com Modelagem Matemática* (3rd ed.. 2ª reimpressão). Contexto.

Biembengut, M. S. (2016). *Modelagem na Educação Matemática e na Ciência*. Editora Livraria da Física.

Bogdan, R., & Biklen, S. (2010). *Investigação Qualitativa em Educação*. Editora Porto.

Cortes, D. P. d. O. (2017). *Re-significando os conceitos de função: Um estudo misto para entender as contribuições da abordagem dialógica da etnomodelagem*. Dissertação (Mestrado em Educação Matemática), Instituto de Ciências Exatas e Biológicas, Universidade Federal de Ouro Preto, Ouro Preto. 226 f.

D'Ambrosio, U. (2013). *Etnomatemática: Elo entre as tradições e a modernidade* (5th ed.). Autêntica Editora.

D'Ambrosio, U. (2020). Sobre as propostas curriculares STEM (ciência, tecnologia, engenharia, matemática) e STEAM (ciência, tecnologia, engenharia, artes, matemática) e o programa etnomatemático. *PARADIGMA*, [S. l.], p. 151–167. Disponível em: http://revistaparadigma.online/ojs/index.php/paradigma/article/view/876. Acesso em: 10 mai. 2022.

D'Ambrosio, B. S., & D'Ambrosio, U. (2006). Formação de professores de matemática: Professor pesquisador. *Atos de Pesquisa em Educação, 1*(1), 75–85.

Dutra, E. D. R. (2020). *Etnomodelagem e café: Propondo uma ação pedagógica para a sala de aula*. Dissertação (Mestrado em Educação Matemática), Instituto de Ciências Exatas e Biológicas, Universidade Federal de Ouro Preto, Ouro Preto. 319 f.

Eça, J. L. M. (2020). *Formação continuada à luz da Etnomodelagem: Implicações para o desenvolvimento profissional do professor que ensina matemática*. Dissertação (Mestrado em Educação em Ciências e Matemática), Universidade Estadual de Santa Cruz, Ilhéus.

Eça, J. L. M. d., & Madruga, Z. E. d. F. (2021). A Etnomodelagem na formação de professores: Uma análise a partir de produções recentes. *Revemop*, (3), e202105. https://doi.org/10.33532/revemop.e202105

Fiorentini, D. (2013). Pesquisar práticas colaborativas ou pesquisar colaborativamente? In M. C. Borba & J. L. Araújo (Eds.), *Pesquisa qualitativa em Educação Matemática* (5th ed.). Autêntica Editora.

Madruga, Z. E. F. (2012). *A criação de alegorias de carnaval: Das relações entre modelagem matemática, etnomatemática e cognição*. Dissertação de Mestrado, Pontifícia Universidade Católica do Rio Grande do Sul, PUCRS, Porto Alegre, RS.

Madruga, Z. E. F. (2021). Concepções de Modelagem Matemática nas Pesquisas em Etnomodelagem. In *VIII SIPEM—Seminário Internacional de Pesquisa em Educação Matemática* (Anais).

Madruga, Z. E. F., & Biembengut, M. S. (2016). *Modelagem & Aleg(o)rias: Um enredo entre cultura e educação*. Appris.

Martins, R. B. G. (2020). *Etnomodelagem: Modelagem matemática no interior de uma comunidade rural sustentável*. Dissertação (Mestrado em Educação), Universidade Federal de São Carlos, São Carlos.

Mesquita, A. P. S. d. S. (2020). *Uma análise sociocrítica da etnomodelagem como uma ação pedagógica para o desenvolvimento de conteúdos matemáticos em uma comunidade periférica*. Dissertação (Mestrado em Educação Matemática), Instituto de Ciências Exatas e Biológicas, Universidade Federal de Ouro Preto, Ouro Preto. 286 f.

Moraes, R., & Galiazzi, M. C. (2013). *Análise Textual Discursiva* (2nd ed.). Editora Unijuí.

Nonaka, I., & Takeuchi, H. (1997). *Criação de conhecimento na empresa: Como as empresas japonesas geram a dinâmica da inovação*. Editora Campus.

Onuchic, L. d. l. R., & Allevato, N. S. G. (2014). Ensino-Aprendizagem-Avaliação de Matemática: Por que através da Resolução de Problemas? In L. d. l. R. de la Onuchic, N. S. G. Allevato, F. C. H. Noguti, & A. M. Justulin (Eds.), *Resolução de Problemas: Teoria e Prática*. Paco Editorial.

Pimentel, C. C. (2019). *Etnomodelagem: Uma abordagem de conceitos geométricos no cemitério de Arraias—TO*. Dissertação (Mestrado Profissional em Matemática), Universidade Federal do Tocantins, Programa de Pós-Graduação em Matemática, Arraias.

Pradhan, J. B. (2020). Artefatos culturais como uma metáfora para comunicação de ideias matemáticas. *Revemop, 2*, e202015.

Reges, A. M. M. (2013). *O ensino da geometria com enfoque na etnomodelagem*. Dissertação (Mestrado), Universidade Federal Rural do Semi-Árido, Mossoró.

Rodrigues, J. (2021). *Explorando a perspectiva de pesquisadores e participantes de trilhas de matemática sobre a (re)descoberta do conhecimento matemático fora da escola: Um estudo qualitativo em etnomodelagem*. Dissertação (Mestrado Profissional em Educação Matemática),

Instituto de Ciências Exatas e Biológicas, Universidade Federal de Ouro Preto, Ouro Preto. 327 f.

Rosa, M., & Orey, D. (2012). A Modelagem como um Ambiente de Aprendizagem para a Conversão do Conhecimento Matemático. *Bolema, 26*(41A), 261–290.

Rosa, M., & Orey, D. (2014). Interlocuções Polissêmicas entre a Etnomatemática e os Distintos Campos de Conhecimento Etno-x. *Educação em Revista, 3*(30), 63–97.

Rosa, M., & Orey, D. (2017). *Etnomodelagem: A arte de traduzir práticas matemática locais.* Editora Livraria da Física.

Rosa, M., & Orey, D. (2018). Etnomatemática: Investigações em etnomodelagem. *Revista de Investigação e Divulgação em Educação Matemática, 2*(1), 111–136.

Santos, J. (2020). *Produção Artesanal de Chocolate e Etnomodelagem: Construção do conceito de função por estudantes do Ensino Fundamental.* Dissertação (Mestrado em Educação Matemática), Universidade Estadual de Santa Cruz.

Santos, J., & Madruga, Z. (2021). Etnomodelagem e produção artesanal de chocolate: Uma investigação no 9° ano do Ensino Fundamental. *Revista de Ensino de Ciências e Matemática, 12*(1), 1–20.

Sonego, G. V. (2009). *As contribuições da etnomodelagem matemática no estudo da geometria.* Dissertação (Mestrado), Centro Universitário franciscano, Santa Maria.

Chapter 4
Ethnomodelling Aspects of Positionality Between Local (Emic) and Global (Etic) Knowledge Through Glocalization (Cultural Dynamism): The Specific Case of a Market-Vendor

Daniel Clark Orey and Diego Pereira de Oliveira Cortes

4.1 Initial Considerations

In the context of academic research, a valid discussion exists in relation to the *positionality* between researchers and the ones being "researched", as well as its in relation to context, fieldwork, and academy (D'Olne Campos, 2000). This concern is apparent in sociocultural research, for example, investigations developed in and around ethnomathematics.

It is important to state here that the concept of positionality in the context of ethnomodelling research presents us with a topic of paramount importance. Positionality is a neologism related to the quality of being positional. It refers to the position of investigators in relation to members of the communities they are studying. In this regard, there is a need for researchers to become aware of their position in relation to the participants of their investigations (Rose, 1997).

This awareness is related to the (re)construction of the *status* of insiders (emic, local) and outsiders (etic, global) in terms of their positionality during the development of their research. This approach allows for an improved, and clearer understanding of the cultural dynamics of members of distinct cultural groups under investigation.

According to Rosa and Orey (2017a), in the ethnomodelling process, the positionality of actual being there in the field can capture the relations between symbolic mathematical practices and the reproduction of a given social context.

D. C. Orey (✉)
Universidade Federal de Ouro Preto, Ouro Preto, Minas Gerais, Brazil
e-mail: oreydc@ufop.edu.br

D. P. d. O. Cortes
Secretaria de Estado de Educação de Minas Gerais, São Joaquim de Bicas, Minas Gerais, Brazil

M. Rosa et al. (eds.), *Mathematical Modelling Programs in Latin America*,
https://doi.org/10.1007/978-3-031-04271-3_4

However, this investigative experience can be considered frustrating, as the fieldwork is composed of a heterogeneous group of spaces, times, and contexts.

In this regard, everyday life is like an arena of sociocultural inquiry through which boundaries of observation, analysis and interpretation are practically limitless. Thus, positionality is one of the main objectives for the successful conduction of research in ethnomodelling because it is necessary to understand the interactions of local (emic), global (etic) and glocal (dialogical) approaches to the development of mathematical knowledge.

It is important to highlight how the aspects related to research conducted was done in a state public school located in a metropolitan region of Belo Horizonte, Minas Gerais, in 2017, and also in a local farmer's market located in the same region. The market in which this research was conducted sells a variety of horticultural products, clothing, food products, and handicrafts. However, in this investigation, only the labor practices of the market in relation to the commercialization of horticultural products were studied.

The main objective of this study was to the identification of how a dialogical approach in ethnomodelling contributes to the process of re-signifying function concepts of 38 students enrolled in the second year of high school during their interaction with a market vendor and his labor practices.

We would like to emphasize here, that we use the term re-signification "as a process of construction of (new) meanings and (new) interpretations of what we know, do and say" (Jiménez-Espinosa, 2002, p. 4). This is mainly in relation to the mathematical knowledge related to the concepts that function and can be reinterpreted through the interaction of students with the members of distinct cultural groups, such as the market vendors.

A second objective was to outline possibilities regarding the positionality of the vendor according to his emic/etic (glocal) perspective in a dialogical way (Rosa & Orey, 2012), which is related to his movement from his cultural environment (the street market) to the school setting and vice versa, in which he develops his labor activities. In addition, the aim of this research was also related to the discussion of the pedagogical potential that contextualized knowledge locally developed by the farmer-vendor in this study, can offer for the study of mathematical concepts, such as functions in a formal school setting.

In this respect, we used empirical information taken from data collected during the conduction of the fieldwork of the research entitled: *Re-signifying Function Concepts—a Mixed Study to Understand the Contributions of the Dialogical Approach of Ethnomodelling*, developed in the master's degree program in Mathematics Education of the Universidade Federal de Ouro Preto (UFOP). As well, we sought to contemplate the conceptual and practical aspects related to culturally rooted mathematical knowledge in the context of the market.

In the context of this research, this work is related to particular aspects of the interpretation of answers given by the vendor-market to the questions of a semi-structured interview conducted by the first author of this chapter. Thus, as the researchers looked for answers to the research question, a diversity of concepts related to functions emerged that offered subsidies for the development of future investigations.

Therefore, reflecting on the findings from this research, it became important to us how this experience is related to the theoretical basis that permeated its methodological development, as well as how practical realities can be used in the pedagogical action in the classrooms.

From this perspective, an aspect of dialogue through a *cultural dynamism*[1] that emerged as a result of this investigation highlighted that ethnomodelling, including its local (emic), global (etic), and glocal (dialogical) approaches, are related to the development of the positionality of the farmer-vendor and the researchers. And finally, this chapter promotes reflections and observations made by us over 3 (three) years of studies after the original research was developed.

4.2 Ethnomodelling and Its Connection to Ethnomathematics and Modelling

In the interrelations between diverse areas of investigations, a diversity of knowledge and culturally rooted practices have been incorporated in distinct research fields. This approach helped the culmination of the development of *ethno-x*. Although various fields of scientific knowledge had advanced rapidly, in the mid-eighteenth-century aspects related to culture were not effectively incorporated into these areas (D'Olne Campos, 2000).

However, D'Ambrosio (2001) and Rosa and Orey (2014) argued that in the early nineteenth century, the development of *ethnosciences*, which sought to understand cultural knowledge and its relationship with scientific knowledge began to merge. This makes sense as through centuries of colonization, travel, trade, and war; scholars from diverse parts of the world began modestly to dialogue and share their knowledge.

In this respect, the growing discussions and rudimentary understanding of cultural knowledge and practices promoted the development of new epistemological fields related to ethnobotany (1896), ethnozoology (1914), ethnogeography (1916), ethnobiology (1935), and ethnoherpetology (1946) (Cardona 1985 apud D'Olne Campos, 2000).

[1]With cultural dynamism, local (emic) mathematical ideas and procedures interact with globally consolidated mathematical practices developed in schools through the development of a reciprocal relationship between emic (local) and etic (global) knowledge. Thus, mathematical practices developed in ethnomodelling investigations value the cultural dimension of emic mathematical knowledge, which is analyzed from the elaboration of ethnomodels that represent daily problems and situations taken from reality and contextualized in the daily life of the members of the school communities. This approach provides the exchange of mathematical knowledge, procedures, and practices between the school and the community contexts by deepening the power of the dynamics of the encounters between these two cultural knowledges (Rosa & Orey, 2017a).

Similarly, reflections on the nature of mathematical cognitive contexts (historical, social, and cultural) can be motivated the by the intention to clarify the understanding from knowledge built by humanity in different sociocultural contexts, by interest groups, communities, societies, peoples, and nations, gave rise to the Ethnomathematics Program as proposed by D'Ambrosio (Rosa & Orey, 2014).

The ethnomathematics program is conceived as an attempt to understand the adventure of humanity in the pursuit of peace, social justice, knowledge, and the adoption of behaviors that can be shared with the members of distinct cultural groups. This perspective presents historical and philosophical aspects that seek to promote the development of pedagogical implications (D'Ambrosio, 2001). Thus, ethnomathematics is associated with:

> (. . .) mathematics found among identifiable cultural groups, such as: national tribal societies, worker groups, children of a certain age, and professional class. Its identity largely depends on the focus of interest, motivation and certain codes and jargons that do not belong to the domain of academic mathematics (D'Ambrosio, 1994, p. 89).

Along with cultural aspects and diverse ways mathematical models can describe phenomena, the first time *ethno/modelling* was cited by Bassanezi (2002), who established that, in certain circumstances, the process of mathematical modelling may be related to and used with the principles of an essentially ethnomathematical nature.

Thus, in seeking a holistic or more comprehensive understanding, Rosa and Orey (2003) discussed the need to use pedagogical approaches that connect the cultural aspects of mathematics (ethnomathematics) with the academic aspects of this knowledge (modelling). This approach is referred by Rosa and Orey (2010) to as *ethnomodelling*.

However, it is important to highlight that mathematical knowledge developed by members of diverse cultural groups has their own interpretation (emic) in opposition to external interpretations of school/academic mathematics (etic). Thus, *ethnomodelling* uses modelling techniques to establish relationships between the structures of local and school/academic mathematics (Rosa & Orey, 2017a) through the elaboration of *ethnomodels*.[2]

In this regard, both ethnomathematics and modelling look at the relations between local and school/academic knowledge (global), thereby promoting cultural dynamism between different mathematical knowledge systems (Rosa & Orey, 2010). This particular research sought to understand how mathematical knowledge

[2]Ethnomodels are descriptions of local (emic), global (etic), and glocal (dialogical) mathematical knowledge through specific methods, procedures, and techniques that help members of distinct cultural groups to develop understandings of their own world by using small units of information that compose its entire representation. Ethnomodels help to link the development of mathematical practices to the cultural heritage of these members, who detain necessary information to solve problems and situations described in systems taken from their own reality. This approach helps the organization of pedagogical actions in classrooms by using emic (local), etic (global), and dialogical (glocal) aspects of mathematical knowledge through the development and elaboration of ethnomodels (Rosa & Orey, 2018).

used in a traditional street market might provide the understanding and link to the mathematical knowledge developed in the classrooms through looking at the mathematical practices used by the street market vendor.

4.3 Contextualizing the Free Market

One of questions discussed here is related to the results of the study conducted by Cortes (2017) in a traditional street market that is located in a metropolitan region of Belo Horizonte, in Brazil. This particular street market has been operating in this same location for over 30 years and runs on Sundays, from 7 am to 3 pm. With, approximately 500 stallholders who show and sell their products that are related to food, jewelry, crafts, clothing, and local grown vegetables and fruits. Currently, according to its administration, approximately 10,000 people attend this street market every Sunday.

4.3.1 Introducing the Market-Vendor

At the time of the conduction of the original research, the vendor-marketeer in this study was 60 years old and studied only until the fourth grade. He has worked at the street market since 1990 and does not have any other profession. He prefers to work at the market, as he has a higher income than if he was working as an employee by receiving only the minimum wage, which would not be enough to support his family. Currently his net income is about three times the Brazilian minimum wage.

Before working at the market, he worked on a farm in a coffee and sugar cane plantation, and in his youth, he also worked for 6 months in a steel mill, returning later to work in the fields and when he developed his own market stall. His family includes three sons and his wife. His eldest son is married and lives in another city, while on certain occasions, his unmarried son helps him sell goods at the market, while his other son accompanies him on his purchases made at *CEASA-MG*,[3] in Contagem, which is a city located in the metropolitan region of Belo Horizonte.

This particular market-vendor sells cassava, tomato, chayote, okra, onion, banana, yams, green corn, lettuce, cabbage, chives, chicory, watercress, spinach, broccoli, cilantro, and mustard in his stall. Of these goods, this vendor himself cultivates cassava, okra, yam, and leaf vegetables on a leased land, and the rest of the products are purchased at CEASA-MG. To purchase the goods at CEASA-MG,

[3]CEASA-MG is the supply center of the state of Minas Gerais. Among the CEASA-MG activities are related to the wholesale trade in grain, meat, processed foods, agricultural inputs, packaging, restaurants, personal financial services, public health agencies, education, and security.

this participant uses a pickup truck, however, when he cannot pick them up, he places orders with his colleagues at the street market to make his purchase.

It is important to emphasize here that the vendor was careful to only buy and sell quality products, despite being more expensive, and having a loss with some of these products, such as tomatoes. In his opinion, the products he cultivates have an excellent quality, as the harvest is developed in a more careful manner and timely way. These products are sold the day after they are harvested, and they are fresher. The street market is held only on Sundays, during the week he works in the cultivation and care of his vegetables on a leased plot of land near where he and his family resides. These weekly activities ensure the livelihood of his family.

4.3.2 Interviewing the Market-Vendor

The interview with the market-vendor was held on May 27th, 2016, and its main objective was to understand some specific aspects and characteristics of his own culture, as well as to understand procedures related to the use of the mathematical practices in his daily life. Another objective was related to the elaboration of *rhetorical emic ethnomodels*[4] developed from the market-vendor's labor practices.

Methodologically, the development of the analysis of the collected data and the subsequent interpretation of the results obtained in the interview was conducted by applying the assumptions of a Mixed Methods Study in its *QUAN* + *QUAL* design in which both quantitative and qualitative data were collected and analyzed. Continuing with this analytical procedure, the qualitative data were quantified, which enabled the interpretation of the results of this study through the elaboration of three categories that allowed the research question of this study to be answered.

One of these categories was identified as "Out-of-school environment of the street market and the market vendor". By analysing the collected data in this interview, the first author chose to prepare a report for this category, in text format, as described in the next paragraphs of this chapter in order to interpret the results of this methodological instrument. It is important to state here, that when analyzing factors related to this investigative context, the researchers particularly focused their attention on:

1. The inclusion of the level of education of the vendor.
2. The working time for marketing the products.
3. The experience of the vendor.
4. The degree of mathematical knowledge used in the market.

[4]The rhetorical emic ethnomodels can be regarded as representations that are developed by the members of a specific cultural group, which are based on mathematical ideas, concepts, and procedures rooted in the cultural aspects of the group, such as religion, clothing, behaviors, ornaments, architecture, and lifestyles. Consequently, ethnomodels are based on ideas, procedures and characteristics that are important for the internal understanding of the sociocultural surroundings of these members (Rosa & Orey, 2017a).

5. The knowledge of addition, subtraction, multiplication, and division that contributed to the satisfactory performance of his work in the market.

At the same time, the activities performed by the vendor were not reduced to only the domain of elementary mathematical operations, and it is important to consider the relevance of the articulation of his mathematical knowledge with other knowledge, such as, for example, reading, writing and, above all, respect for the social roles played by members of this particular cultural group and context.

Accordingly, it is important to highlight that the commercial activities developed in the street markets in this region revealed an environment full of possibilities, ideas, procedures, techniques, and mathematical practices that can be translated by the actions of "comparing, classifying, quantifying, measuring, explaining, generalizing, inferring, modelling, and diverse ways used to evaluate, are all ways of thinking, and are present in all human beings" (D'Ambrosio, 2005, p. 30).

For example, through the analysis of the answers given by the market-vendor to the open questions of the interview, it is inferred that there is evidence of the application of local (emic) mathematical practices, which emerged when the vendor mentioned that he sells his products by using plastic packages or small packages by stating that:

> They make calculation easier. I make smaller packages, and I make a package of (. . .) three, five, and fifty *reais*.[5] There are some products that I do pack from three to fifty because the customer wants a smaller or bigger package. Depending on the product I sell, the price goes up because the product is heavy, and the pack is bigger. Thus, I gain profit on weight, in kilos or in grams, right!

For sales, the market-vendor sells his products in packages or per kilo according to the choice of the customers. Packages typically weigh half a kilo and were weighed on a balance scale. According to the market-vendor, this type of weighing makes it possible to obtain a profit from sales, as "I put half a kilo in the package to make a profit and I weigh everything on the scale".

Thus, this *knowhow* is related to the development of mathematical thinking that seeks to explain and understand the diverse ways that the market-vendor has to deal with the work environment in which he is inserted (Cortes, 2017). Then, the vendor makes available a product that can be purchased by customers in packages of a variety of sizes, with the aim of speeding up the purchase process by customers. In this sense, the vendor commented that: "I make a package to make it easier for the customer, as they are in a hurry, and I need to help them quickly".

The way the market-vendor uses to quantify his goods through different sized packages denotes the use of a certain set of skills such as estimation and calculation, as it enables the understanding of how local (emic) knowledge is impregnated in the mathematical procedures and *practices* that are developed in the street market environment. As well, the packaging of products can be considered as an artifice

[5] The Brazilian real or reais (R$) is the official currency of Brazil that is subdivided into 100 cents (Orey & Cortes, 2020).

shared in the street market as one of the "*ad hoc*[6] practices to deal with problems and situations arising from reality" (D'Ambrosio, 2012, p. 16).

Nevertheless, according to Rosa (2010), diverse *ad hoc* practices are aimed at temporary situations or instant solutions that are developed and disseminated in different contexts, such as the street market, for the resolution of problems arising in these environments.

The analysis of the answers given to the questions in the interview showed that the market-vendor used several variables to calculate his expenses, losses, and profits in the market, such as fuel expenses to pick up products at CEASA-MG, fertilizers, and packaging. Thus, the market-vendor commented that "costs were spent on compost, manure, gasoline, and plastic (packaging)" (Cortes, 2017, p. 160).

However, though it was not possible to identify the mathematical practices he used for resolving this particular everyday problem, it was inferred that the monthly amount collected by the market-vendor was sufficient for the support of his family and, also, to cover daily household expenses. For example, the market-vendor argued that "I work as a vendor because I earn more, because if I work as someone's employee, I will only earn the minimum wage, which is not enough for me to support my family, so, I prefer to work in the market" (Cortes, 2017, p. 176).

This analysis also showed that when the market-vendor was asked about how he calculates the price of his products in order to avoid losses, he argued that "if I pay for the merchandise from forty *contos*[7] [reais] upwards, I sell my products it in the range of about five contos [reais] a kilo to six contos [reais] a kilo, if I pay from sixty contos [reais] upwards, then I'll put it in the range of seven contos [reais] upwards". The market-vendor also develops the sales and promotions prices of his products by stating that:

> I do promotion with vegetables and lettuce, in this case of three contos [reais], then I put a promotion on these products, two for five contos, it's the promotion of the street market, at the moment, and it can be three packages for five packages in this case. There are promotions of three for five and two for five, and we do it, we sell it early and then we lower the price. Then, it goes until the end of the day in the market, because my products have good quality, they are first class products, and then, I don't sell for less than that (Cortes, 2017, p. 170).

The answers given by the market-vendor, gave evidence also of an informal use of his mathematical knowledge, by which his commercial and financial labor activities are conducted with the use of personal techniques and strategies (mental calculation and measurements). Table 4.1 shows an excerpt taken from the interview between the first author and the market-vendor.

This excerpt from the interview with the vendor-market shows that the use of *emic* (local) and *etic* (global) knowledge through the application of a *dialogic*

[6] *Ad hoc* is a Latin expression that means *for this purpose*. It generally means a solution designed for specific problems or tasks, non-generalizable, and which cannot be adapted to other purposes (Rosa & Orey, 2010).

[7] Conto is a former Brazilian currency unit used until 1942. Currently, it is a jargon employed by the market-vendor when he wants to talk about the monetary unit of reais.

Table 4.1 Excerpt from the interview conducted by the first author with the market-vendor

Researcher: Explain how you calculate the price of your products that are sold at the fair. Can you use okra as an example? How do you calculate its sale price?
Market-Vendor: In grams?
Teacher-researcher: In kilograms.
Market-Vendor: How I calculate this price (...). If it is ten contos a kilo (...). If it's one hundred grams, then, I calculate the price of how much a kilo and two hundred grams of okra will be. And also, one kilo and two hundred and fifty grams, one kilo three hundred and fifty grams, then, I calculate each price, I already have everything here in my head.
Researcher: Let's use some examples just to understand it and see your strategy that you use. For example, imagine that the product okra is at four contos [reais] a kilo, then, you weigh it and see that the customer took one kilo and four hundred grams.
Market-Vendor: One kilo and four hundred grams.
Researcher: How do you think, how do you do calculate its price?
Market-Vendor: If the price of okra is four reais (...) then, one hundred grams will forty cents. It's four times, then, it is one conto [real] and sixty cents, and it is three contos and twenty cents. It's seven contos [reais] and twenty cents (...). Is that okay?
Researcher: But it's four reais a kilo, right? Could you please rephrase it, as its weight is one kilogram and four hundred grams?
Market-Vendor: One kilo and four hundred grams, how much is that? (...) [thinking] (...) forty cents per gram, eighty cents, one conto [real] and twenty cents (...) [thinking] (...) it's two reais, right? (...) it gives two reais to four hundred grams.
Researcher: Going back to the question of kilograms, can you explain again how do you calculate the price to be paid? Let's see the example, at four reais, the person takes one kilo and eight hundred grams, how do you calculate its price?
Market-Vendor: One kilo and eight hundred grams? (...). It will be (...) [thinking] (...) minus eighty cents, then, it will be seven and twenty. Four reais per kilo, right? Did I do it right?
Researcher: Yes! So, is that how do you thought about it? Can you explain it?
Market-Vendor: I thought from the top to the bottom.
Researcher: But, please, can you explain how and why do you calculate it? [Laughter].
Market-Vendor: Because it's easier, right? I added eight hundred grams with one kilo, and then, I did eight contos [reais] minus eighty cents, right? So, the value decreased.
Researcher: Could you, please, explain it in more detail? I didn't get it yet.
Market-Vendor: You didn't understand it (...) [laughs] (...) the thing is that I take it (...) how is it? (...) one kilo and eight hundred grams, (...), then, one kilo is four reais and one and half kilos is six reais, then, there is another three hundred grams, which is one another one real and twenty cents. It's seven reais and twenty cents.
Researcher: Yes! But now, as I understand it, you did the calculation, the second time, in a unique way.
Market-Vendor: It's a different way.
Researcher: But could you explain how you did it in the first time and then you explain how you did it at this time.
Market-Vendor: The first time I decreased it, right?
Researcher: But can you explain how did you calculate it?
Market-Vendor: I don't know how to explain it (...). I did the calculation according to what is easier for me, I do the calculations at the moment in the market, at the time in my stall and, sometimes, and my stand is full of customers, and I have to do the calculations quickly.

(continued)

Table 4.1 (continued)

Researcher: I understand (...) and just for me to reflect here (...) because I think I did understand what you just explained (...), because I think you did it like that (...) and if I'm wrong, you can correct me (...), I think you thought like this: a kilo costs four reais, and two kilos cost eight reais, but since I asked you to calculate one kilo and eight hundred grams, you did two hundred grams less, so you calculated that two hundred grams would be eighty cents.
Market-Vendor: Eighty cents!
Researcher: Isn't that, right? Then, you took the eight reais minus the eighty cents and gave you got seven reais and twenty cents. Is that right?
Market-Vendor: *Hum-hum* (...) [confirming the assertation].
Researcher: So, that's it. And now, in the second time, how did you calculate it?
Market-Vendor: In the second time I put it together (...), let's assume that one and half kilos cost six reais and that three hundred grams times forty cents is one real and twenty cents, right?
Researcher: So, what did you do next?
Market-Vendor: Then, the total cost is seven reais and twenty cents, isn't it?
Researcher: Yes!
Market-Vendor: It's easier!
Researcher: Let's just use one more example.
Market-Vendor: It's because, as a simple and humble person, sometimes I have no reading and I have no writing practice, and I did the calculation to find the price of my products like this (...). Have you seen how I did the calculation? If I work more, I do the calculations faster. Have you noticed that?
Researcher: Yes! That's what I was going to ask you now because I can see that when you are there, at the market, you would have already done that same calculation much faster.
Market-Vendor: Yes, it is.
Researcher: Could you explain about your feeling of being there in the market doing the calculations to determine the price of your products and by being here doing this interview.
Market-Vendor: Yes! My feeling is that when I am there in the market, I am with a warm body and I have an active memory, and I am connected to my products and to the place I work and, thus, I already have the prices of my products in my head, which makes easy for to find the price for each product. Here, I am not in my working place and the calculations were done in a different context that is an environment that I'm not used to experience.
Researcher: I understand!
Market-Vendor: Let's suppose that a product costs one real and ninety cents per kilo or two reais and ninety cents per kilo, then, it's difficult to calculate its price. For me, it's more difficult to do this kind of calculation, and I can do it only by using a calculator. And let's suppose that one kilo of a product costs five reais and then, I know that a hundred grams cost fifty cents. Then, it's much easier, right?

Source: Adapted from Cortes (2017, p. 151/152)

(glocal) approach enables members of distinct cultural groups to better understand how mathematical practices are locally developed and how they are used in the daily lives of these members, which are contextualized in environments defined according to the diverse and unique customs, history, language, and culture of these members, in this case, the vendor.

Yet, the activities and tasks performed by the market-vendor in his daily labor practices include skills of mental calculation and estimation, which can be considered as a starting point for the elaboration of problems and situations for the implementation of a pedagogical action that can dialogue with local (emic), and school/academic (global) mathematical knowledge used in classrooms.

4.4 Positionality Between the *Going* and *Coming* of the Market-Vendor

The possible duality between *research field* and *school/academic* contexts can foster the development of *strangeness* and *familiarity* (D'Olne Campos, 2000), which is related to *etic* (global) and *emic* (local) approaches as proposed by Rosa and Orey (2012) as a theoretical foundation for the development of ethnomodelling. This context promotes a unique dialogue (glocal) between diverse kinds of knowledge (local and global) in which exchange of information improve a deeper understanding of mathematical procedures and practices developed in diverse contexts (Rosa & Orey, 2017a).

This approach enables the development of the re-signification of mathematical concepts such as functions in a more meaningful way, which shows the connection between local and academic learning (Cortes, 2017). In this regard, the *research field* can be characterized as the *locus of* the collection of empirical data such as indigenous communities, school settings, and markets. On the other hand, *academy* can be considered as the starting point of the researchers who may be in universities, schools, or laboratories, searching for empirical data by comprising research practices (Rosa & Orey, 2017a).

Therefore, in this research, we continue to use emic (local), etic (global), and dialogical (glocal) approaches so that we can understand the theoretical and practical knowledge that is developed respectively, in relation to the *research field* and the *academy*. We emphasize that both *emic* (local) and *etic* (global) concepts have been proposed as an analogy to the terms Phon-***emic*** and Phon-***etic***, which were first used and studied by the linguist Pike (1954).

In this sense, these terms were used in correspondence to the sounds used by a given language, for example, the phonemic is associated with the study of specific sounds used in a given language, while phonetics studies the general aspects of vocal sounds and production of sounds in different languages (Rosa & Orey, 2017a).

In this correspondence between emic (local) and etic (global) approaches, Sturtevant (1974) cited by D'Olne Campos (2000) highlights that the terms "Etic: refers to real-world characteristics independent of culture. Emic: an attempt to discover and describe the behavioral system of a given culture in its own terms, identifying not only the structural units but also the structural classes to which they belong" (p. 121).

According to this assertion, the emic (local) approach can be understood as the perception that members of a given cultural group have in relation to their customs, traditions, and beliefs. Thus, this approach can be considered as the "view of the self towards our own[8]" (Rosa, 2015, p. 333). Here, the emic (local) approach is related to the mathematical perception of the market-vendor regarding his labor practices.

The etic (global) approach can be understood as the mathematical understanding of members who do not belong to a particular cultural group, but who seek to interpret the mathematical knowledge developed by members of that group through an external view. This approach can be considered as the "view of the self towards the other[9]" (Rosa, 2015, p. 333). In this study the etic (global) perspective is related to the view of the researchers and the students in relation to the labor practices of the market-vendor.

On the other hand, a dialogical (glocal) approach can be understood as the *symmetrical dialogue*[10] developed with *alterity*[11] between the emic (local) and etic (global) approaches, as no perspective is more important than the other because they are complementary in the search for a mutual understanding of the mathematical knowledge involved in the mathematical practices developed by the members of distinct cultures (Rosa & Orey, 2017b).

Therefore, the movement of *coming* and *going* of the market-farmer between the *research field* and the *academy* and, consequently, between the *emic* (local) and *etic* (global) approaches, may assume a dichotomous position regarding to the observation of members of a given cultural group through an exclusively internal (emic) or external (etic) postures (Rosa & Orey, 2017b). However, this dichotomy should not be seen as an obstacle for conducting research in ethnomodelling, as the dialogue between both the emic (local) and etic (global) approaches enables *translation*[12] between understandings developed in the *field* and in the *academy* (Rosa & Orey, 2012).

[8]This is how we do it.

[9]This is how they do it.

[10]A symmetrical dialogue is a type of bidirectional communication in which members of distinct cultural groups have the right to speak symmetrically. It means that this dialogue is developed without the predominance of members of a particular cultural group over others. In this type of dialogue, ideas and previously acquired knowledge are socialized by generating behavioral change in these members through the development of transformative actions in society (Freire, 1996).

[11]According to Levinas (1970), alterity derives from the Latin word *alter*, which is a philosophical term related to *otherness*. It is generally taken as the philosophical principle of exchanging one's own perspective for that of the *others*. In this regard, alterity refers to the state of being that of the *others* and *diversity*. It contains concepts like difference and *otherness* within itself. Hence, it is important that difference and *otherness* are unpacked to begin understanding *alterity* and the cluster of meanings associated with *otherness*.

[12]Translation is defined by Miremadi (1993) as a reciprocal process from one culture to the other and from other cultures into one culture. According to Rosa and Orey (2017a), this translational process implies a holistic performance that incorporates globalization and localization by expanding the intracultural flow, which seeks to value and respect the mathematical knowledge developed by members of distinct cultural groups. Therefore, this translational process implies in using alternative ways of expressing cultural meanings, which aims to allow investigators to perceive and experience other realities, cosmologies, and worldviews in an interactional process that mutually influences local (emic) and global (etic) mathematical practices though cultural dynamism.

In this direction, the emic approach is related to observations of mathematical knowledge from an internal perspective of the members of a specific cultural group, while the etic approach is associated with external observations, which are conducted outside the context of theses members by an outside observers and investigators. On the other hand, the dialogical approach is related to the reciprocal complementarity between the knowledge developed emically and etically (Rosa & Orey, 2012).

For example, it is important to note that the excerpt of the interview, shown in Table 4.1, evidenced the pedagogical richness that the use of emic and etic forms of knowledge, complemented by the application of the dialogical approach, can offer to the understanding of mathematical concepts developed by the market-vendor.

However, in the course of this interview, it quickly became evident that when the market-vendor left his workplace to meet the researcher in a school setting (academic). The data interpretation shows that the market-vendor was slightly uncomfortable in this unique environment and with questions asked by the researcher (first author), which caused a certain discomfort that led him to make some mistakes when performing his calculations, which would probably not occur if he was at the street market.

By justifying his discomfort, the vendor originally explained that: “It’s because there, at the street market, my memory is active and I already have the prices of the products in my head” (Cortes, 2017, p. 153). This justification incurs in a kind of positionality in which the market-vendor left his typical cultural environment (emic), which is internal, within the culture, to an external cultural environment (etic) from outside of his culture (Rosa & Orey, 2017a).

Because positionality is a necessary condition for the dialogical interaction to manifest itself in the fieldwork conducted in research in ethnomodelling, it is necessary for the researchers and researched to recognize this movement of *coming* and *going*, in this study, between the street market and the school environment, as well as the approximation and/or distance between the researcher (first author) and the researched (market-vendor) (Rosa & Orey, 2017b). This movement of *coming* and *going* of the market-vendor, between the street market and the school (academic environment), was related to the process involved in human relationships and interactions that take place during the conduction of research in ethnomodelling.

In the context of the street market, mathematical knowledge is not always developed in the school context. In this specific study, local mathematical knowledge was used for contextualizing and enriching school mathematical concepts and content, which increased the power of students’ reasoning in a holistic way. In this regard, Rosa and Orey (2006) argue that there is a need to value social, political, economic, and cultural knowledge linked to daily life of the students. For example, in this study the market-vendor participated in a seminar at the school by positioning himself in the academic context.

The seminar occurred at school, with the participation of all 38 students, the researcher (first author) and the market-vendor. In this seminar, which was essential for the development of the analytical and interpretative process of this research, the researcher (first author) observed the interactions and collaborations between the students and the market-vendor by recording them in his field diary.

However, the market-vendor seemed to be comfortable with his position in the school environment as he explained to the students about the daily labor practices he developed in his own workplace. According to D'Olne Campos (2002), the understanding of this inseparable movement of *coming* and *going* enabled the conversation, the mathematics, and demonstrated symmetrical dialogues or an *otherness*[13] that permeated the dynamics of encounters between the researched (emic) and researchers (etic).

Thus, students were encouraged to question about the mathematical procedures developed by the market-vendor during the deployment of his labor practices, as well, they were able to clarify issues related to the elaborations of ethnomodels related to the re-signification of function concepts (Cortes, 2017). In this regard, during the development of the seminar, the market-vendor shared the mental and physical skills and abilities that he had performed during the deployment of his labor activities at the street market.

For example, during the seminar, several questions were raised in relation to his knowledge and experience related to the determination of the price of his products. In this context, students also requested that the market-vendor explained about the operation of the manual weight scale, as well as how discounts are determined and how he used mental calculations related to the determination of the price regarding to his expenses so that he can make profit.

Therefore, the results of this study show how the market-vendor, in this study, continuously developed his mathematical ideas and procedures that directed him to the development of his own critical and reflective capacity, which is related to knowledge sharing and also with the compatibilization of mathematical practices that are in correspondence with his daily labor practices (Cortes, 2017). This is evident now, as on a recent visit to the street market, the first author observed that the market-vendor has moved towards electronic tools and is no longer using the manual weight scale, which can be considered as a transcendence of his labor practices.

According to this context, market-vendors are considered members of a specific cultural group, whose daily activities are intrinsic to the development of pedagogical implications in the elaboration of mathematics curricular tasks that can contribute to the implementation of an ethnomodelling perspective in the classrooms.

[13] Otherness can be considered as the quality of being different, which enables members of distinct cultural groups to perceive diverse sociocultural features and characteristics in which the main objective is to contemplate diversity. Then, otherness is a situation, state or quality that is constituted through relations of difference, contrast, and distinction. The practice of otherness is linked to relationships among members of one's own group or of distinct cultures. In this way, otherness is also recognized as the estrangement and detachment of the investigators who are there in the research field and who are here in the academic environment in this back-and-forth movement of *coming* and *going* that is mediated by dialogicity (Rosa & Orey, 2017a).

4.5 Implications of the Pedagogical Action of Ethnomodelling Regarding to Positionality of a Market-Vendor

It is important to outline here some of the pedagogical aspects that were revealed during the conduction of the fieldwork of this study, as well as our thoughts about some implications for research after observing the positionality of a market-vendor and his *coming* and *going* movement between his workplace and the school setting.

For example, Cortes (2017) argued for and described the street market as an informal multicultural setting through which students and the market-vendor came together and shared mathematical knowledge in a social, economic, and educational environments that had relevance in a transdisciplinary extra-escolar fashion.

We state here that, during the development of this study, local (emic) mathematical procedures and techniques developed by the market-vendor in his workplace (street market) and mathematical knowledge applied in schools (etic, academic context) were observed, analyzed, and interpreted. This approach offered pedagogical opportunities for the elaboration of ethnomodels by applying emic (local) knowledge related to the labor practices of the market-vendor, which showed the existence of specific mathematical concepts used in everyday situations present in the street market.

For example, according to Cortes (2017), the market-vendor explained how he uses mathematical operations to determine the price of these products:

> *Researcher*: Going back to the question about kilograms, can you explain again how do you calculate the price to be paid for the product? For example, if a person wants to buy one kilo and eight hundred grams of tomatoes, how do you calculate that if its price is 4 reais each kilo?
>
> *Market-Vendor*: One kilo and eight hundred grams? (...) it will be (...) [thinking] (...) it is 8 reais minus eighty cents, then the prices are seven reais and twenty cents. Did I do it right?
>
> *Researcher*: Yes! So, is that how did you think about it? Can you explain it to me?
>
> *Market-Vendor*: I did it from the top down (p. 155).

We observed that the market-vendor performed this operation in a different way from that commonly used in the classrooms. For example, the market-vendor affirmed that: ""I did it from the top down" and instead of calculating the product of 1.8 kg by R$4.00, he calculated the difference of 200 grams to complete it from 1.8 kg to 2.0 kg" (Cortes, 2017, p. 160).

Then, first, the market-vendor calculated the multiplication of 2.0 kg by $4.00, whose result is R$8.00 and then, he subtracted R$0.80, which was equivalent to the value of R$7.20. In this perspective, Biembengut (2000) argued that it is necessary to:

> (...) know, understand, and explain a model [from an etic perspective] or even as certain people or social groups have used or use it [from an emic perspective], can be significant, mainly because it offers us an opportunity to "penetrate the thought" of a culture and obtain a better understanding of its values, its material and social base, among other advantages (Biembengut, 2000, p. 137).

According to this assertion, Rosa (2010) argued that it is from everyday knowledge that learners are enabled to unlock mathematical meanings implicit in culturally specific contexts as, for example, the street market, through the approximation of mathematical knowledge mathematician developed in other cultures or diverse contexts (street market) with the mathematical practices used in the school environment.

Thus, the market-vendor replied that, early in the day, at the beginning of the street market working day, the sale of products is "more expensive due to expenses with gasoline, plastic (packaging), snacks, and the helper". This argument is complemented by his comment on the following example:

> Let's say I am going buy tomatoes, and it will cost 40 reais a box, which is 40 cents per 100 grams. Thus, I cannot sell the tomatoes at that price because of my expenses. In this way, I put the price at five reais a kilo. The price of this product should be more expensive because I don't go to CEASA to buy and sell tomatoes at the same price. So, the price increases 100 percent, 60 percent, and 50 percent, depending on the price I buy the product. This system is used on any product. If the product is priced at 80 or 100 reais a box, then, its price must be 10 reais ou 12 reais (Cortes, 2017, p. 160).

According to the perspective provided by the market-vendor, it is inferred that his emic ethnomodel is related to the multiplicative thinking, in which the sale price is approximately a tenth part of the purchased price. Moreover, the vendor adds another value to this price, which aims to cover the costs and surcharges for the products sold in his stall.

Although the market-vendor did not develop formal academic (school) knowledge in relation to the study of functions and their main mathematical characteristics, he has developed his own mathematical procedures to determine the price of his products, which deals with the labor experience and observation of the world around him. This context allowed him to apply an intuitive concept of function he developed during his work in the street market.

Therefore, through these emic (local) constructs, which are subject to adjustments, the market-vendor can add charges or admit discounts that are inherent to the labor needs and the sale conditions his products. These constructs are typical of the market-vendor's culture, which are comprised in "several ways, techniques, and skills (*tics*) of explaining, understanding, dealing, and living with (*mathema*) different natural and socioeconomic contexts (*etnos*)" (D'Ambrosio, 2001, p. 63) that are present in his own cultural environment.

It is important to emphasize that the market-vendor's mathematical knowledge can be considered as the development of his own ethnomathematics since culture is not considered as a construct apart from and causing the development of mathematical practices because it is not inseparable from the development of mathematical knowledge represented by the members of distinct cultures. According to D'Ambrosio (2001), this happens when members of distinct cultural groups develop their own strategies and techniques in order to calculate, quantify and, also in this case, to work with money and pricing products.

According to these results, it was inferred that the market-vendor performed his duties so that, when communicating his survival strategies, he proposed new ways of relating to mathematics and his own reality. Consequently, in this context, ethnomodelling provided an opportunity for the market-vendor to strengthen his cultural roots through his mathematical knowledge and the peculiarity of his labor practices. This opportunity also provided for students, the development of an understanding of daily activities in their own reality in the context of the street market.

Hence, students observed mathematical strategies, which are used by the market-vendor and associated with the development of his own *tics* and/or techniques. When students developed their ethnomodels related to the market-vendor's mathematical ideas, they experienced a direct connection between what members of distinct cultural groups do with mathematics, and what they are learning in the classrooms.

The implication of this pedagogical action is to show the relevance of the use of diverse ways to solve problems that are inherent to the process of pricing and selling products in the street market by the market-vendor. Accordingly, D'Ambrosio (1990) argues that these *ad hoc* practices develop mathematical procedures used by the members of distinct cultural groups in order to help them to deal with phenomena and daily activities from their own reality.

Therefore, D'Ambrosio (1997) has affirmed that in order to solve specific problems, members of these cultural groups create *ad hoc* solutions, and methods that are generalized to solve similar situations. Thus, members of distinct cultural groups come to *know* mathematics in ways that are quite different from academic-western mathematics taught in schools.

Thus, the analysis of this study on the daily life of the market-vendor showed to the students how the market-vendor's quest for survival can be transformed into transcendence by the development of his unique mathematical knowledge and talents. This approach enabled students to strengthen their own culture through the use of mathematical knowledge developed by the market-vendor in his workplace at the street market.

4.6 Final Considerations

In the everyday environment of the street market, we can easily recognize local practices, such as the mathematical procedures and techniques developed by the market-vendor. This approach provides a study of practical mathematical content that involves fast mental calculations to resolve problems and situations related to discounts, profits, and losses, as well as notions of proportional thinking related to functions. Thus, the daily life of the market-vendors is impregnated with the mathematical *knowledge* and *practices* found in a typical urban street market culture, evidencing the quantifications, measurements, classifications, and comparisons with the knowledge and instruments that are available in this context.

These street markets provide a rich context for study and interactions for students in local schools. In this sense, the daily use, and the commercialization of products in the street market, are ripe for the study and inclusion of teaching and learning process in mathematics revealed mathematical practices learned outside of the school environment, which can be considered as the: *Ethnomathematics of a Street Market through Ethnomodelling*. Therefore, a key component of ethnomodelling is to enable a critical view of reality by using mathematical tools that provide techniques for the pricing of products, profit, and loss and, also, allowed for a look at making the analyses of comparative prices, accounts and budget making, which enables the development of curricular mathematical activities.

On the other hand, after a few years during the pandemic, on a visit to the market, the first researcher noticed that the vendor was now using electronic scales and in the face of the change, was positioned to reflect on what might be the effects of the innovative technologies on the work practices and culture of the market-vendor. This is a question that may lead to a more in-depth study in the future. In this sense, in common cultural settings such as these, one easily sees aspects of mathematics and interactions worthy of study, based on a particular cultural dynamism of the market and the community it serves, which is developed with an emphasis on dialogical communication that is characterized by mutual respect.

In accordance with this context, the initial investigation conducted in this study showed how ethnomodelling is based on the understanding of the mathematical labor practices developed by the market-vendor and its connections with the re-signification of the function concepts. Thus, it would be good to organize ongoing studies that revisit the street market on various occasions over time. It is important to state here that, currently, we are looking at long-term funding of research that would allow a longitudinal growth study.

Thus, in this research, one of the main contributions of ethnomodelling was to organize and present the market-vendor's mathematical practices (emic approach, local) to facilitate their communication, transmission, and dissemination in the school environment. Thus, the representation of the vendor's local mathematical knowledge was translated through scientific methods (etic approach, global) that were related to the re-signification of the concept of function (dialogue, glocalization). Consequently, we have seen how ethnomodelling contributes to the appreciation of the ways of *knowing/doing* of the market-vendor who developed his own mathematical practices, such as counting, measuring, comparing, classifying, and modelling.

The activities carried out at the street market unveiled an environment full of ideas, procedures, and mathematical practices, inherent to the product marketing process, which were implicit in this context, and which are different from those practiced in the school environment. What is powerful as well, is how we showed how to look at ethnomodelling in contexts that are more *day to day* in their cultural context.

Therefore, ethnomodelling enabled the insertion of the re-signification of function in the mathematical curriculum through the elaboration of mathematical activities that originated in the sociocultural context of the school community. This approach enabled the dialogical development between ideas, procedures, and intrinsic mathematical practices to the vendor marketing process (local, emic approach) and school mathematical content (global, etic approach) with the use of problem situations that emerged in the context of the street market.

In this regard, "the use of everyday purchases to teach mathematics reveals practices learned outside the school environment, which is a true ethnomathematics of commerce" (D'Ambrosio, 2001, p. 23). In this way, it is important to emphasize that the school/academic mathematical knowledge related to the concept of functions was adjusted to include the market-vendor's daily life for the elaboration of the activities proposed in the mathematics curriculum.

Thus, an important contribution of ethnomodelling to the process of re-signifying the concept of functions was to provide an analysis of the informal and labor strategies used by the market-vendor, as well as the formal techniques used by students in each cultural context, as these environments constitute spaces for the effective exchange of mathematical knowledge, which is essential for the constitution of mathematical *knowledge*.

In this regard, ethnomodelling is understood as a teaching and learning process that favors a critical analysis of the multiple sources of mathematical knowledge used by students in carrying out proposed activities in the classrooms. This approach corroborated the point of view of Rosa and Orey (2012) who argue that mathematical knowledge can be entered, localized, guided, and grounded in the cultural profile of students and their community.

According to this context, in the dialogical approach, the emic observation sought to understand the mathematical practices developed from the internal cultural dynamics and the relationship between the vendor and the market, which is the cultural environment in which he is inserted. On the other hand, the etic approach provided interculturality, as it employed comparative perspectives with the use of academic mathematical concepts, such as, for example, the concept of functions. Again, we are curious to see, if after the Covid (when it is safe) and with innovative technologies if these results have changed.

In conclusion, the data from this study, showed us how dialogue between mathematical knowledge is an inherent commodity, as it were, of the marketing process developed by the vendor by developing activities based on problem situations that emerged from a street market. This dialogue enabled a sharing of experiences and knowledge that generated innovative mathematical knowledge, the re-signification of the concept of function for students.

The insertion of the dialogical approach to ethnomodelling and the school curriculum of students furthered the valuation of other epistemologies (knowledge) and cosmologies that were related to the mathematical knowledge of the vendor. This is why Rosa and Orey (2012) have argued that a mathematical curriculum based on the perspective of ethnomodelling can provide a theoretical basis, which is a context, for learning, as it uses diverse cultural and linguistic elements of members of distinct cultural groups in the pedagogical action for the process of teaching and learning mathematics.

It is important to note that ethnomodelling as evidenced here, provided the recognition that mental calculations, strategies, and instruments of non-standard measures reveal specific dynamics of knowledge, *to do*, to *understand* and *comprehend* the ideas, the procedures and mathematical practices used in the daily life of the market, in a way learners will revisit over and over in the course of their lives.

This approach can be considered as a peculiar territoriality of the market-vendor that made possible the development of perspectives and a relationship with the students in this study and also with the mathematical practices taught at school. Thus, the dialogue between the *coming and going* through the dialogical activity enabled the *emic* (local) connection and narratives with descriptions and related ideas with *etic* (global) knowledge by enhancing reciprocity.

In this process, the positionality of the market-vendor in being there in the academic setting captured the relationship between his symbolic mathematical knowledge and practices in his own sociocultural context (street market) and its connection to the school environment.

Consequently, it is necessary that students participate in extramural or extracurricular activities because they can observe the outside world to find out diverse ways in solving problem situations they face in their daily life (Monteiro & Pompeu Jr., 2001).

In this regard, Rosa (2010) argued that there is a need to articulate both academic/school knowledge and everyday knowledge, as this path enables the development of a teaching and learning process in contextualized and meaningful mathematics.

Therefore, respect and attention to the daily experiences of students are relevant to the promotion of a meaningful relationship between everyday local mathematical knowledge and that systematized in schools is vital now (post-pandemic), more than ever.

Thus, we can both conclude and agree with Grijó (2011), that it is important to consider work with mathematics in a holistic way by valuing knowledge brought *outside the school walls* such as a street market, which stimulates reflective and creative thinking, and critical students, as well as celebrating and valuing the existing mathematical diversity.

References

Bassanezi, R. C. (2002). *Ensino-aprendizagem com modelagem matemática*. Editora Contexto.

Biembengut, M. S. (2000). Modelagem e etnomatemática: Pontos (In) comuns. In M. C. S. Domite (Ed.), *Anais do Primeiro Congresso Brasileiro de Etnomatemática—CBEm1* (pp. 131–141). Faculdade de Educação, USP.

Cortes, D. P. O. (2017). *Re-significando os conceitos de função: Um estudo misto para entender as contribuições da abordagem dialógica da Etnomodelagem*. Dissertação (Mestrado Profissional em Educação Matemática), Instituto de Ciências Exatas e Biológicas—ICEB, Departamento de Educação Matemática- DEEMA, UFOP, Ouro Preto, MG.

D'Ambrosio, U. (1990). *Etnomatemática: Arte ou técnica de explicar e conhecer*. Editora Ática.

D'Ambrosio, U. (1994). Ação pedagógica e etnomatemática como marcos conceituais para o ensino de matemática. In M. A. V. Bicudo (Ed.), *Educação Matemática* (pp. 14–22). Moraes.

D'Ambrosio, U. (1997). Ethnomathematics and its place in the history and pedagogy of mathematics. In A. B. Powell & Frankenstein (Eds.), *Ethnomathematics: Challenging eurocentrism in mathematics education* (pp. 13–24). State University of New York Press.

D'Ambrosio, U. (2001). *Etnomatemática: Elo entre as tradições e a modernidade*. Autêntica.

D'Ambrosio, U. (2005). Etnomatemática e educação. In G. Knijnik, C. J. Oliveira, & F. Wanderer (Eds.), *Etnomatemática: Currículo e formação de professores* (pp. 39–52). Editora EDUNISC.

D'Ambrosio, U. (2012). Um enfoque transdisciplinar à educação e à história da matemática. In M. A. V. Bicudo & M. C. Borba (Eds.), *Educação matemática: Pesquisa em movimento* (4th ed., pp. 13–29). Cortez.

D'Olne Campos, M. (2000). "Estar aqui" e "Estar lá": Tensões e intersecções com o trabalho de Campo. In M. C. S. Domite (Ed.), *Anais do Primeiro Congresso Brasileiro de Etnomatemática—CBEm1* (pp. 117–131). FE-USP.

D'Olne Campos, M. (2002). Etnociência ou etnografia de saberes, práticas e técnicas? In M. C. M. Amorozo, L. C. Ming, & S. M. P. S. Silva (Eds.), *Métodos de coleta e análise de dados em etnobiologia e etnoecologia e disciplinas correlatas* (pp. 47–91). UNESP.

Freire, P. (1996). *Pedagogia da autonomia: Saberes necessários à prática educativa*. Paz e Terra.

Grijó, D. R. (2011). *Práticas sociais e matemática: Uma proposta transdisciplinar em uma escola rural de Paty do Alferes*. Dissertação de Mestrado, Programa de Pós-Graduação Stricto Sensu Mestrado Profissional em Educação Matemática, Universidade Severino Sombra, Vassouras, RJ.

Jiménez-Espinosa, A. (2002). *Quando professores de matemática da escola e da universidade se encontram: Resignificação e reciprocidade de saberes*. Tese de Doutorado, Faculdade de Educação, UNICAMP, Campinas, SP.

Levinas, E. (1970). *Alterity and transcendence*. (M. B. Smith, Trans.). Continuum International Publishing Group.

Miremadi, A. (1993). *Theories of translation and interpretation*. SAMT.

Monteiro, A., & Pompeu, G., Jr. (2001). *A matemática e os temas transversais*. Editora Moderna.

Orey, D. C., & Cortes, D. P. O. (2020). Unity in differences: Reflections on the doing of ethnomodelling and dialogue through connecting ethnomathematics and modelling. In M. Rosa & C. C. Oliveira (Eds.), *Ethnomathematics in action: Mathematical practices in Brazilian indigenous, urban, and afro communities* (pp. 197–209). Springer.

Pike, K. L. (1954). *Emic and etic standpoints for the description of behavior*. Summer Institute of Linguistics.

Rosa, M. (2010). *A mixed-methods study to understand the perceptions of high-school leaders about ELL students: The case of mathematics*. Tese de Doutorado não publicada, College of Education, California State University, Sacramento (CSUS), Sacramento, CA.

Rosa, M. (2015). Aspectos de insubordinação criativa nas pesquisas em etnomatemática. In B. S. D'Ambrosio & C. S. Lopes (Eds.), *Vertentes da subversão na produção científica em educação matemática* (pp. 325–346). Mercado de Letras.

Rosa, M., & Orey, D. C. (2003). Vinho e queijo: Etnomatemática e modelagem! *BOLEMA, 16*(20), 1–16.

Rosa, M., & Orey, D. C. (2006). Abordagens atuais do programa etnomatemática: Delineando um caminho para a ação pedagógica. *BOLEMA, 19*(26), 19–48.

Rosa, M., & Orey, D. C. (2010). Alho e sal: Etnomatemática com modelagem. *Perspectivas da Educação Matemática, 2*(4), 149–162.

Rosa, M., & Orey, D. C. (2012). O campo de pesquisa em etnomodelagem: As abordagens êmica, ética e dialética. *Educação e Pesquisa, 38*(4), 865–879.

Rosa, M., & Orey, D. C. (2014). Interlocuções polissêmicas entre a etnomatemática e os distintos campos de conhecimento etno-x. *Educação em Revista, 30*(3), 63–97.

Rosa, M., & Orey, D. C. (2017a). *Etnomodelagem: A arte de traduzir práticas matemáticas locais*. Editora Livraria de Física.

Rosa, M., & Orey, D. C. (2017b). Posicionalidade em pesquisas em etnomatemática: Discutindo os movimentos de ir e vir entre o campo e a academia. *Revista Latinoamericana de Etnomatemática, 10*(3), 233–255.

Rosa, M., & Orey, D. C. (2018). Reflecting on glocalization in the contexts of local and global approaches through ethnomodelling. *Educação Matemática Pesquisa, 20*(2), 171–201.

Rose, G. (1997). Situating knowledges: Positionality, reflexivities and other tactics. *Progress in Human Geography, 21*(3), 305–320.

Chapter 5
Ethnomodelling as a Pedagogical Action in Diverse Contexts by Using a Dialogical Knowledge

Érika Dagnoni Ruggiero Dutra, Jéssica Rodrigues, Ana Paula Santos de Sousa Mesquita, and Milton Rosa

5.1 Initial Considerations

In 1996, Professor Ubiratan D'Ambrosio gave a plenary talk at the *Annual Meeting of the California Mathematics Council*, at the Asilomar Conference Center, in California. After he descended the stage, Professor Daniel Clark Orey asked to talk with him, and their initial conversation became 48 h that changed Orey's life forever. Together they walked the grounds, shared lunch and dinner, and before he returned to Brazil, Professor D'Ambrosio suggested Professor Orey to apply for a *Fulbright Scholarship* to work with mathematical modelling and ethnomathematics at the Pontifícia Universidade Católica de Campinas (PUCC), in the state of São Paulo, Brazil.

In 1998, Professor Orey spent a semester in the modelling/ethnomathematics specialization program with Professors D'Ambrosio, Rodney Bassanezi, and approximately 40 mathematics teachers from diverse states in Brazil who worked in groups of 5 or 6 participants to develop their research projects. One of those teachers who participated in these projects would become Professor Milton Rosa. The PUCC modelling groups developed a broad and diverse selection of themes, such as HIV, sex education, beer, esoterics, games, and coffee. Themes were very

É. D. R. Dutra (✉)
Colégio América do Norte, Manhuaçu, Minas Gerais, Brazil

J. Rodrigues
Universidade Federal de Ouro Preto, Belo Horizonte, Minas Gerais, Brazil

A. P. S. de Sousa Mesquita
Universidade Federal de Ouro Preto, Ribeirão das Neves, Minas Gerais, Brazil

M. Rosa
Universidade Federal de Ouro Preto, Ouro Preto, Minas Gerais, Brazil
e-mail: milton.rosa@ufop.edu.br

M. Rosa et al. (eds.), *Mathematical Modelling Programs in Latin America*,
https://doi.org/10.1007/978-3-031-04271-3_5

much based on culture, mathematics, and modelling, which enabled participants in these projects to develop models along these themes.

Professor Orey soon was able to see how modelling was used to bring out aspects of the emerging themes of ethnomathematics, and how modelling is used by thousands of Brazilian teachers to bring-out aspects of, indeed translate how, ethnomathematical ideas, procedures, and practices that are present in diverse contexts. Professor Orey began spending most of his time with the coffee group and doing field research in Professor Rosa's classrooms in Amparo, in the São Paulo state, Brazil, and numerous visits to various coffee plantations and production facilities in the region.

Fast forward 22 years later, both Professors Rosa and Orey are now professors at the Universidade Federal de Ouro Preto (UFOP), in Ouro Preto, Minas Gerais, Brazil, where they have been developing investigations with numerous students in ethnomathematics, mathematical modelling, and ethnomodelling. In this regard, Rosa and Orey (2010) proposed the notion of ethnomodelling, which is the study of mathematical practices developed by members of distinct cultural groups through modelling.

In 2014, researchers Neil da Rocha Canedo and Marco Aurélio Kistemann published a theoretical essay based on a study related to scientific production in Mathematical Modelling within the scope of Mathematical Education in Minas Gerais. In this bibliographic research, they investigated theses, dissertations, and articles in books and journals authored by professors, researchers, master and doctoral students from a diversity of educational institutions and universities in the state of Minas Gerais.

By considering their research objectives, Canedo and Kistemann (2014) presented results they found by conducting an investigative paradigm related to the *State of the Art* by including a methodological trend concerned with developing studies aimed at systematizing published investigations through cross-sectional studies. This approach enables researchers to unveil prevailing perspectives as well as theoretical frameworks used in studies with the highest level of general development, such as techniques, strategies, and/or a scientific field achieved at a particular time and place.

Thus, Canedo and Kistemann (2014) investigated studies conducted in this line of inquiry to systematize the accumulated theoretical knowledge and methodological procedures by presenting a portrait of the development of mathematical modelling in the state of Minas Gerais. As well, they showed the contribution to the development of ethnomodelling by researchers Daniel Orey and Milton Rosa, both professors at the Universidade Federal de Ouro Preto (UFOP).

Hence, it is relevant to acknowledge that in any city or region of the country, there is at least one deep-rooted tradition or culture, whose knowledge, procedures, and practices are spread from generation to generation. Thus, when teachers develop pedagogical actions based on the sociocultural context of their students and community by using ethnomathematical influences (local, emic) that are present in classrooms together with the techniques and strategies of mathematical modelling (global, etic), these professionals use ideas, procedures and mathematical practices that are different from the mathematical contents studied academically in the school environment.

However, the pedagogical action of Ethnomodelling enables the connection between local (emic) mathematical knowledge and practices with global (etic) mathematical knowledge through its dialogical (glocal[1]) approach. According to Ruggiero (2020), these practices are found beyond the physical space of the school environment, and it can be used in the development of activities based on the sociocultural context of the students.

Therefore, we highlight that the way students learn must be linked to their own sociocultural context, however, it is important that teachers understand the social and cultural aspects that are present in everyday life of the students by aiming to provide a teaching and learning process in mathematics that is motivating, contextualized, and meaningful, which enables the development of ethnomodelling (Ruggiero et al., 2019).

In this context, we point out here that the first empirical study developed on ethnomodelling was conducted from 2015 to 2017, by Cortes (2017), at the Universidade Federal de Ouro Preto. Subsequently, from 2017 to the present date, 5 (five) investigations were developed at UFOP and there are 7 (seven) research projects in process on ethnomodelling, which are related to its emic (local), etic (global), and dialogical (glocal) approaches studied in distinct sociocultural contexts.

Thus, the main objective of this chapter is to share data gleaned from three studies conducted by our students in the development of their masters' degree research on diverse ethnomodels they found in their own communities under the ethnomodelling perspective, which deal with local, global, and dialogical mathematical knowledge found on coffee production, peripheral communities, and math trails. However, despite the diversity of these themes, each study has built on the last, as ethnomodelling as a theoretical basis, grows and matures and grows in its theoretical and methodological applications in diverse sociocultural contexts.

5.2 Ethnomathematics, Modelling, Ethnomodelling, and Ethnomodels

In 2010, Professors Daniel Clark Orey and Milton Rosa developed the concept of ethnomodelling in order to establish mathematical modelling as a relevant pedagogical action for the development of an ethnomathematics program, which helps members of distinct cultures to understand how cultural origins and linguistic, social values, morals and lifestyles influence the evolution of mathematical knowledge developed in diverse context.

[1]Glocal (dialogical) approach mixes the local and global forms of knowledge, experience, cosmologies, paradigms, and worldviews (Orey & Cortes, 2020).

Similarly, Shockey and Mitchell (2016) claim that ethnomodelling is a research field that has matured in Brazil and has spread globally. They also state that the theoretical and methodological foundations of ethnomodelling initiated an international discussion about the interdisciplinary role of this research field as a potential contribution to the ethnomathematics program.

According to this context, D'Ambrosio (2017) states that the introduction of the concept of ethnomodelling can be considered as the recognition that modelling is the cognitive strategy *per excellence* for these members to deal with situations and problems present in their own *ethnos*, not only in everyday life, but also in the imaginary world.

In this regard, Lachney et al. (2016) emphasize that researchers in mathematics education can use the insights of ethnomodelling to describe an engaging dynamic of the cultural encounters that promotes interactions between emic (local) and etic (global) knowledges through the development of a dialogical approach.

The etic approach corresponds to the perceptions of external observers towards others whereas the emic (local) approach corresponds to a vision of internal observers towards their own cultural context. Thus, ethnomodelling seeks to connect emic (local) and etic (global) mathematical knowledge by offering a dialogical approach as an alternative point of view based on the dialogue between these two perspectives (Rosa & Orey, 2017b).

Thus, ethnomodelling can be defined as a practical application of ethnomathematics that adds cultural perspectives to the concepts of mathematical modelling, which provides a theoretical basis for the development of pedagogical action that values and respects the diverse cultural and linguistic elements of members of distinct cultures for teaching mathematics through the elaboration of ethnomodels (Rosa & Orey, 2010).

In this context, D'Ambrosio (1993) describes ethnomathematics as the art or technique (*techné* = *tics*) used by members of distinct cultures to explain, know, and understand the problems of reality (*mathema*) in the environment in which these members of these (*ethnos*) live. Thus, ethnomathematics can be understood as the ways in which members of distinct cultures develop ideas, procedures, and techniques so that they can use them to solve problems they face daily.

For example, *tics* are related to measurements, inferences, and performing calculations, comparisons, and classifications, which are different ways of modelling the social, cultural, political, economic, and environmental contexts. Therefore, ethnomathematics proposes the study of sociocultural aspects of mathematical knowledge (Rosa & Orey, 2006). There is need for effective teaching and learning processes in mathematics to be shared in classrooms, and it is important to understand how culture is present in students' learning and, also, how cultural knowledge can be used as a pedagogical resource in schools (Rosa & Orey, 2017a).

In this regard, Orey and Cortes (2020) argued that ethnomodelling is a teaching and learning strategy linked to ethnomathematics, in which students solve everyday problem situations by understanding mathematical practices used in their own sociocultural contexts. In this process, Ruggiero et al. (2021) stated that ethnomodelling is related to the daily activities carried out by members of different cultural groups (*ethnos*) by valuing and respecting the development of their techniques (*tics*) so that these members can deal with the phenomena, situations and problems that make up the different *mathemas* of their daily life.

In this aspect, ethnomodelling connects diverse cultural characteristics of mathematics to school/academic aspects of modelling through the use of a set of features that can be translated between different systems of mathematical knowledge (Rosa & Orey, 2017b). Hence, ethnomodelling is understood as a pedagogical action that connects cultural forms of mathematical development with school curriculum through the modelling process, which results in the exchange of mathematical ideas, procedures, and practices that are historically and dialogically shared among members of distinct cultural groups (Mesquita, 2020).

For example, the former promotes an etic approach (from the outsiders of the culture) and the latter develops an emic (local) approach (from the insiders of the culture), which deal with the elaboration of ethnomodels produced by the members of distinct cultural groups (Canedo & Kistemann, 2014).

Consequently, ethnomodelling seeks to connect mathematical knowledge with culture by using emic (local) and etic (global) approaches through its dialogic (glocal) approach and thus enable a holistic understanding and a broad understanding of mathematical knowledge developed locally by members of different cultural groups. In this perspective, emic (local) and etic (global) approaches can complement each other through cultural dynamism provided by the dialogicity between these two kinds of knowledge (Orey & Cortes, 2020).

This occurs when the emic approach is associated with the point of view of the members of cultural groups, who are the observers from within the culture. In this approach, emic mathematical knowledge is acquired through observation and dissemination of locally developed mathematical ideas, procedures and practices (Rosa & Orey, 2017b).

The etic approach is related to the point of view of external observers (researchers, teachers, and educators) in relation to customs, beliefs, and the mathematical knowledge of members of a given cultural group, who are observers from outside the culture, who develop concepts, theories and hypotheses of local knowledge, which is important and meaningful to internal observers of culture (Rosa & Orey, 2017b).

The dialogical approach highlights the interdependence and complementarity between emic and etic approaches as it provides a balanced or symmetrical dialogue with otherness (alterity). By establishing that none of these approaches is more important than the other, ethnomodelers come to see how they complement each other in the search for mutual understanding and holistic approach to mathematical knowledge involved in mathematical practices developed in different contexts (Rosa & Orey, 2017a).

Therefore, this approach enables researchers, teachers, and educators to become aware of their own prejudices and become familiar with cultural differences that are relevant to members of different cultures. Consequently, Ethnomodelling can be considered as a practical application of ethnomathematics that adds cultural characteristics to the modelling process (Rosa & Orey, 2017b).

Therefore, it is important to state here that ethnomodelling investigators use *translation* to describe the process of modelling emic (local) mathematical systems, as well as to enable the development of etic (global) forms of school/academic representations and vice-versa. These characteristics are related to the ideas, procedures, and mathematical practices developed by members of distinct cultural groups, which are *translated* through the process of ethnomodelling, which depends on acts of translation between local (emic) and global (etic) approaches and vice versa (Rosa & Orey, 2019).

These translations are related to the interrelations between the local (emic) and school/academic (etic) knowledge, which are addressed to help members of distinct cultural groups to understand specific mathematical procedures, techniques, and practices acquired in diverse contexts. This translational process is used to describe the process of modelling local (emic) and global (etic) cultural by applying a dialogical approach through cultural dynamism (Orey & Cortes, 2020).

Thus, the use of ethnomathematics and the application of mathematical modelling helps members of distinct cultures, such as students and educators, to move away from exotic interpretations of mathematical ideas, procedures, and practices by enabling the establishment of relations between local conceptual frameworks and mathematical concepts embedded in global practices through dialogical processes (Lachney et al., 2016).

In this context, Canedo and Kistemann (2014) stated that ethnomodelling was used as an educational practice in a non-school context related to a math trail, which was composed by activities such as the elaboration of ethnomodels related to the curve on a school wall and the slope of the streets in Ouro Preto, as developed by Professor Orey with university students and learners from the municipal school system in the city of Ouro Preto. In these activities, students developed ethnomodels to show the potentiality of ethnomodelling as a pedagogical approach to mathematics education curriculum at all educational levels.

In another research, Pinto and Orey (2018) conducted a project related to the math trail that was aimed to encourage students to develop a sense of awareness and importance towards the conservation and preservation of cultural heritage as well as the identification of the presence of mathematics and modelling in the sociocultural contexts of their own towns. This investigation was developed with 29 seventh grade students in a municipal school in Itabirito, in the state of Minas Gerais, Brazil.

According to Pinto and Orey (2018), this study was developed from the mathematical point of view of elements of historical heritage, based on increasing awareness and sensitivity of students to this theme based on the perspective of ethnomodelling. Data were collected through photographs taken by the students and notes and operations carried out in their field notebook.

During the fieldwork, students were able to observe various geometric patterns, in the lines of houses and churches, which drew their attention because of their high degree of complexity. Results showed that students learned about their own local history and the architectural importance of the town by examining the existing geometric richness found in their walks by elaborating ethnomodels (Pinto & Orey, 2018).

In this regard, ethnomodels establish relations between local (emic) mathematical knowledge and the procedures and practices used in the school and/or academic contexts. Ethnomodels are developed by the members of distinct cultural groups were elaborated in order to instigate a greater potential for association between local communities (emic) and school and/or academic (etic) mathematical knowledge, which is one of the most important aims of the development of these representations (Rosa & Orey, 2017b).

In accordance with this context, ethnomodels are consistent representations of the knowledge socially constructed and shared by the members of distinct cultural groups. Thus, ethnomodels help to link the development of mathematical practices developed by these members with their cultural heritage (Rosa & Orey, 2010). In the ethnomodelling process, ethnomodels can be classified as emic (local), etic (global), and dialogical (glocal).

Emic (local) ethnomodels are representations developed by the members of distinct cultural groups taken from their own reality as they are based on mathematical ideas, procedures, and practices rooted in their own cultural contexts, such as their own science, religion, clothing, ornaments, architecture, and lifestyles.

Etic (global) ethnomodels are elaborated according to the view of the external observers in relation to the systems taken from reality. In this regard, ethnomodelers use techniques to study mathematical practices developed by members of different cultural groups by using common definitions and metric categories.

Dialogical (glocal) ethnomodels are based on the shared understanding that complexity of mathematical phenomena is only verified in the context of cultural groups in which they are developed. In these ethnomodels, the emic approach seeks to understand a particular mathematical procedure based on the observation of the local internal dynamics while the etic approach provides a cross-cultural understanding of these practices.

Consequently, Albanese and Perales (2014) affirmed that ethnomodelling promotes an understanding of the ways in which members of distinct cultural groups communicate, think, diffuse, and disseminate mathematical ideas, procedures, and practices developed locally. This approach is related to the elaboration of ethnomodels that are understood as cultural artifacts that can be used as pedagogical tools that enable the understanding and comprehension of systems taken from the daily lives of these members.

In this process, ethnomodelling is related to the daily activities carried out by members of different cultural groups (*ethno*) by valuing and respecting the development of their techniques (*tics*) so that these members can deal with the phenomena, situations and problems that make up the different *mathemas* of their daily life (Ruggiero et al., 2021).

According to this context, Ethnomodelling seeks to connect the cultural characteristics of mathematics with its school/academic aspects of modelling through the use of a set of characteristics that can be translated between different systems of mathematical knowledge (Rosa & Orey, 2017b).

Thus, ethnomodelling can be understood as a pedagogical action that aims to connect cultural forms of mathematical development with the school curriculum, as one of its main objectives is to have a cultural view of the modelling process, which can result in the exchange of ideas, mathematical notions, procedures and practices, which are historically and dialogically shared among members of different cultural groups, aiming at the transcendence of these knowledge and practices (Rosa & Orey, 2017b).

In the pedagogical action of ethnomodelling, it is important to develop ethnomodels, which are defined as cultural artifacts, which are the tools used to enable and facilitate the understanding and comprehension of systems taken from the daily lives of members of different cultural groups (Rosa & Orey, 2017b). Therefore, the use of ethnomodelling as a pedagogical action for the ethnomathematics as a program values the *tacit knowledge*[2] of the members of distinct cultural groups.

According to Ruggiero (2020), this approach enables the development of students' ability to develop and elaborate ethnomodels for the different applications and contexts from their interests in the sociocultural reality in which they are inserted and, not only, by imposing a curriculum without context or meaning for their learning.

5.3 Research Methodological Design: An Adaptation of Grounded Theory

Grounded Theory is an inductive, qualitative, and exploratory methodology that emerged in the social sciences with sociologists Barney G. Glaser and Anselm L. Strauss, in 1967, having as one of its objectives the validation of qualitative research as methods for the elaboration of an emerging theory. This theory enables a detailed analysis and in-depth interpretation of information through encodings, which allows the identification of concepts based on data analysis. Thus, the data are collected and systematically analyzed through the identification of preliminary codes in the open coding process so that the results obtained are then interpreted through the elaboration of categories identified in the axial coding (Gasque, 2007).

In this codification process, the methodological steps are pre-established to enable the possible elaboration of a theory that emerges from the data from its qualitative analysis. Thus, in this inductive theory, the data are collected, analyzed, and interpreted systematically by enabling the development of a methodological model theoretically based on the information obtained during the analytical phase of the study, which is organized, analyzed, and prepared to conduct the open and axial coding process that is proposed by the Grounded Theory.

---

[2]Tacit knowledge is embedded in personal experiences, it is subjective, contextualized, and analogous. For example, members of distinct cultural groups do not learn how to ride a bicycle by reading a manual, as they need personal experimentation and practices to acquire the skills necessary to learn this action. Therefore, this knowledge is acquired and accumulated through individuals' experience, as it involves intangible factors such as beliefs, perspectives, perceptions, value systems, ideas, emotions, norms, hunches, and intuitions (Rosa & Orey, 2012).

According to Strauss and Corbin (1990), to start an analytical process, first by making use of direct quotes from the participants, which are fragmented in order to be analyzed line by line and sentence by sentence. This procedure allows the identification of preliminary codes through common characteristics and concepts related to a studied problem situation.

Continuing with this analytical process, axial coding is initiated by developing a detailed analysis of the preliminary codes obtained while performing the open coding. In this phase, preliminary codes are grouped according to similar properties and concepts, which helps to identify the conceptual categories (Strauss & Corbin, 1990).

Thus, there is a need to emphasize that this interpretive process can be conducted through a dense description of the conceptual categories that are identified in the analytical process. For the writing of these categories, direct quotations from the students are also used to provide a trustworthy context of the studied problem, thus enabling a detailed interpretation of the results obtained in the conduction of the research.

It is important to highlight that in the 3 (three) investigations described in this chapter, the authors developed and adapted Grounded Theory, as selective coding that enabled the development of the central category was not used as a methodological procedure in these studies given that there was no writing of an emerging theory of the data, since the main objective of the investigators was to answer the proposed research question.

5.4 Ethnomodelling Investigations

This section outlines 3 (three) investigations that focus on how we can use their results to emphasize and introduce ethnomodelling by applying common day-to-day activities and phenomena. Then, we show the connections of these projects to daily events that enable us to demonstrate how they form interesting, practical, and useful pedagogical actions to be developed by educators for the process of teaching and learning mathematics conducted for students in their own sociocultural contexts.

In these projects we selected activities developed with students inside and outside of school contexts so that ethnomodelling could provide an integrative approach to the school mathematics curriculum that considers emic (local), etic (global), and dialogical (glocal) mathematical knowledge origins, which helps educators, teachers, and learners to understand its main theoretical foundations and methodological assumptions based on an adaptation of grounded theory.

This approach helps educators and student to understand ethnomathematics in realistic, holistic, and a more comprehensive way, while looking at diverse mathematical procedures and practices developed by members of distinct cultural groups that make up diverse students' population in the schools. These three studies present the possibility of using daily practices as pedagogical actions in the process of teaching and learning mathematics through the elaboration of ethnomodels during the development of ethnomodelling.

5.4.1 Investigation 1: Ethnomodelling and Coffee: Proposing a Pedagogical Action for the Classrooms

According to Ruggiero (2020), the main objective of this qualitative research was to assist members of distinct cultural groups in developing perspectives related to both emic (local) and etic (global) mathematical knowledge through a dialogical approach (Rosa & Orey, 2017b). This research activity also enables a sociocultural appreciation of the members of different cultural groups (coffee and school cultures) by students through an understanding of cultural dynamism. In this regard, emic (local) mathematical knowledge was used holistically for the development and formulation of etic (global) curricular mathematical activities, in a dialogical manner, which enabled the understanding of mathematical processes developed by members inserted in the coffee culture.

This study was conducted in a private school, located near Manhuaçu, in the Zona da Mata Mineira, Minas Gerais, with 35 students from the second year of a high school. The following research question was defined to guide the researcher in the conduction of this investigation: *How does an application of ethnomathematics, together with modelling tools, contribute to the development of a broader understanding of mathematical and geometric contents for students of the second year of high school, through a pedagogical action based on ethnomodelling, relating to coffee culture of a city near of Manhuaçu, in Minas Gerais?*

To support the development of this research, the theoretical basis of ethnomodelling, mathematical modelling, and ethnomathematics were used. Data collecting was conducted by using two questionnaires (both an initial and a final one), four blocks of activities, a seminar, notes in the teacher-researcher's field diary, and three semi-structured interviews. The analysis of the collected data was conducted according to the methodological design adapted from Grounded Theory (Gasque, 2007) through both open and axial codings, which provided, respectively, the development of preliminary codes and the conceptual categories. In this study, the researcher did not use selective coding because her intention was not to writing an emerging theory, but to answer its research question.

In the context of coffee culture, Rosa (2010) highlights that student involvement with the communities and cultures in which they are a part of, is an important tool to support the teaching and learning process in mathematics. Thus, Ruggiero (2020) states that it is necessary to consider the development of a pedagogical action for mathematics in a holistic way through the valorization of knowledge and actions brought from outside school environments. This pedagogical action stimulates the development of creative, critical, and reflective thinking of the students, as well it provides the use and appreciation of various mathematical practices existing in everyday life.

In this regard, baskets used in the coffee harvest made it possible to develop curricular activities through which the emic (local) mathematical knowledge developed by the members of the coffee culture met with other mathematical knowledge systems, such as the school and academic contexts (etic) through cultural dynamism provided by the ethnomodelling dialogical approach. This approach is related to the dynamics of the encounters between different cultures (Rosa, 2010).

Regarding coffee harvesting, at the site visited by some researchers, workers use artisanal baskets to transport this product. Thus, they receive payment for all the coffee they can harvest on a working day. When these workers were asked about the form of payment, the investigators were informed that the farmer used the basket they made as a unit of measure for payment. Thus, Ruggiero (2020) asked students to discuss and write their conclusions about how to know if the farmer was making the correct payment for the coffee workers.

The analysis of answers given to this problem-situation show that students from 3 (three) groups answered this question positively while students from the other 3 (three) groups answered it negatively. For example, students in *Group C* replied that "Yes, because we think the payment method is fair despite the fact that some baskets have different volumes, they just approximate them, not making payment so uneven". On the other hand, students in *Group B* stated that "No, he is paying less than he should because the farmer is paying 60 liters while coffee workers are harvesting 64 liters", which was the volume calculated by these students previously (Ruggiero, 2020).

The researcher also argued that the analysis of these responses also shows that these students identified the need to standardize the size of the basket used to harvest coffee. For example, students in *Group A* stated that "the farmer should make the payment based on the weight of the coffee bag or by using a basket with standard measures, since the handmade basket for each picker could have different measures". On the other hand, students in *Group C* and *Group D* highlighted the importance of initially calculating the volume of the basket when commenting that the harvesters "should calculate the volume of the basket and then stipulate a value for this measure". Figure 5.1 shows a basket used for harvesting coffee that was shown to students during their visit to the coffee farm.

The next proposed activity was related to the standardization of coffee baskets because during a visit to the coffee farm, students identified that these cultural artifacts have no standardized size for their measurements. This question was related to the etic (global) school knowledge of the students with the emic knowledge (local) of the coffee workers through the dialogical approach of ethnomodelling. As the students had not yet studied the contents of Spatial Geometry, the researcher explained the main characteristics of geometric solids before they solved the following question: *Which geometric solid do you think is more suitable to represent this basket? Determine its volume from an approximate representation of the basket used in coffee harvesting.*

Subsequently, the researcher drew specific geometric solids on the blackboard and asked the students to determine what would be the best approximation for the geometric shape of the coffee basket, which has a circular mouth and a quadrilateral-shaped bottom. Figure 5.2 shows the geometric solids (etic knowledge) represented by the researcher on the blackboard in the classroom.

The answers given to this question show that students in Group B, Group D and Group E chose the truncated cone, students in Group C and Group F chose the

Fig. 5.1 Basket used to harvest coffee. (Source: Ruggiero, 2020, p. 205)

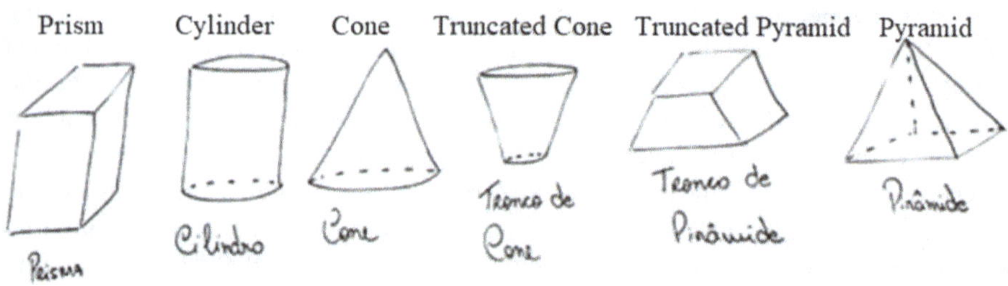

Fig. 5.2 Geometric solids represented by the researcher on the blackboard. (Source: Ruggiero, 2020, p. 205)

cylinder and students in Group A chose the truncated pyramid as the geometric solids to represent the basket used in the coffee harvest. The researcher also asked students to represent through a process of drawing the basket used in the coffee harvest.

The answers given by the students show that the etic representations (ethnomodels) of the coffee basket were developed by using emic information, for example, baskets were made by using real measurements (emic) and geometric shapes learned in the school (etic) in order to verify the necessity of the standardization of this cultural artifact. Figure 5.3 shows the dialogical representation of the baskets used to harvest coffee and, also, their measurements.

Table 5.1 shows the volume determined by the students for the baskets through a dialogical representation that simultaneously used emic and etic knowledge related to this cultural mathematical practice.

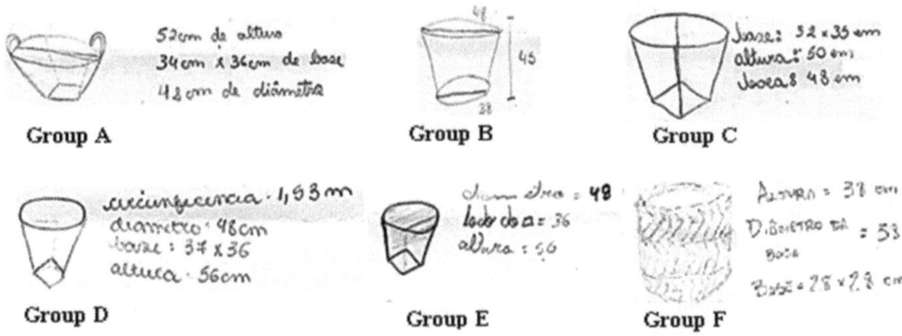

Fig. 5.3 Dialogical representations of the baskets used to harvest coffee. (Source: Ruggiero, 2020, p. 205)

Table 5.1 Volume of the baskets by using dialogical representations

Groups	Basket measurements	Geometric solid	Volume
A	Height = 52 cm; square base side = 34 cm; square base side = 48 cm	Truncated pyramid	V ≅ 88.1 L
B	Height = 45 cm; smaller base radius = 19 cm; largest base radius = 24 cm	Truncated cone	V ≅ 64.7 L
C	Height = 50 cm; base radius = 24 cm	Cylinder	V = 86.4 L
D	Height = 56 cm; smaller base radius = 24 cm; largest base radius = 25.8 cm	Truncated cone	V ≅ 107.3 L
E	Height = 56 cm; smaller base radius = 18 cm; largest base radius = 24 cm	Truncated cone	V ≅ 78 L
F	Height = 38 cm; base radius = 19 cm	Cylinder	V ≅ 42.5 L

Source: Ruggiero (2020, p. 206)

Figure 5.4 shows the elaboration of a dialogical ethnomodel developed by the students of *Group C* in relation to the approximate calculation of the volume of the coffee basket.

The researcher inferred that these ethnomodels were elaborated according to the view of external observers (etic/students) who represented the baskets in the way that they imagine how mathematical developed internally and locally (emic/coffee producers) work. According to Rosa and Orey (2017b), these ethnomodels are related to school and academic mathematical knowledge that predominate in curricular activities developed in schools, however, these knowledges and practices are also rooted in the local coffee culture.

The results of this study demonstrated that the students participating in this investigation developed the necessary abilities to understand their own reality in order to improve the quality of life of the members of their own community, such as when students designed a vehicle prototype to assist coffee workers to harvest this

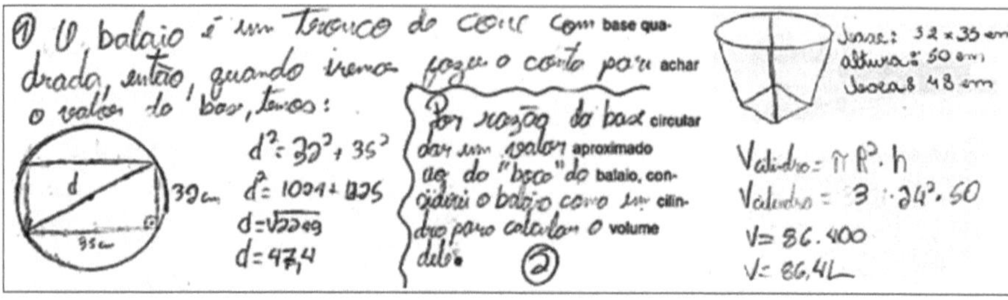

Fig. 5.4 Dialogical ethnomodel elaborated by students in *Group C*. (Source: Ruggiero, 2020, p. 209)

product in mountainous terrains. This perspective shows the possibility of using daily practices developed by members of a certain cultural group (coffee culture) in the teaching and learning process in mathematics through ethnomodelling.

5.4.2 *Investigation 2: A Sociocritical Analysis of Ethnomodelling as a Pedagogical Action for the Development of Mathematical Content in a Peripheral Community*

This study was conducted by Mesquita (2020) in a public state school, located in a peripheral community in the metropolitan region of Belo Horizonte, Minas Gerais. Collected data originated with for the study originated from 6 (six) students in the eighth grade of middle school and with a collector of recycled materials. Its main objective was related to conducting a sociocritical analysis of ethnomodelling as a pedagogical action in the development of mathematical content in a peripheral community.

Other objectives are related to: (a) the discussion and problematization of the relations between mathematical modelling and ethnomathematics that culminated with the concept of ethnomodelling, (b) identification cultural characteristics of peripheral communities and (c) investigation of mathematical practice through ethnomodelling in the classroom.

The following research question was elaborated: *How can ethnomodelling as a pedagogical action constitute a critical environment for the development of mathematical content of eighth grade students in a peripheral community in the metropolitan region of Belo Horizonte?* The theoretical basis of this study was related to concepts of ethnomathematics (D'Ambrosio, 2001), mathematical modelling and its sociocritical perspective, ethnomodelling (Rosa & Orey, 2017b), and peripheral communities (Lana, 2016).

As in the above research, the data were collected, analyzed and coded according to the assumptions of the *Grounded Theory* (Glaser & Strauss, 1967). For the data analysis, both open and axial coding were used, which allowed for the elaboration of conceptual categories that provided the interpretation of results obtained in this study.

In order to collect data, Mesquita (2020) elaborated three blocks of activities based on ethnomathematics (emic), mathematical modelling (etic), and ethnomodelling (dialogical) approaches; two questionnaires (initial and final), field diary of the researcher and a semi-structured interview conducted with a recycled material collector were developed, which were used as instruments of data collection. For example, one of the activities of the fieldwork developed in this research was related to the interview with Mr. João, who is a collector of recycled materials, whose interview excerpt is available below:

Researcher: How does the sale of recycled materials work? Does each material have a different price? How is it?

Mr. João: Yes, they have a different price. One kilo of cardboard costs 30 cents of real.

Researcher: So you need to collect a lot of cardboard to sell, right?

Mr. João: Yes, but a kilo of cardboard is fast to collect. I also collect pet bottles and sell them by kilo. However, there are different types of pet bottles. For example, pet soda and mineral water bottles are different categories. So, a kilo of pet soda bottle costs 1 real and an aluminum can cost 4 reais a kilo. But, the price can go up if a lot of people show up to buy them. This is a question of demand and supply. Each one can costs 5 cents of real and I know that 80 cans weights approximately 1 kilo and, more or less, 1 can costs 5 cents of real.

Researcher: And how do you that? When you sell the cans, how do you know that that amount of cans gives you that value?

Mr. João: I do this work by myself, and I know it because of my experience in collecting cans and approximate their weight in my hands in order to compare the weight of the cans and calculate each price (Mesquita, 2020, p. 135/136).

Then, Mesquita (2020) asked students to read and discuss this interview excerpt in order to solve questions 1, 2, 3, and 4. Hence, the answers given to Question 1: *According to Mr. João, a can costs 5 cents of real when sold in junkyard. So, how much would Mr. João earn if he sells 5 cans?*, show that 6 (six) students answered it correctly by stating that the total value was 25 cents (p. 155). These students performed the multiplicative calculation between the number of cans (5) and its unit value (R \$0.05) to determine the total value. For example, student *F12* affirmed that "he would earn R \$0.25 if he sold 5 cans" (Mesquita, 2020, p. 156).

The analysis of the answers given to Question 2: *By following the same logic of question A, how much would Mr. João earn if he sells 20 cans?* shows that 6 (six) correctly answered this question by stating that the cost of 20 cans are equivalent to R \$1.00 because they multiplied the unit value of each can by the amount of cans sold by Mr. João.

The analysis of the answers given to Question 3: *Now calculate how much would Mr. João earn if he sold 32 cans? And 240 cans?* shows that 6 (six) students correctly answered this question by stating that this amount would be R \$1.60 for 32 cans. Again, these students applied the multiplicative process to solve this question, however, *F8* used multiplication by decomposing the number $32 = 30 + 2$ (Mesquita, 2020, p. 180). Figure 5.5 shows the emic ethnomodel elaborated by student *F8*.

This analysis also shows that when these participants were asked to determine the price of 240 cans, 4 (four) students correctly answered that the value was R \$12.00 by using the multiplication algorithm as performed in the previous questions while

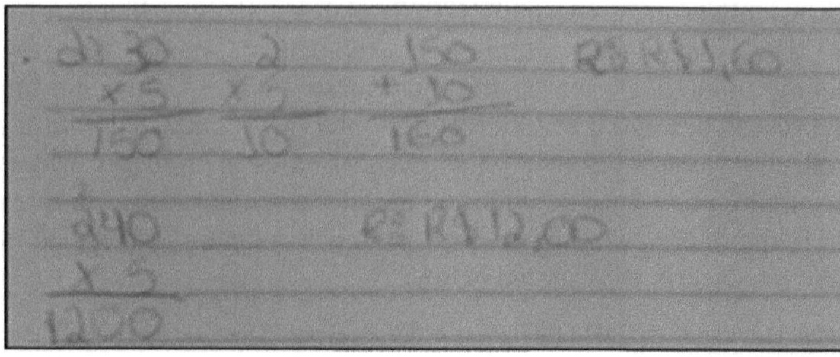

Fig. 5.5 Emic ethnomodel elaborated by student *F8*. (Source: Mesquita, 2020, p. 189)

I would do that by weighing the cans and according to their weight, it would be easier to calculate the total value. For example, if I have 800 cans and 80 cans is equal to 1 (one) kilo, then I have that:

80 = 1 kg	1kg = R$ 4,00	10kg = R$ 40,00
800 = x	10 (kilos)	
80x = 800	x 4 (reais)	
x = 10 kg	40 reais	

Fig. 5.6 Resolution of question 4 by student *F8*. (Source: Adapted from Mesquita, 2020, p. 188)

2 (two) students did not answer this question. It is important to emphasize here that these students solved questions B and C of this activity by elaborating emic and rhetorical ethnomodels, which were prepared according to the information provided by Mr. João.

The analysis of the answers given to Question 4: *How could we calculate the total value that Mr. João would earn if the quantity of cans was very high.*, shows that 2 (two) students did not answer this question while 4 (four) students exemplified this how to determine this value by using larger quantities of cans. For example, student *M7* argued that "I just multiplied the value of each can by the total number of cans". Figure 5.6 shows how student *F8* exemplified this problem-situation by using 800 cans in order to determine the total value of these cans.

This example showed that participant *F8* elaborated an emic ethnomodel to solve the proposed question that shows its generalization to determine the value to be obtained for any quantity of cans. This mathematical thinking shows evidence of the dialogue between distinct mathematical knowledge related to the school community through Mr. João with the school through the students via the development of curricular mathematical content.

The results or this study showed that ethnomodelling contributed to the development of mathematical content of the students, who are residents of a peripheral community, which enabled them to discuss about the lack of adequate basic sanitation that make up the daily lives of its members. These results also show that ethnomodelling provided students with the development of a critical reflection in relation to their space itself.

In addition, these students reflected about multicultural and interdisciplinary experiences from the development and presentation of the project: *Projeto Wall-e: Repensando a Produção de Lixo em Comunidades Periféricas por meio da Matemática* (Wall-e File: rethinking production of garbage in peripheral communities through mathematics), at the 3rd Minas Gerais Scientific Initiation Fair—FEMIC, which culminated with a 1st place in the category of Exact and Earth Sciences and the accreditation for the 18th Brazilian Fair of Sciences and Engineering—FEBRACE.

5.4.3 Investigation 3: A Mathematical Trail and the (Re)Discovery of Mathematical Knowledge Outside of the School

In this investigation, Rodrigues (2021) conducted research entitled: *Exploring the Perspective of Researchers and Participants of the Maths Trails about the (Re)-Discovery of Mathematical Knowledge Outside of School: A Qualitative Study in Ethnomodelling*, which sought forms and contributions of ethnomathematics and its connection with mathematical modelling in the context of the sociocultural perspective of ethnomodelling.

This research was developed with input from two national and three international researchers, who investigate themes related to *Math Trails* and also with six former students of an ethnomathematics course that was offered in a Mathematics Education Master's degree program. All of these participants had experiences with Math Trails.

In this regard, D'Ambrosio (2001) stated that the ethnomathematics program aims to discover and analyze processes of origin, transmission, diffusion, and institutionalization of mathematical knowledge developed by the members of distinct cultural groups. Hence, Rosa (2010) states that this approach enables the promotion of a teaching and learning process in mathematics that respects the values and knowledge of members of different cultural groups, which are brought to the classroom. In this context, Rosa (2005) states that mathematical modelling is considered as a set of procedures whose objective is to build a parallel to try to explain, mathematically, the phenomena present in the human being's daily life, helping him to make predictions and take decisions.

According to these assertions, Rosa and Orey (2017b) argue that ethnomodelling can be used when ethnomathematics is actively used as a system based on a theoretical basis that can solve everyday problems related to the contexts: social, cultural, economic, political and environmental through the mathematical modelling procedures, as it considers the knowledge acquired from cultural practices developed in the communities with the use of locally developed procedures and techniques.

One of the main objectives of this study was to identify local mathematical and geometric concepts that are relevant to the development of awareness and appreciation of mathematical procedures and practices that emerge from different sociocultural contexts through the development of math trails according to the perspective of these participants. In addition, this study also made it possible to understand the contribution of the Ethnomathematics Program and its connection with mathematical modelling in its sociocultural perspective through ethnomodelling in the context of Math Trails.

In this regard, the dialogicity between these theoretical bases enabled the understanding of the mathematical and geometric processes developed by the members inserted in their own reality in a holistic way. In this way, the following research question was elaborated: *How can the perspective of researchers and participants of math trails can contribute to the development of mathematical modelling activities in an ethnomathematics perspective through ethnomodelling*?

In order to support this research, the theoretical basis of ethnomathematics, mathematical modelling, ethnomodelling and math trails were used in its conduction. Data collection was carried out through one questionnaire, five semi-structured interviews, one focus group, and also through the notes in the teacher-researcher's field diary. The data analysis and the interpretation of results of this study were conducted according to a methodological design adapted from the Grounded Theory, through the open and axial codifications, which made it possible, respectively, to identify the preliminary codes and conceptual categories.

The results of this investigation show that math trails present a good form of pedagogical action, as they consist of a sequence of designated places or stations along a planned route, in which students stop to explore the mathematical content contextualized in everyday situations (Richardson, 2004). In this context, this study identified local mathematical and geometric concepts relevant to the development of awareness and appreciation of mathematical procedures and practices that emerge from different sociocultural contexts through the mathematical trails. For example, Fig. 5.7 shows an activity related to the Column Fountain that was performed by the students in one of the math trails be developed in Ouro Preto.

In this context, some students from the discipline Ethnomathematics in the Professional Masters' degree in Mathematics Education, from UFOP that was taught by Professors Milton Rosa and Daniel Orey, in which students explored the concepts of spatial geometry, as well as historical, social and cultural aspects of Column Fountain, on Rua Alvarenga, in Ouro Preto. Figure 5.8 shows the exploration of the concepts of symmetry and spirals involved in the construction of fountains in Ouro Preto, which demonstrates the geometric curves concepts at Contos Fountain that were determined by the students.

These results also show that math trails operate and connect historical, social, and culturally relevant contexts for problem solving, as well as providing pedagogical potential for the development of mathematical content in a creative way through the contextualization of teaching and learning processes in mathematics.

According to Orey (2011), although educators can present examples of school mathematical content originating from their sociocultural experience, it is important that they make a connection between community mathematical knowledge/doing with school mathematical thoughts through dialogicity. Thus, teachers can use

Fig. 5.7 Column Fountain on Rua Alvarenga. (Source: Orey, 2011, p. 10)

Fig. 5.8 Contos Fountain in Ouro Preto. (Source: Orey, 2011, p. 11)

cultural artifacts developed locally to contextualize everyday problem situations to involve students in the teaching and learning process in mathematics.

This approach is similar to the development of Math Trails with respect to cultural artifacts found during the course of the previously determined path (Owens et al., 2003). In this direction, these results also show that these math trails enabled the humanization of mathematics, as the mathematical contents become alive for students by engaging them cognitively, physically, and emotionally (Kenderov et al., 2009) in the activities proposed inside and outside the classrooms.

The interpretation of these results show that there is a need for students to become aware of the relevance of cultural aspects of mathematics in the process of building ethnomodels that are related to the activities performed in the daily life of the Mathematical Trails. This interpretation also showed that there is a need for members of the school community to develop differentiated pedagogical actions so that enable students to understand that mathematical knowledge is related to the sociocultural aspects of ideas, procedures, and mathematical practices developed by the member of local communities.

5.5 Final Considerations

In the three investigations presented here, we were interested in discussing mathematical knowledge intrinsic to the development of different activities, communities and cultures. This aspect is not always valued nor perceived as important by scholars in the academia and, consequently, may not be considered nor explored in school contexts and/or in formal academic pedagogical practices. In this regard, ethnomodelling seeks to connect the cultural aspects of mathematics (ethnomathematics) to school and/or academic mathematics (modelling) (Rosa & Orey, 2017b) in a dialogical manner.

Ethnomodelling also values the relevance of and for the elaboration of ethnomodels developed by the members of distinct cultural groups who translate problem-situations and phenomena taken from their own realities to other mathematical knowledge systems. In order to enable the understanding and comprehension of these systems, the students in these investigations developed ethnomodels that are considered accurate and consistent representations of mathematical knowledge that is socially constructed in their own terms.

Hence, the main objective of this chapter was to show how a group of investigators conduct research based on an ethnomodelling perspective within the scope of mathematics education, and how it is beginning to construct understanding of the mathematical knowledge developed by the members of distinct cultural groups through the development of ethnomodels, which help to give voice to, revitalize, and enhance the identity of community members and, at the same time, enhance the importance of the acquisition of mathematical knowledge developed in their daily lives.

In this context, an ethnomodelling approach was developed to help investigators, educators, and students to comprehend how cultural aspects of mathematics might play a role in the development of the modelling process through the lens of the members of local communities and researchers who develop investigations in this field through the development of important epistemological, theoretical, and methodological applied in their investigations.

It is important to emphasize here that these studies revealed that the conduction of our investigations within the ethnomodelling perspective in the state of Minas Gerais is still evolving and represents a work in progress. Yet we can envision that this research area has a promising future in developing a concrete understanding of different mathematical practices originating in the sociocultural contexts of distinct communities.

For example, ongoing investigations on ethnomodelling related to the mathematical knowledge developed in math trails, soap-stone artifacts, abandoned gold mines, deaf culture, weaving and tapestry, local communities, rural and urban schools, traditional handmade lace work, and the architecture of the historic towns in our region are taking place in the realm of our investigations regarding to the elaboration of emic, etic, and dialogical ethnomodels.

This approach has allowed for a way for us to dialogue and listen to the *voices* of those who are actually doing mathematics in the context of their own daily activities that form a part of diverse ethnomathematical contexts in our region. For example, for far too long, *outsiders*, *external observers* have been describing how the *others* (insiders) think or how they develop their own mathematical ideas, procedures, and practices without giving a strong voice to, or direction about the questions of those that actually are observed to be considered ethnomathematically.

In this context, we are trying to develop a platform in which the *others (insiders)* show us, the outsiders, how they foster mathematical knowledge and do mathematics on their own terms. Thus, it is necessary that we respectfully listen, value, and respect mathematical experiences of diverse peoples, not traditionally studied or given a voice. This context also enables students, educators, and researchers in our region, to make an effort to translate school/academic mathematics on other terms or mathematical knowledge systems and eventually tell their own story.

It is important to emphasize that members of distinct cultural groups form the essence for the development of an ethnomodelling process, which contemplates, inherently, the complementarity of ethnomathematics and modelling. In this process, ethnomodelling discusses the evolution of mathematical knowledge through the history of humanity as a response to a variety of situations and problems originated in and by the distinct contexts.

In our opinion it is a pleasure to be able to share these possibilities to this community of educators and researchers in mathematics education, especially, to those who, like us want to give voice to the people that have often been marginalized and forgotten. We hope that this chapter encourages *other voices* as well who hope to develop ethnomodelling as an investigative program that values and respects different mathematical worldviews, cosmologies, and paradigms in their own contexts. It is our invitation for future investigations in this research field.

There is nothing more rewarding than to have locals walk across a street and engage in sharing the context, history, and background of objects being modelled and being able to mathematize these perspectives for *others* to consider!

References

Albanese, V., & Perales, F. J. (2014). Pensar matemáticamente: una visión etnomatemática de la práctica artesanal soguera. *Relime, 17*(3), 261–288.

Canedo, N. R., Jr., & Kistemann, M. A., Jr. (2014). Movimento das pesquisas em modelagem matemática em Minas Gerais—Brasil. Edição Temática. *Revemat, 9*, 100–123.

Cortes, D. P. O. (2017). *Resignificando os conceitos de função: um estudo misto para entender as contribuições da abordagem dialógica da etnomodelagem*. Dissertação de Mestrado, Departamento de Educação Matemática, Universidade Federal de Ouro Preto, Ouro Preto, MG.

D'Ambrosio, U. (1993). Etnomatemática: um programa. *A Educação Matemática em Revista, 1*(1), 5–11.

D'Ambrosio, U. (2001). *Etnomatemática: elo entre as tradições e a modernidade*. Autêntica.

D'Ambrosio, U. (2017). Prefácio. In M. Rosa & D. C. Orey (Eds.), *Etnomodelagem: a arte de traduzir práticas matemática locais* (pp. 13–16). Editora Livraria da Física.

Gasque, K. C. G. D. (2007). Teoria fundamentada: nova perspectiva à pesquisa exploratória. In S. P. M. Mueller (Org.), *Métodos para a pesquisa em ciência da informação* (pp. 83–118). Thesaurus.

Glaser, B. G., & Strauss, A. L. (1967). *The discovery of grounded theory: Strategies for qualitative research*. Aldine.

Kenderov, P., Rejali, A., Bartolini Bussi, M., Pandelieva, V., Richter, K., Maschietto, M., Kadijevich, D., & Taylor, P. (2009). Challenges beyond the classroom sources and organizational issues. In E. Barbeau & P. Taylor (Eds.), *Challenging mathematics in and beyond the classroom—New ICMI study series 12* (pp. 53–96). Springer.

Lachney, M., Bennett, A., Appiah, J., & Eglash, R. (2016). Modelling in ethnocomputing: Replacing bidirectional flows with recursive emergence. *International Journal for Research in Mathematics Education, 6*(1), 219–243.

Lana, H. F. A. M. M. (2016). *Uma experiência de produção de espaço coletivo na Ocupação Dandara: a comunidade real como horizonte teórico de uma assessorial*. Dissertação de Mestrado em Arquitetura e Urbanismo. Escola de Arquitetura, Universidade Federal de Minas Gerais—UFMG, Belo Horizonte, MG.

Mesquita, A. P. S. S. (2020). *Uma ação pedagógica para o desenvolvimento de conteúdos matemáticos em uma comunidade periférica*. Dissertação de Mestrado em Educação Matemática, Departamento de Educação Matemática, Universidade Federal de Ouro Preto—UFOP, Ouro Preto, MG.

Orey, D. C. (2011). *Projeto trilha de matemática de Ouro Preto: TRIMOP*. UFOP.

Orey, D. C., & Cortes, D. P. O. (2020). Unity in differences: Reflections on the doing of ethnomodelling and dialogue through connecting ethnomathematics and modelling. In M. Rosa & C. C. Oliveira (Eds.), *Ethnomathematics in action: Practices in Brazilian indigenous, urban and Afro communities* (pp. 197–209). Springer.

Owens, K., Pattison, J., & Lewis, E. (2003). Organising maths trails. *Reflections, 28*(1), 27–30.

Pinto, R. A., & Orey, D. C. (2018). *The historical patrimony of Itabirito (Minas Gerais) under the view of ethnomodelling*. Math Trail project. UFOP, Ouro Preto, MG.

Richardson, K. M. (2004). Designing math trails for the elementary school. *Teaching Children Mathematics, 11*(1), 8–14.

Rodrigues, J. (2021). *Explorando a perspectiva de pesquisadores e participantes de trilhas de matemática sobre a (re)descoberta do conhecimento matemático fora da escola: um estudo qualitativo em etnomodelagem*. Dissertação de Mestrado em Educação Matemática, Departamento de Educação Matemática, Universidade Federal de Ouro Preto—UFOP, Ouro Preto, MG.

Rosa, M. (2005). Curriculum e matemática: algumas considerações na perspectiva Etnomatemática. *PLURES Humanidades, 6*(6), 81–96.

Rosa, M. (2010). *A mixed-methods study to understand the perceptions of high school leader about English language learners (ELL): The case of mathematics*. Educational Leadership Program. Unpublished doctorate dissertation, California State University, Sacramento.

Rosa, M., & Orey, D. C. (2006). Abordagens atuais do programa etnomatemática: delineando um caminho para a ação pedagógica. *Boletim de Educação Matemática, 19*(26), 1–26.

Rosa, M., & Orey, D. C. (2010). Ethnomodelling: A pedagogical action for uncovering ethnomathematical practices. *Journal of Mathematical Modelling and Application, 1*(3), 58–67.

Rosa, M., & Orey, D. C. (2012). A Modelagem como um ambiente de aprendizagem para a conversão do conhecimento matemático. *Boletim de Educação Matemática, 26*(41A), 261–290.

Rosa, M., & Orey, D. C. (2017a). *Influências etnomatemáticas em salas de aula: caminhando para a ação pedagógica*. Editora Appris.

Rosa, M., & Orey, D. C. (2017b). *Etnomodelagem: a arte de traduzir práticas matemáticas locais [Ethnomodelling: The art of translating local mathematical practices]*. Editora Livraria da Física.

Rosa, M., & Orey, D. C. (2019). Ethnomodelling as the art of translating mathematical practices. *For the Learning of Mathematics, 39*(2), 19–24.

Ruggiero, E. D. D. (2020). *Etnomodelagem e café: propondo uma ação pedagógica para a sala de aula*. Dissertação de Mestrado em Educação Matemática, Departamento de Educação Matemática, Universidade Federal de Ouro Preto—UFOP, Ouro Preto, MG.

Ruggiero, E. D. D., Orey, D. C., & Rosa, M. (2019). Conexões entre a modelagem matemática e a etnomatemática por meio da cultura cafeeira: uma perspectiva da etnomodelagem. In *Anais da XI Conferência Nacional sobre Modelagem na Educação Matemática—XI CNMEM* (pp. 1–13). UFMG.

Ruggiero, E. D. D., Orey, D. C., & Rosa, M. (2021). Etnomodelando artefatos (balaios) da cultura cafeeira. *Revista de Ensino de Ciências e Matemática—RENCIMA, 12*(2), 1–20.

Shockey, T. L., & Mitchell, J. B. (2016). An ethnomodel of a traditional penobscot summer dwelling. *International Journal for Research in Mathematics Education, 6*(1), 178–195.

Strauss, A., & Corbin, J. M. (1990). *Basics of qualitative research: Grounded theory procedures and techniques*. Sage Publications.

Chapter 6
Ethnomodelling: Weaving Networks Between Academic Mathematical Knowledge and Cultural Knowledge in the Tocantins Southeast Region

Alcione Marques Fernandes, Cristiane Castro Pimentel, and Nayane Rodrigues de Deus

6.1 Introduction

The study and research group in Mathematics and Mathematics Education as part of the *Mestrado Profissional em Matemática em Rede Nacional*[1] (PROFMAT) at the Univ*ersidade Federal de Tocantins*, Campus de Arraias, in the state of Tocantins, Brazil, was created in 2018 with the objective of aggregating the research developed by the teachers participating in this Program. Together with their advisees, the researchers establish the necessary reflective dimension for a graduate program as a professional master's degree program.

Within the Study and Research Group, the line of study related to Etnomatemática e Formação de professores[2] proposes to discuss the Ethnomathematics Program as a research area, establishing its foundations in the mathematics teachers' education, both at undergraduate and graduate levels. Thus, this line of research brings together students from the Teacher Education Programs in Mathematics, as well as students from the PROFMAT Program.

[1] National Network Professional Master's Degree Program.

[2] Ethnomathematics and Teacher Education.

A. M. Fernandes (✉)
Universidade Federal do Tocantins, Arraias, Tocantins, Brazil
e-mail: alcione@uft.edu.br

C. C. Pimentel
Escola Estadual Brigadeiro Felipe, Arraias, Tocantins, Brazil

N. R. de Deus
Colégio Estadual Odolfo Soares, Ponte Alta do Tocantins, Tocantins, Brazil
e-mail: nayanerodrigues@uft.edu.br

M. Rosa et al. (eds.), *Mathematical Modelling Programs in Latin America*,
https://doi.org/10.1007/978-3-031-04271-3_6

This chapter presents and discuses 2 (two) investigations conducted by the students: Cristiane Castro Pimentel (Master's Degree Thesis) and Nayane Rodrigues de Deus (Completion of Course Work), both completed in 2019, and whose theoretical scope is based on the concepts of ethnomodelling.

Currently, investigations related to research in Ethnomathematics Research and Teacher Education Programs have been developed within the Lagoa da Pedra Quilombola Community, located in the municipality of Arraias, in the state of Tocantins, Brazil, with a view to providing guidance for the development of students' project related to the completion of their final course work.

6.2 Ethnomathematics and Ethnomodelling: The Faces of Mathematics Seen Through a Cultural Mirror

Ethnomathematics was constituted as a research field in the Mathematics Education and advanced to its establishment as an investigation area after the *5th International Congress on Mathematics Education (ICME-5)*, held in Adelaide, Australia, in 1984, where Ubiratan D'Ambrosio chaired the opening conference entitled: *The sociocultural bases of Mathematics Education* (D'Ambrosio, 2018).

However, discussions about the cultural influences on mathematical thought organization are not restricted to this historical period; previously there was some sparse research on mathematical practices related to culture (Rosa & Orey, 2014). However, in terms of historical landmarks, we can consider the year 1984 as the birth of Ethnomathematics as a program. The Ethnomathematics word was created from an etymological game established by the mathematics educator D'Ambrosio (2018) who stated that:

> Why not *ethno* [for a commonly accepted group of myths and compatible values and behaviors] + *techné* [for manners, arts, techniques] + *mathema* [for explaining, understanding, learning]. My proposal is a research program to understand *tics* of *mathema* in different *ethnos*. The three Greek roots together form ethno+mathema+tics or, as it would sound better, ethnomathematics (p. 28).

This perspective of Ethnomathematics adopted by D'Ambrosio allows us to investigate different cultural groups based on their ideas, their history, their comparisons and inferences that are used to understand the phenomena and solve problems in the daily lives of the members of these group (D'Ambrosio & Rosa, 2016).

In this regard, Barton (2006) stated that ethnomathematics can be defined as "a research program on the way cultural groups understand, articulate and use concepts and practices that we describe as mathematic, whether or not the cultural group has a concept of mathematics" (p. 53). We understand that Ethnomathematics makes it possible to understand local knowledge and practices by using concepts and methodologies specific related to mathematical thinking.

According to D'Ambrosio (2009), when entering the universe of local mathematical knowledge, whether from a community, a professional group, or any ethnic group, we come across different ways and procedures of living with reality and everyday life. Thus, it is important to highlight that the:

> Daily life is impregnated with the knowledge and practices typical of culture. At all times, individuals are comparing, classifying, quantifying, measuring, explaining, generalizing, inferring and, in some way, evaluating using the material and intellectual instruments that are typical of culture (D'Ambrosio, 2009, p. 22).

Thus, when translating these daily activities into mathematical language, we are considering the dialogue between diverse sociocultural practices[3] and school/academic mathematical knowledge through ethnomathematical and modelling perspectives (Rosa & Orey, 2017).

In this pedagogical action, ethnomodelling points in the direction of establishing a methodological studies for comprehension of knowledge/tasks from the perspective of school/academic mathematics and vice-versa, that is, "ethnomodelling can be considered as the study of ideas and procedures used in mathematical practices developed by the members of different cultural groups" (Rosa & Orey, 2016, p. 57).

The study and research of practices developed by the members of different cultural groups enable the establishment of a dialogue between such mathematical practices with geometric, architectural, and artistic concepts, which are typically related to school/academic mathematics (Rosa & Orey, 2017).

The convergence between these cultural practices and the elaboration of mathematical models is addressed in Ethnomodelling, which is a research area that falls at the intersection between Ethnomathematics, Cultural Anthropology, and Mathematical Modelling (Rosa & Orey, 2017). Figure 6.1 shows ethnomodelling as an intersection area of these three knowledge fields.

In the intersection between 3 (three) areas of investigation: Ethnomathematics, Cultural Anthropology and Mathematical Modelling, we can establish that the research process of sociocultural practices is developed from the translation of these practices into other mathematical systems, such as school/academic environments and vice-versa (Rosa & Orey, 2017).

For example, in the "process of translating locally developed systems the elaboration of ethnomodels takes place through the use of culturally mediated tools, which seek to bring local practices closer to those used in academia" (Rosa & Orey, 2016, p. 59).

For Rosa and Orey (2016), in establishing this dialogue between sociocultural practices and academic mathematics, it is possible to consider that observers, who are members of a given cultural group, share the same concepts and ideas (emic approach) and, in counterpoint, the observation of researchers and educators of these sociocultural practices is considered as an etic approach.

[3]It is necessary to think about ways to conceive and practice an exercise in mathematics education that signals to ways of reading, understanding, and explaining the world in order to make sense of the paths of mathematical construction in different cultural contexts, through a cultural learning process (Farias & Mendes, 2014, p. 38).

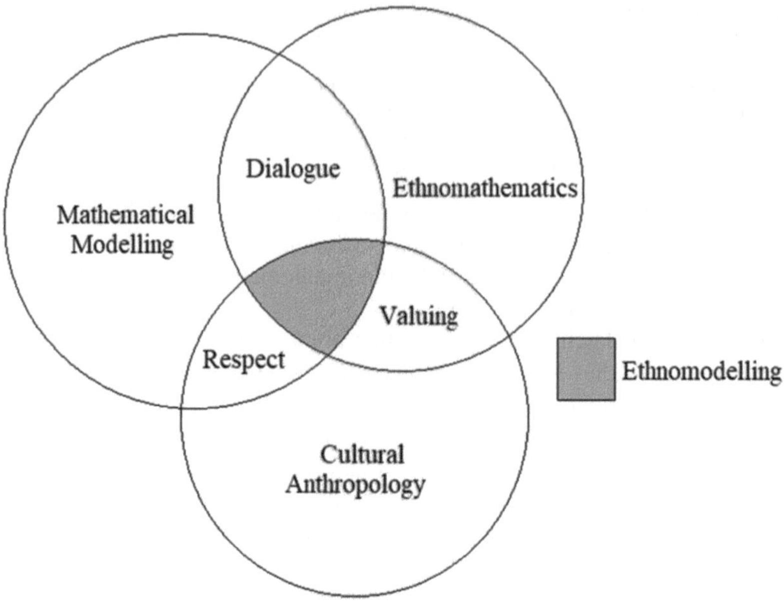

Fig. 6.1 Ethnomodelling as an intersection of areas. (Source: Adapted from Rosa & Orey, 2017, p. 36)

According to Rosa and Orey (2017), the search for the interpretation of the members of a specific cultural group's knowledge and practices leads to a dialogical approach that can be understood as the bridges built between their own mathematical conceptions regarding to these practices and the translations conducted by the external observers, such as researchers and educators.

6.3 Ethnomodels

Throughout history, members of different cultural groups have developed techniques, tools, and artifacts that can help them to solve daily problems faced in their own communities. For example, it is important to consider this quest for survival and transcendence as inherent elements to humanity, because:

> As I said earlier, I see survival and transcendence as the essence of being human [verb]. The human [noun] being, like all living species, seeks only their survival. The will to transcend is something more distinctive of our species (D'Ambrosio, 2009, p. 50).

In this search through different problem-solving techniques, Rosa and Orey (2016) state that ethnomodels can be described as "cultural instruments or artifacts used to provide comprehension and understanding of systems that are taken from the reality of members of different cultural groups" (p. 61). They also affirm that "from this perspective, the primary objective for the elaboration of ethnomodels is to translate the mathematical ideas, concepts, and practices developed by the members of distinct and diverse cultural groups" (p. 66).

This context enabled Rosa and Orey (2021) to state that ethnomodels are based on three different characteristics: emic, etic, and dialogical. Emic ethnomodels are constructed and manipulated within a given cultural group by its members. These models are based on mathematical procedures that are directly related to religious, artistic, and mythological principles, which means that:

> Local or emic ethnomodels are representations developed by members of distinct cultural groups taken from their own reality as they are based on mathematical ideas, procedures, and practices rooted in their own cultural contexts, such as religion, clothing, ornaments, architecture, and lifestyles (Rosa & Orey, 2021, p. 447).

Hence, ethnomodels can be considered as the translation developed by researchers and educators to describe processes used in mathematical practices by the members of a certain community or cultural group. This translational process is conducted between two complementary systems, such as local (emic. communities) and school/academic (educational institutions) mathematical language (Rosa & Orey, 2017). In this way, we can establish that etic ethnomodels use concepts, methods and theories distinct from the members of other cultural groups from which ethnomodels are derived (Rosa & Orey, 2021).

Finally, according to Rosa and Orey (2016), dialogical ethnomodels allow for a holistic approach to be created between the two forms presented above: emic and etic ethnomodels. It is necessary to state that "this approach is necessary to understand and explain this mathematical practice in its entirety, from the point of view of external observers regarding the perception of mathematical knowledge developed by members of this local cultural group" (p. 72).

Below, we present the results from two investigations that were conducted at the Research and Studies Group in Ethnomathematics and Teacher Education at Profmat/Arraias, from the perspective of Ethnomodelling.

6.4 Ethnomodelling in the Manufacturing of Jewelry at Mestre Juvenal's Goldsmithery

The monography, which is the final paper presented in the Teacher Education Course in Mathematics was developed by the student Nayan Rodrigues de Deus, in 2019, under the guidance of the first author/researcher, focused on identifying mathematical practices developed by goldsmiths in the manufacture of jewelry at the Mestre Juvenal workshop, located in Natividade, Tocantins.

In this chapter, we emphasize that the elaboration of ethnomodels practiced by goldsmiths in their work of making jewelry is an important cultural component of the members of this specific cultural group. For example, Bonfim (2019) highlights that jewelry produced in Natividade more than a century ago is a tradition that has been

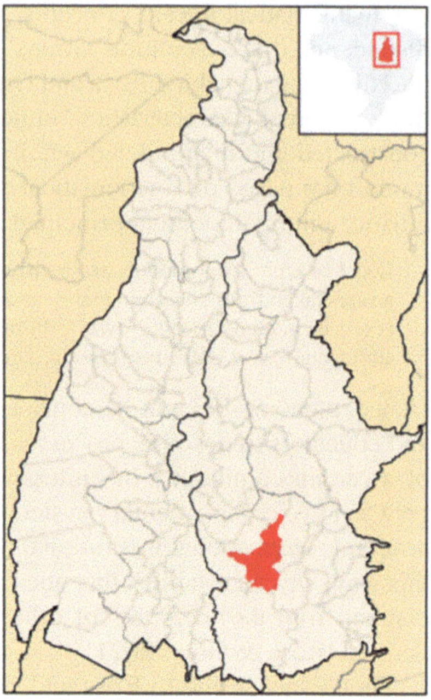

Fig. 6.2 Location of Natividade in Tocantins state, Brazil. (Source: https://upload.wikimedia.org/wikipedia/commons/5/50/Tocantins_Municip_Natividade.svg)

passed down from generation to generation by the first masters who applied this local knowledge.

In this context, it is important to contextualize the relevance of Natividade for the development of this research. Natividade is a town located in the Southeast region of Tocantins state, Brazil, on the right bank of the Tocantins River; listed by the *Instituto do Patrimônio Histórico e Artístico Nacional*[4] (IPHAN) as a national historic heritage since 1987, preserving in its streets, churches, alleys, and squares, original traces of the state's colonial period. Figure 6.2 shows the location of Natividade town.

According to IPHAN (2007), Natividade is significantly influenced by the past that goes back to around the eighteenth century in which the ruins of the beginning of the town demonstrate the relevance of these characteristics. This municipality is the oldest in the state of Tocantins since it was founded in 1734 during the discovery of the gold and, currently, this town still has mineral extraction as one of its main sources of income.

The population of Natividade cultivates different ways of knowing and doing their daily practices that arise from orality and, among of them, there is the knowledge developed by the artisans and their artisanal production of jewelry in gold and silver. In order to develop these artifacts, they use an ancient technique known as

[4] National Historical and Artistic Patrimony Institute.

Fig. 6.3 Filigree technique. (Source: Araújo, 2006)

filigree that can be considered as a fundamental and representative part of the process of the historical constitution of this town.

Filigree is an art of working with metal, which is essentially a popular type of goldsmithing technique passed from generation to generation. We emphasize that, according with IPHAN (2007), the jewelry is handcrafted by using the filigree technique. In this regard:

> (...) technique of using gold or silver threads [filigrees] as fine as those of a hair, which, intertwined and welded, form a delicate lace, transforming into whole pieces or being able to be applied as details to other objects. This technique also refers to the ethno-linguistic origins of slavery, which went there from the foundation of the town. This slavery revealed an important number of black mines, from the Gold Coast and who brought important mining techniques and possibly jewelry (as was customary in West Africa) to the village of Natividade (p. 57).

Therefore, the filigree technique used by the artisans in Natividade has its origins developed by both the Portuguese influence and the African roots. In addition, it has undergone a process of resignification in the elaboration of techniques in a South American environment. Figure 6.3 shows the filigree technique used by the artisans in the Natividade community.

In accordance with Bonfim (2019), currently in Natividade, the master goldsmiths and/or filigree artisans pass on their knowledge to apprentices, who collectively mark their geography and history, which are permeated with senses and meanings that help them to build the cultural identity of their people.

In this context, Wall and Araújo (2015) affirmed that there are 3 (three) goldsmiths in this town: Mestre Juvenal and João Bosco. It is necessary to emphasize that the only artisan who is interested to discuss the educational purpose of this approach is Mestre Juvenal, who aims to train new goldsmiths and, in addition, introduce them to the production of handcrafted jewelry. Currently, Mestre Juvenal goldsmithing has 12 members and during the conduction of this interview, he stated that "I have passed on my teachings on the art of the knowing/doing of handcrafted jewelry to 42 young *nativitanos*[5]".

The research presented in this chapter was conducted in 2019 by using primary sources: semi-structured interviews with three goldsmiths and three apprentices. In addition, documentary sources that are part of the *Associação Comunitária Cultural de Natividade*[6] (ASCCUNA) collection and the IPHAN Monument Project. In this research, we did not deal with the investigation conducted with gold miners in the Príncipe district, which is 35 km away from Natividade, and who were interviewed about their craft.

The results of this study showed that the masters and his apprentices develop their own works from an ethnomodelling perspective, bearing in mind that jewelry manufacture in Natividade originates from a centuries-old tradition, and which was developed from Portuguese and African roots. Thus, Deus (2019) states that:

> We can see through the explanations given by the goldsmiths that the preservation of the ancestry jewelry confection, highly values the tradition of this local knowing/doing. In this sense, it is worth emphasizing that the approximation of the natives with their adornments is presented as something familiar and representative to the community (p. 39).

The handcrafted jewelry manufactured in Natividade can be considered as representative symbols of the history and culture of this town. Figure 6.4 shows examples of important manufactured jewelry.

As these pieces are among the most manufactured by the goldsmiths and can be considered as cultural symbols of the town of Natividade. It is important to discuss the ethnomodel elaborated in the making process of the Native Heart jewelry.

During the interviews with the 3 (three) goldsmiths and 3 (three) apprentices, we inferred that this cultural artifact is considered by the majority of interviewees as the town's cultural symbol by standing out as a favorite among tourists (Deus, 2019). For example, one of the participants stated that "the jewelry most requested by tourists is the Native Heart". Figure 6.5 shows the technical description of the confections of the Native Heart pendant.

It is necessary to point out that the Native Heart represents the origin of the town, as it has the classic meaning of the symbol of love, which is linked to religiosity. In this context, the yellow heart represents pure, sincere, and true love and, in addition to love, it also represents strength, truth, wisdom, justice, the divine, spirit, birth, intuition, and regeneration (Wall & Araújo, 2015).

According to Wall and Araújo (2015), 4 (four) stages are necessary to make the Native Heart: (a) preparation of the frame, (b) filigree making, (c) finishing, and (d) cleaning, which can be considered as the necessary steps for the elaboration of ethnomodels.

---

[5]Nativitano is a demonym or gentilic given to people born in the town of Natividade, in the state of Tocantins, Brazil.

[6]Natividade Cultural Community Association.

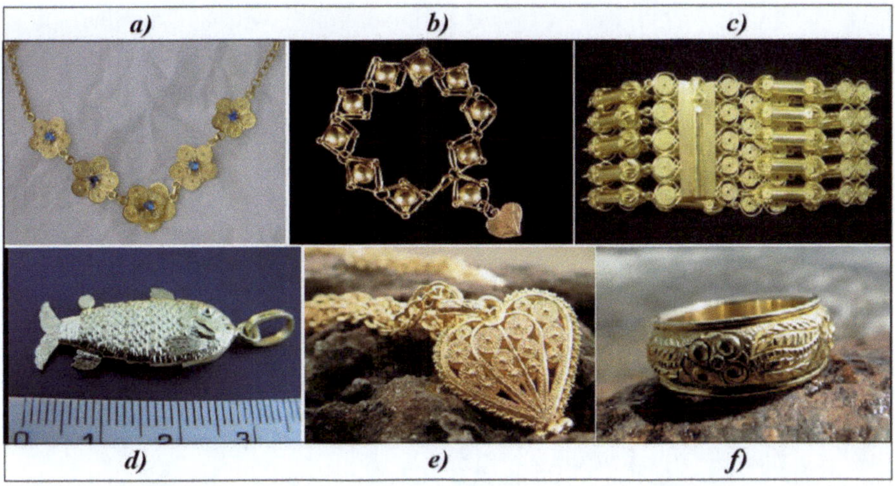

Fig. 6.4 Examples of important manufactured jewelry of Natividade. (**a**) Passionflower earrings in gold, (**b**) bead bracelet in gold, (**c**) slave bracelet in gold, (**d**) fish pendant in gold, (**e**) native heart pendant in gold, and (**f**) slave ring in gold. (Source: Araújo, 2006)

6.4.1 Dialogical Ethnomodels of Native Heart Jewelry

Below, we briefly describe each of the 4 (four) stages related to the putting together of the Native Heart through the elaboration of dialogical ethnomodels.

6.4.1.1 Preparation of the Frame

In this first step, Wall and Araújo (2015) comment that it is necessary to define the size of the piece to be developed. Usually, the diameter of the jewelry is between 64 and 76 mm and its weigh is from 7.5 to 8.5 g. In defining the process to determine the diameter of the frame, a caliper is used to measure its diameter and, after taking its measurements, the gold is melted with an oxygen torch at a temperature between 900 and 1000 °C. In sequence, the gold is melted and transformed into an ingot and then undergoes to lamination processing in order to become a fine wire (filigree). In this case, three wires are used for each face of the heart and the wire thickness is determined as a function of the width of the diameter. After the completion of these processes the piece is welded.

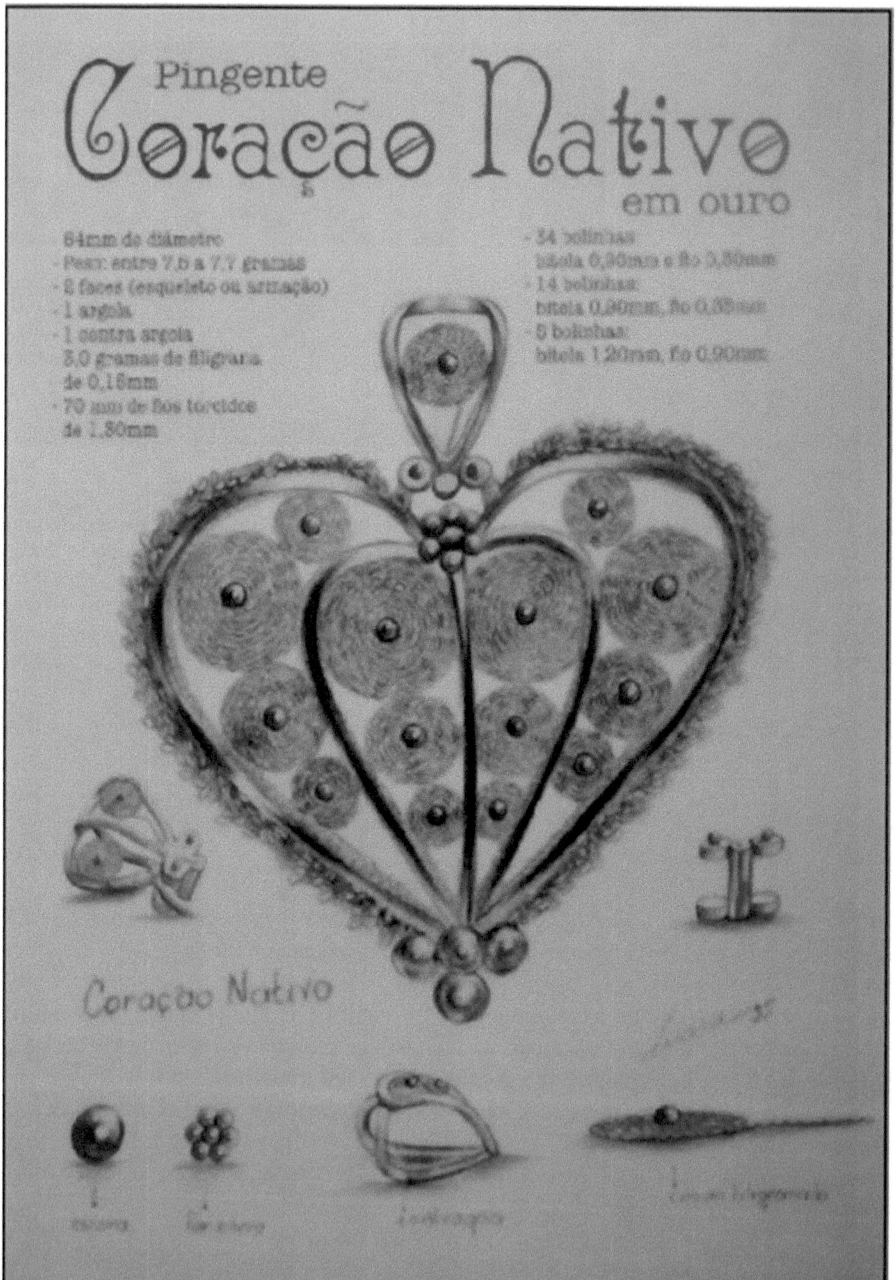

Fig. 6.5 Technical description of the confection of the Native Heart jewelry. (Source: Wall & Araújo, 2015)

6.4.1.2 Filigree Making

In this second step, Wall and Araújo (2015) affirm that the filigree is worked intensively to achieve the ideal filigree thickness between 0.20 to 0.25 mm. To start, make a sharp point at one end of the wire, using pliers to pull it and then anneal it. The next process is to bend the filigree and braid it. Use the drift motor with mandrel to braid the filigree. It is with the filigree that you fill in the frames or the skeletons of this jewelry. It is with filigrees that frames in the jewelry model are filled in. To get a good frame fill, start the process with the inner details of the top of the heart. Then flatten both sides of the heart, using the inlay and always annealing. In summary, the filigree is braided, annealed, and placed on the frame produced in the previous step.

6.4.1.3 Finishing

In order to finish the previous process, Wall and Araújo (2015) state that a total of 53 units (little balls) are produced to compose the Native Heart. Of these 53, 28 little balls are placed in the body of the heart; 6 on its back ring; 14 are part of the finishing in the top of the heart (little flower) and another 5 are placed inside of the heart. The filigrees in the laterals of the heart give the finishing on its sides. Then, they make the ring that is attached to the top of the heart to join its two sides. Then they make the counter ring with a drop shape that are the support for the chain. Fill the counter ring with filigree and weld the little balls. To finish the heart, weld the little flowers (balls) to the top of the heart and, finally, weld the back ring.

6.4.1.4 Cleaning

At the end of the previous stage, Wall and Araújo (2015) highlight that for the cleaning step, the jewelry is washed with water and neutral soap, lathering and cleaning with a brass brush, and quickly dipped in a muriatic acid. The jewelry is ready for use. Figure 6.6 shows a finished Native Heart pendant.

The filigree process used to the confection of the Native Heart production in Natividade, refers to the cultural identity of this town, it is also noticed that the jewelry is related to religious beliefs, deities, and local traditions. Frequently, the production of Nativitan jewelry is wrapped in a twisted gold thread that forms filigree circles and flowers, which are three-dimensional figures placed at the center of the heart as a decoration, as well as a way of bringing the two bodies together. Oftentimes, hearts can appear made with half a gold cap in the center of the flowers (Deus, 2019).

It is important to highlight here that the elaboration of a dialogical ethnomodels related to the production of the artisanal jewelry is developed by using the measures

Fig. 6.6 Finished Native Heart pendant. (Source: Araújo, 2006)

of the diameter of this cultural artifact in order to define its size, which is established at the beginning of this process (Deus, 2019).

During the elaboration of this ethnomodel, the goldsmiths determine the amount of gold needed to produce the jewelry, the percentage of metal casting, the temperature in degrees for the casting of these metals, as well as the circumferences formed by the filigree wires that are inserted inside of the piece, they also place small balls inside of each circle, which remind us of the spheres.

All stages of casting, baking, annealing, as well as lamination, understood as the transformation of the gold ingot into increasingly finer threads, have inherent ethnomodelling characteristics. In this context, the goldsmiths in this study use filigree as a delicate jewelry embellishment in which fine, pliable threads of precious metal are twisted or curled into a design and then soldered onto a piece of jewelry.

In the end of this process, each handcrafted piece possesses mathematical concepts of symmetry and harmony directly related to the concept of beauty. In this regard, the creation of "beauty represents the realization of the highest spiritual potential of human beings, in the manifestation of their sensitive consciousness. The forms of beauty are powerful" (Ostrower, 1998, pp. 280–281).

Therefore, in addition to the cultural identity of the members of this community, jewelry manufactured in Natividade has several mathematical elements that constitute it, such as mathematical procedures, techniques, and concepts of symmetry that can be observed in the double face of the Native Heart.

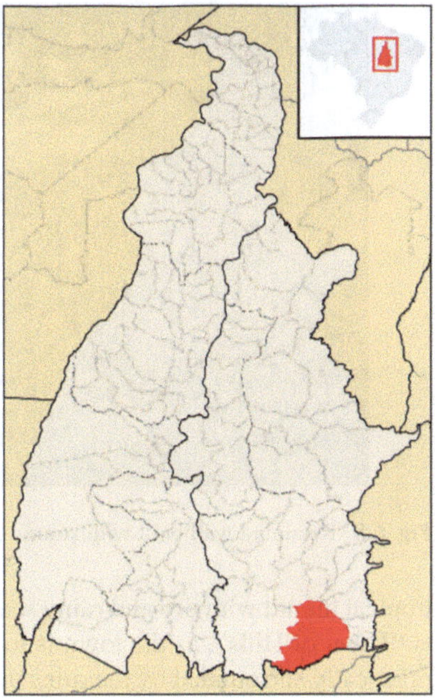

Fig. 6.7 Arraias location in the state of Tocantins. (Source: https://pt.wikipedia.org/wiki/Arraias)

6.4.2 Ethnomodelling of the Walls and Reception Square of the Arraias Cemetery, in the State of Tocantins, Brazil

The dissertation defended in the PROFMAT, in 2019, addresses the geometric elements of the cemetery in the town of Arraias, in the state of Tocantins, Brazil, in order to describe the ethnomodels present in the stone walls that limit its borders and in the reception square in its entrance (Pimentel, 2019).

This qualitative research was based on testimonies of former inhabitants of the town and artisans with the purpose of unveiling the ethnomodels used in the process of using stones to develop their craft work. The development of this investigation aimed to answer the following research question: "What are the geometric ethnomodels present in the wall of the Arraias cemetery and in its reception square?" (Pimentel, 2019, p. 20).

Arraias is a Tocantins secular municipality with an estimated population of 10,525 inhabitants,[7] located in the south part of the state. Figure 6.7 shows the location of Arraias town in the state of Tocantins, Brazil.

The town of Arraias is located in the southeast region of the state of Tocantins, Brazil, which is 420 km away from the state capital named Palmas. Its climate is

[7] For more information, please, access: https://www.ibge.gov.br/cidades-e-estados/to/arraias.html. Retrieved on November 7, 2021.

Fig. 6.8 Retaining wall built with stones in the town center. (Source: Pimentel, 2019, p. 68)

tropical humid with dry and rainy seasons divided throughout the year. Its codename is "Town of Hills", as the town is located in the middle of many hills, and it is also practically surrounded by the ruins of stone walls built by the slaves in the eighteenth century (Fernandes, 2016).

The use of stones to build the walls has been constant in the town' s history, dating back to the time of slavery. Because of this tradition, there are still a few artisans dedicated to this ancient technique of fitting stones by considering its different configurations in the shape of building walls. Figure 6.8 shows a retaining wall built in the town center by using stones.

For the construction of the cemetery walls and its reception square, we assume that the same traditional technique used in the stone walls built by slaves around the town was also used to build the walls in this graveyard, as well as in other old buildings in the Arraias town. The conversations held with Mr. Domingos de Souza, known as Dominguinho, a 95-year-old man, born and raised in Arraias, shared his active participation in the process of building the cemetery walls, since its initial construction. His learning took place through contact with two stone construction masters from Natividade, with the aim of constructing the first wall of the Arraias cemetery.

The results of the interview conducted with Mr. Domingos, on January 20th, 2019; about the process of building the walls, he reported that "in the past there was no cement, so they used lime made by them that was mixed with clay to build the wall with stones". According to him "we take the stone row and burn it in order to have the dust, then we mix the dust with the clay to build the wall" (Pimentel, 2019, p. 72).

In this interview, Mr. Domingos commented about the technique he used in the execution of the stone works by stating that this practice is derived from observations and practices, which are characteristics inherent to traditional knowledge. In this regard, Almeida (2010) highlights that this work is:

> Strengthened by creative adaptation to the ecological environment from which they emerge. Since this knowledge is passed on orally and experimentally, it is responsible for maintaining hundreds of cultural groups spread across places and not yet co-opted by the logic of the market system that levels and standardizes everything (p. 63).

It is necessary to emphasize that, during this investigation, Mr. Domingos was unable to report details about the local procedures (emic knowledge) used by him to carry out the development of his stone works. Therefore, he only described that this work is a lengthy process and difficult to be expressed in words. For example, Pimentel (2019) affirms that the "technique of working with stones is difficult because it takes time, as Mr. Domingos often has to do and undo his work in order to find a perfect fit between the stones (p. 73).

In relation to the research question of this study, we chose to present and describe etic ethnomodels observed in the constructions of both the stone wall and the reception square at the Arraias cemetery entrance.

6.4.2.1 Etic Ethnomodels on Stone Walls and in the Reception Square of the Cemetery

At the Arraias cemetery entrance, we come across a reception square, built in stone, as well as the walls and floor. This construction dates from the early 1980s and it was planned by the city government, yet we did not have access to its initial project. We were also able to identify that Mr. Domingos was the project's executor by using the techniques related to the handicraft of stone construction. According to Rosa and Orey (2016), etic ethnomodels present the view of external observers regarding to cultural practices developed by the members of a particular cultural group.

Based on this assumption, we analyze the shapes present in the reception square and on the cemetery walls from the Euclidean geometric perspective through the elaboration of etic ethnomodels by establishing several parallels in the artisanal construction with academic geometric patterns. Figure 6.9 shows geometric ethnomodels of the reception square at the Arraias cemetery entrance.

Fig. 6.9 Ethnomodel of the reception square at the Arraias cemetery entrance. (Source: Pimentel, 2019, p. 88)

Fig. 6.10 Etic ethnomodel of the columns at the entrance of the cemetery. (Source: Pimentel, 2019, p. 89)

Fig. 6.11 Ethnomodel of the rectangular prism. (Source: Pimentel, 2019, p. 90)

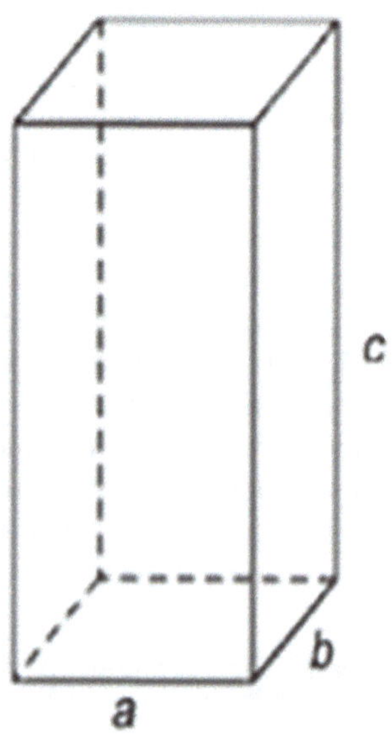

According to Fig. 6.9, we identified the shape of a pentagon at the entrance of the reception square, as well as this construction presents elements of spatial geometry. Figure 6.10 shows the next ethnomodel related to the columns at the entrance of the cemetery, which can be considered as a rectangular prism in a etic ethnomodel as shown in Fig. 6.11.

For the artisan Mr. Domingos, the process of accommodation of the stones in order to produce the geometric shapes showed the possibilities of the elaboration of etic ethnomodels that represent his mathematical knowledge that he learned by observing other master craftsmen as they developed their work.

Finally, we can consider that the stone constructions in the Arraias cemetery follow an emic ethnomodel developed over the years by artisans such as Mr. Domingos, who established appropriate criteria for laying the stones, in order to constitute a firm, solid and visually symmetrical construction by reproducing various geometric shapes.

6.5 Final Considerations

The ethnomodels presented in this chapter are the result of two investigations conducted, in 2019, within the *Study and Research Group in Mathematics and Teaching of Mathematics* in accordance with Ethnomathematical perspective and Teacher Education.

The research with the handcrafted jewelry of Natividade, Tocantins, resulted in a final monograph as part of the required course work that made use of the methodological procedures focusing on observations and interviews with artisans and apprentices in order to identify their perceptions about the mathematics present in their handcrafted jewelry work. We presented here ethnomodels of the Native Heart jewelry, all the other forms of jewelry have geometric elements that can be translated into school/academic mathematical language and, thus, be described as ethnomodels.

The research about the walls and the reception square at the cemetery entrance in Arraias, which was developed as a Master's thesis at PROFMAT. The results showed the application of a proposal for pedagogical action composed of activities that used geometrical elements present in the construction, and architecture, enabled the study of geometric shapes. The etic ethnomodel described in the study represents the mathematical knowledge of the artisan Sr. Dominguinhos' memories that emerged from the answers of this interview.

The Ethnomathematics and Teacher Education research line continues to develop studies in the area of Ethnomathematics and Ethnomodelling at Tocantins southeast region. The result of one of these investigations is related to the recent publication of a chapter in partnership with the student Jeferson Dias dos Santos in relation to the production of cassava flour in the Lagoa da Pedra Quilombola Community. In this work, Fernandes and Santos (2021) described an ethnomodel developed in this Community for the production of the flour, with the objective of improving the quality of production and its reduction preparation time.

We understand that research and discussions about Ethnomodelling in the Tocantins southeast region will continue to bear results, given the vast research field that is presented. The region has several quilombola communities and other traditional communities that develop, in their daily lives, ethnomodels inherited from their ancestry that shape the field of Ethnomodelling. In this regard, Rosa and Orey (2021), affirm that:

> These tools [ethnomodel] enable them [member of distinct cultural groups] to identify and describe the beautiful and often very unique mathematical ideas, procedures, and practices developed in diverse cultural contexts by using the processes of schematizing, formulating, and visualizing problems in truly different ways, as well as by discovering relations and regularities in order to translate real-world problems through mathematization processes (p. 446).

In this context, the group of studies and investigations regarding ethnomodelling, and teacher education has contributed to observing and recording emic, etic, and dialogical ethnomodels developed by the members of traditional communities at Tocantins state in Brazil.This context shows that mathematical knowledge impregnated in the daily activities of these members must considered and respected, through the elaboration of ethnomodels. This process can help the development of a process related to the recovery of the importance of the knowledge local developed in their communities by valuing and legitimizing their knowledge that is transmitted and diffused from generation to generation.

References

Almeida, M. (2010). *Complexidade, saberes científicos, saberes da tradição*. Livraria da Física.

Araújo, S. C. (2006). *Fotos da Ourivesaria Mestre Juvenal de Natividade [CD-ROM]*. ASCUNA.

Barton, B. (2006). Dando sentido à etnomatemática: etnomatemática fazendo sentido. In J. M. Ribeiro, M. S. Domite, & R. Ferreira (Eds.), *Etnomatemática: papel, valor e significado* (2nd ed., pp. 39–74). Zouk.

Bonfim, W. F. (2019). *Os filigraneiros de Natividade, Tocantins: Patrimônio imaterial, identidade e turismo*. Dissertação de mestrado, Universidade Federal do Tocantins, Programa de Pós-Graduação em Geografia, Porto Nacional, TO.

D'Ambrosio, U. (2009). *Etnomatemática: elo entre as tradições e a modernidade* (3rd ed.). Autêntica.

D'Ambrosio, U. (2018). Como foi gerado o nome etnomatemática ou alustapasivistykselitys. In M. Fantinato & A. V. Freitas (Eds.), *Etnomatemática: concepções, dinâmicas e desafios* (pp. 21–30). Paco.

D'Ambrosio, U., & Rosa, M. (2016). Um diálogo com Ubiratan D'Ambrosio: uma conversa brasileira sobre etnomatemática. In F. D. Bandeira & P. G. Gonçalves (Eds.), *Etnomatemáticas pelo Brasil: aspectos teóricos, ticas de matema e práticas escolares* (pp. 13–37). CRV.

Deus, N. R. (2019). *Etnomatemática na Ourivesaria Mestre Juvenal: um estudo sobre o saber/fazer matemático na fabricação de joias em Natividade-TO*. Trabalho de conclusão de curso, Universidade Federal do Tocantins. Retrieved from http://hdl.handle.net/11612/2120

Farias, C. A., & Mendes, I. A. (2014). As culturas são marcas das sociedades humanas. In I. A. Mendes & C. A. Farias (Eds.), *Práticas socioculturais e Educação Matemática* (pp. 15–48). Livraria da Física.

Fernandes, A. M. (2016). *Louceiras de Arraias: do olhar etnomatemático à Ecologia de Saberes na Universidade Federal do Tocantins*. Tese de Doutorado, Universidade de Brasília, Faculdade de Educação, Brasília.

Fernandes, A. M., & Santos, J. D. (2021). A Criatividade como Destino na Comunidade Quilombola Lagoa da Pedra. *Journal of Mathematics and Culture*, 48–62. Retrieved from https://journalofmathematicsandculture.files.wordpress.com/2021/06/152-article_3-1.pdf

Instituto do Patrimônio Histórico e Artístico Nacional. (2007). *INRC de Natividade: estudos para o registro da ourivesaria em filigrana de Natividade, Tocantins como patrimônio cultural imaterial brasileiro*, Brasília, DF, Brasil.

Ostrower, F. (1998). *A sensibilidade do intelecto*. Campus.

Pimentel, C. C. (2019). *Etnomodelagem: uma abordagem de conceitos geométricos no Cemitério de Arraias—TO*. Dissertação de Mestrado, Universidade Federal do Tocantins—Campus Universitário de Arraias, Programa de Mestrado Profissional em Matemática—Profmat, Arraias. Retrieved from http://hdl.handle.net/11612/1372

Rosa, M., & Orey, D. C. (2014). Fragmentos históricos do programa etnomatemática: como tudo começou? In *Anais do 6° Encontro Luso-Brasileiro de História da Matemática—6° ELBHM* (pp. 535–558). SBHMat.

Rosa, M., & Orey, D. C. (2016). Etnomodelagem: uma relação dialógica entre Etnomatemática e a Modelagem. In F. Bandeira & P. F. Gonçalves (Eds.), *Etnomatemáticas pelo Brasil: aspectos teóricos, ticas de matema e práticas escolares* (pp. 55–76). CRV.

Rosa, M., & Orey, D. C. (2017). *Etnomodelagem: a arte de traduzir práticas matemáticas locais*. Livraria Editora da Física.

Rosa, M., & Orey, D. C. (2021). Ethnomodelling as a glocalization process of mathematical practices through cultural dynamism. *The Mathematics Enthusiast, 18*(3), 438–468.

Wall, M., & Araújo, S. C. (2015). *Manual das 7 joias de Natividade—Tocantins por Mestre Wall*. 10-Empresa de Comunicação.

Chapter 7
Mathematical Analysis of the Ceramic Designs of the Pre-Columbian Culture of Ecuador Trough Ethnomodelling with a Sociocultural Approach

Juan Ramón Cadena and Ronald Patricio Chasiloa Llumiquinga

7.1 Introduction

This chapter proposes mathematical modelling as a teaching-learning methodology based on the conceptual framework of ethnomathematics that integrates a historical, semiotic and epistemological approach. For this proposal, some examples of pre-Columbian ceramics from Ecuador are taken as an object of study.

The learning of mathematics in the educational system of Ecuador has always represented a problem of importance and evident implications. According to the analysis of the latest results of the exam that assesses the development of aptitudes and skills that students must achieve upon completion of intermediate education and that are necessary for successful development as citizens and to be able to access higher education studies, the subject of mathematics is the one with the lowest quantitative results within the basic subjects.

It can be seen in Graph 7.1 that their average barely exceeds the elementary level. In the framework of the Pisa Tests, in which the country participates for the first and only time in 2017, 70% of students did not reach the basic level of mathematical mastery skills, which reflects the low level of international competitiveness in this science (Ineval, 2018).

In our country, teaching schemes are still maintained in which traditional school pedagogies predominate, memorization of standard algorithms, non-contemporary approaches and decontextualization with the environment, prioritizing of results

J. R. Cadena (✉)
Central University of Ecuador, Quito, Ecuador
e-mail: jrcadena@uce.edu.ec

R. P. C. Llumiquinga
Unidad Educativa Píntag General, Quito, Ecuador
e-mail: ronald.chasiloa@educacion.gob.ec

M. Rosa et al. (eds.), *Mathematical Modelling Programs in Latin America*,
https://doi.org/10.1007/978-3-031-04271-3_7

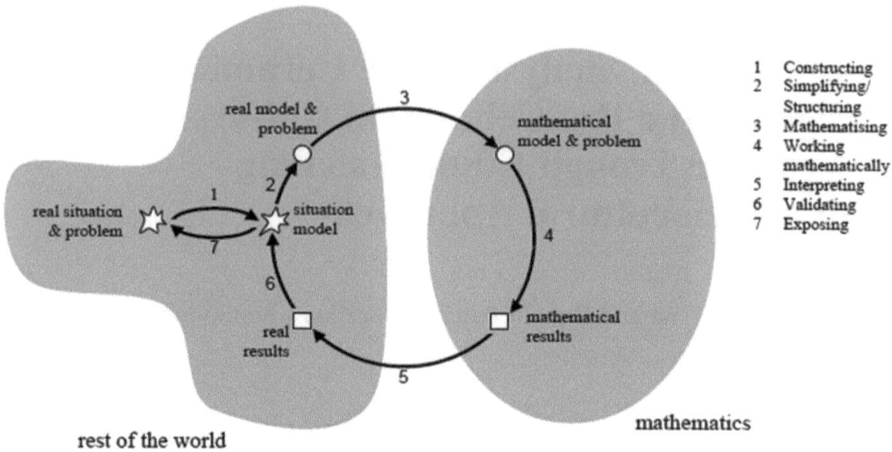

Graph 7.1 Average mathematics proficiency (Ecuador). (Source: Ministry of Education of Ecuador)

before processes, memory before reasoning and, especially, teaching attached to forced abstraction, all combine to limit the potential of this science in the development of cognitive types of reasoning, proposition, and criticality.

We understand that professional training of mathematics teachers must go through processes that consider reflection as a constant didactic practice (Huincahue et al., 2018). Mathematics education in Ecuador presents a dialectical contradiction between Western European rationality, which is reduced to instrumental reason vis-à-vis nature, centered on the individual human being, and Andean rationality that recognizes the duality between nature and human beings (Cadena & Trujillo, 2016).

As Lévi-Strauss (1977) said, in the scenario of the Newtonian model of formal science, it is not exempt from myth and with other forms of sensitivity in understanding the world, such as symmetry (reciprocity that is reflected in a dualistic worldview of reality) and non-arbitrariness (complementarity), the question then arises: How can we improve the learning of mathematics in Ecuador from a cultural, anthropological and historical perspective?

The following section presents an educational proposal from the ethnomodelling perspective, which aims to improve the learning of mathematics in Ecuador from a cultural context, considering the archaeological wealth of the region. The ceramic design of the Andean region offers the possibility of semiotic, geometric, and arithmetic interpretations.

7.2 An Educational Proposal for the Development of Ethnomodelling

This work, under the ethnomodelling scheme, according with Rosa and Orey (2016), theoretically aims to carry out three differentiated and interpretive visions of the phenomenon of mathematics teaching.

(a) *Global vision*: learn about the conceptions of contemporary culture and education in a dialogue with anthropology and history, trying to explain from outside, or from an academic perspective, the way in which these currents converge in the pedagogy of mathematics.
(b) *Local vision*: investigate the aspects related to the characteristics of each cultural sector, considering the conditions of the generation of cognition and conceptualization of mathematical elements, such as counting, measurement, classification, and generalization (Rosa & Orey, 2016). In this context, a study of the ceramics of the pre-Columbian cultures of the northern region of Ecuador presents with a valuable research opportunity, from the perspective of mathematics.
(c) *Glocal vision*: is used to inquire about the aspects that connect the two visions, the external in symbiosis with the internal, the objective analysis of the different sources emanating from the West and that of the Andean onto semiotic, through the generation of flexible methodologies that incorporate and allow for self-study and to rediscover our own episteme.

There is already a route started a few years ago by researchers determined to find innovative methods for teaching mathematics in multicultural contexts. In addition, as an expectation and valuation of the knowledge of the indigenous peoples in many South American countries (Cadena & Trujillo, 2016).

Mathematics contributes to the achievement of competences that make it possible to solve problems outside of it, such as enhancing capacities for discernment, systematization, creativity, criticality, and crucial decision-making. Inherent in this categorization of mathematics, teacher training plays a very important role (Cadena & Trujillo, 2016).

According to Llinares (2012), there must be a sustainable interrelation of the level of knowledge of mathematics and the associated pedagogical load through the didactic transposition. Furthermore, Blanco-Álvarez et al. (2017) state that, in another context, the disciplinary knowledge must be complemented with an inclusive vision of the cultural environment that produces significant learning.

Then, to the extent of the problem raised, a proposal is suggested that seeks to reduce school failure in learning mathematics from a study that has ethnomathematics as its central nucleus, as a new look at education from anthropological, historical, and educational perspectives (D'Ambrosio, 1990).

It is important while considering the mathematical knowledge accumulated through generations is based, among others, on ideas concerning comparison, classification, measurement, operability, and generalization. Therefore, these must be explained under the phenomenological characteristics of a historical, cultural, social, political, and ideological context (Rosa & Orey, 2016).

We must also consider that the view of ethnomathematics conceives a very particular episteme, with an *Andean ethos*[1] that measures the dimensionality of time and space in contrasting contexts with the perspective of mathematics coming from the European, Arab, Babylonian, and Egyptian sources. Then, it is necessary to generate research that allows us to recover, recreate, and reinvent the contributions of the native American peoples (Cadena & Trujillo, 2016).

In this regard, from an Ecuadorian perspective, diverse mathematical concepts or ideas are best studied by looking at its content in the timeline, and beyond the historical niche, in the conditions of society, in its reality as an ontological construct and its relationship with nature (Cadena & Trujillo, 2016). In this direction, we emphasize that the teaching-learning process of mathematics in Ecuador has been based mainly on the acceptance and uncritical reproduction of the western rationality model.

As mentioned earlier here, the Pisa data suggest what has been done in the past does not work. A new perspective that respects history, culture, technology, and the diversity of the Andean region is called for. For its part, Andean rationality recognizes otherness (difference), as something essential that admits and enriches.

As well, with the introduction of reason and myth used together with other forms of understanding the world, symmetry (reciprocity, which is reflected in a dualistic vision of reality) and non-arbitrariness (complementarity). All form a context for understanding and learning. In this context, the Andean person builds a collective identity in relation to the human being (Lizcano, 2006).

Considering that the notion of *culture* is difficult to demarcate and apply in the field of education, this implies both historical and political contexts, along with different meanings in terms of language, the topics to be discussed should have characteristics that allow an epistemological model, which combines the universal with the particular thought (Kragh Sørensen, 2014).

Finally, with the union of the two approaches: global and local, referred to by Rosa and Orey (2016), as glocal, the topics under study are viewed according to a more integrative and disruptive model, in such a way that as the contents are taken into account, the dialogue between knowledge developed by the insiders and outsiders, which makes them acquire an epistemological and otherness.

That is, the recognition of the knowledge of others and how it interacts with local knowledge as a practice of knowledge creates a clearer understanding. In this sense, Cadena and Trujillo (2016) state that it is also important to recognize the different ethical conceptions of cultures and tend towards a universal ethos.

Research can then be carried out under the following theoretical assumptions: mathematical modelling as a didactic strategy used to produce meaningful learning in mathematics with unique methodological and strategic implications, and

[1] Andean ethos does not refer to a reflection on the normativity of human behavior, rather it refers to the harmony of the human being with nature (Sobrevilla, 2008).

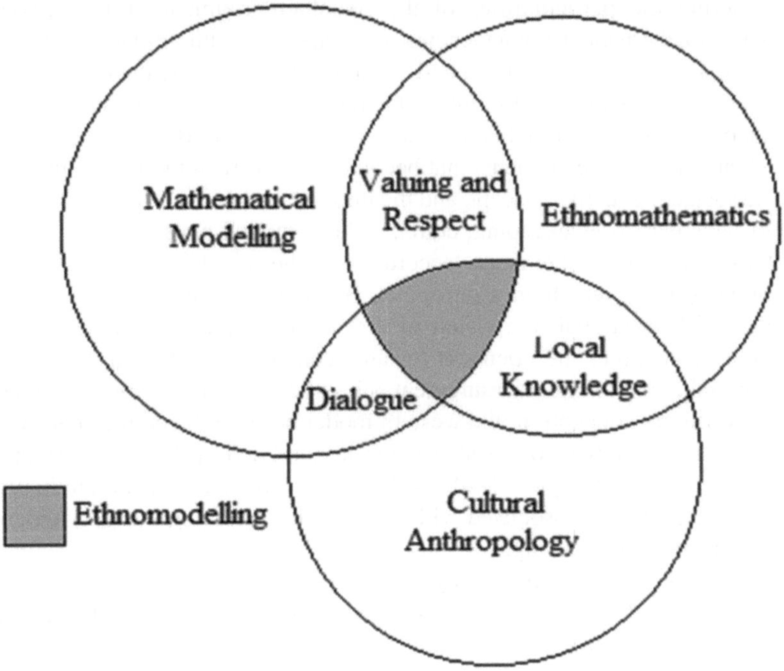

Fig. 7.1 Model sets of modelling. (Source: Blum & Borromeo Ferri, 2009, p. 46)

ethnomodelling, which incorporates sociocultural elements to the mathematical modeling approach, taking into account the aspects that promote an interaction between the interpretation of the mathematical phenomenon and the objective reality (Orey & Rosa, 2021), in our case the ethnomodelling contributes to the understanding of the ceramic designs from the perspective semiotics and mathematics.

Regarding mathematical modelling, we start from the general vision of this process as a construct for the interpretation of reality based on approximations of mathematical processes in their broadest spectrums.

Mathematical modelling elaborates schemes that allows for solving a problem in the areas of human knowledge, such as the social, biological, and epistemological, by means and elaboration of models that identifies its characteristics to discover, improve or propose a medium alternative to solve and verify a solution. Modelling as a method of teaching mathematics presupposes some interesting stages: integration of mathematics into other areas of knowledge, stimulating the student's interest in solving everyday problems, developing group work skills, by using technological, utilitarian, and communicational resources (Biembengut & Hein, 2004).

Mathematical modelling needs to be considered first as a mathematics teaching strategy. According to Blum and Borromeo Ferri (2009), this scheme consists of a Cantor-type diagram, in which two disjoint sets relate to interaction loops. Figure 7.1 shows model sets of modelling.

Considering the particularities of the social environment, it is necessary to address the mathematical competencies associated with this didactic scheme, the main thing, according to Greefrath and Vorhölter (2016) is the development of the ability to transfer a real model to the mathematical model in both directions.

When referring to the initial contact with a real situation, Bassanezi (2002) states that students incorporate elements that become familiar with their surroundings, the history associated with the concept, and the implications of everyday life, such as the shape of wine barrels, beekeeping, crafts, and ceremonial plates.

In terms of ethnomathematics, according to the six dimensions proposed by D'Ambrosio (1990), which are Conceptual, Historical, Cognitive, Epistemological, Political, and Educational; the vision of the socio-cultural environment, ideology and identity constitutes the openness towards concrete activities in the classroom, which are inherent to the measurement, classification, hierarchy and inference, provide a different conception of a western model because the ethnic characteristics referring to the members of local groups are outlined with their own cultural, historical, and epistemological load. In the case of diverse Andean environments, some real situations are associated with engineering, terracing bridges, agriculture, artisan production, and fishing.

In this chapter, we present a study related to ceramic designs developed by the members of pre-Columbian cultures in Ecuador. A process is induced that allows the mathematical modelling by applying the conditionalities inherent to prior knowledge.

In this context, Rosa and Orey (2016) emphasize that when referring to mathematical modelling with the additivity of the cultural, ethnographic, social, and anthropological approach. We work with ethnomodelling, which can be considered as a set of techniques and strategies that allow researchers to look at how members of specific cultural groups solve problems faced in their own cultural environment and context.

Furthermore, with reference to Rosa and Orey (2016), it can be said that, if ethnomathematics emphasizes the production of mathematical knowledge in specific cultural environments, ethnomodelling seeks to academically catalog cognitive mathematical processes in distinct sociocultural contexts by using modelling procedures.

According to Rosa and Orey (2018b), ethnomodelling originates from the union between ethnomathematics and mathematical modelling that has a much deeper conception in which reference is made to connections between emic (local) and the etic (global) mathematical knowledge through dialogue (glocal).

It is important to understand that mathematical experiences and knowledge developed from generation to generation are important characteristics of people, societies, and cultures, which are contained within the emic points of view while the etic perspectives are presented as academic scientific knowledge that has the capacity to be reproduced anywhere. The combination of the emic and the etic perspectives aims to generate a dialogical mathematical knowledge. Figure 7.2 shows ethnomathematics as an intersection between three fields of study.

Fig. 7.2 Ethnomathematics as an intersection between three fields of study. (Source: Rosa & Orey, 2018a, p. 2)

The application of ethnomodelling in problem solving does not preclude mathematical representations of the real in the modelling process (Saxena et al., 2016). It also seeks to develop forms of conceptual interpretation. In this context, ethnomathematics through ethnomodelling connects the historical, cultural, and anthropological versions of mathematical phenomena in environments that are meaningful for student understanding (Rosa & Orey, 2017).

In Latin America, the new educational currents propose interculturality as the basis for dialogue between the different cultures on the continent and outside it, rejecting the incommensurability between cultures and the differentiation of social behavior between them. This includes ethnomathematics, ethnoengineering and other ethnosciences (Esterman, 2015).

This philosophy raises interculturality as the basis of dialogue between the different cultures of this region and outside of it by rejecting the incommensurability between cultures and the differentiation of social behaviors and ongoing exchanges between them. Western formal philosophy draws on other philosophies. If we consider Western science as hegemonic, it is necessary to consider other views of it, the so-called ethnosciences that are structured in differently from the Cartesian model (Cadena & Trujillo, 2016).

Next, we present several cases of pre-Columbian plates from the northern highlands of Ecuador that were analyzed from the perspective of the practical application of ethnomodelling.

7.3 Case Analysis: Ethnomodelling in Pre-Columbian Plates from Ecuador

This study is based on the mathematical analysis of the ceramic plates found in the northern highlands of Ecuador dated between 200 BC to 630 AD. Thanks to archaeological exploration and ethnomodelling methodology; it has been possible to identify underlying mathematical elements.

In such a way that it allows for it to address several aspects: semiotics, mathematical modelling, and the didactic transpositions of knowledge from ancestral cultures to new generations of students to achieve an effective learning of mathematics through identity elements with culture, the history, and the appropriation of ethos characteristics of the Andean peoples.

This research is part of the search and analysis of the semiotics and mathematics of pre-Hispanic societies that inhabited the territory that includes the provinces of Pichincha from the Rumipamba ravine (Quito), Imbabura and Carchi in Ecuador, and the Altiplano of Ipiales in Colombia (northern highlands) Fig. 7.3 show the map of the northern highlands of Ecuador.

In order not to limit ourselves to the descriptive aspect of this research, which could raise arbitrary theoretical assumptions, or force explanations to adapt to our own logic and point of view, it is necessary to be willing to meet the unexpected. The designs found in the pre-Hispanic containers of the Sierra Norte represent symbolic shapes and colors that express the finite manifestation of a reality that unfolds beyond the sensible.

It often happens that, when entering the depths of the pre-Hispanic heritage, and when compared with other cultures, whether Western or Asian, certain implications arise not only of conceptual structures, but of traditional history and cultural contexts. In these societies, the vehicle for the transmission of ideas and concepts was the formulation of conventional symbols embodied in different materials that were often taken from nature and the animal world that surrounded them (Molestina Zaldumbide, 2020).

Fig. 7.3 Map of the northern highlands of Ecuador. (Source: Quito Geographical Institute)

Among the archaeological remains recovered and preserved by the Zaldumbide Rosales Foundation, in the Sierra Norte area of Ecuador, are ceramics with anthropomorphic shapes that are reflected in plates, which contain geometric figures such as triangles, rhombuses, squares, and circles. There are also lines, parabolas, hyperbolas, and catenaries. The colors in the designs are red, brown, and black (Molestina Zaldumbide, 2020).

Each one of the plates is represented with anthropomorphic, zoomorphic, and phytomorphic motifs, generally applied to the internal part of the ring-based plates, representing men and women, animals such as birds, deer, monkeys, round bats, felines, among others. There are also drawings of people in the form of dance with the same ceremonial representations.

These peoples worshiped the *Sun of the Pastos*, an eight-pointed star that has become the insignia of these cultures and has been used to represent them as their own in our country (Albis, 1987). The painter Guayasamín used it as a central element in the mural on a wall of the Congress Building of Ecuador (Fig. 7.4).

With the help of computer software, Fig. 7.5 shows the designs embodied in ceramics that have been modeled to improve the understanding of geometric shapes and implicit mathematical concepts.

For example, the plate F436 comes from the funerary equipment of the Tuza culture located in the north of Ecuador, the red color represents life, there are three concentric circles, in the inner circle a quadripartition is made: two opposite triangles through the vertex triangles and another two towards the circumference, in the triangle's figures of snakes, are observed considered as messengers of the mountain gods. There are also two circular sectors with multiple crossed lines that represent seeding furrows. These figures induce us to think about the complementary duality, which is typical of Andean philosophy or Pachasofía. Figure 7.6 shows plate F436 coiled snakes.

Figure 7.7 shows plate F837 elaborated by using the GeoGebra program.

Figure 7.8 shows a plate from the same culture, but with a 7-pointed star inscribed in a circle. Inside of the star are 13 circles, inside which are three small circles that appear to represent human heads. It is complemented by a filling of sectors with grooves. In plate F120, Fig. 7.8 shows an anthropomorphic semiotics, axial, and central symmetries, but this plate also incorporates a human figure, a fisherman, who indicates the geometric relationship with daily activities.

Next, the mathematical analysis is presented through ethnomodelling with specific cases of ceramics from the northern highlands of Ecuador.

7.3.1 Conics: Hyperbola and Ellipse

Figure 7.9 shows that there are two conics on an ellipse located in the center and a hyperbola with a vertical transverse axis on the outer contour of the plate F837.

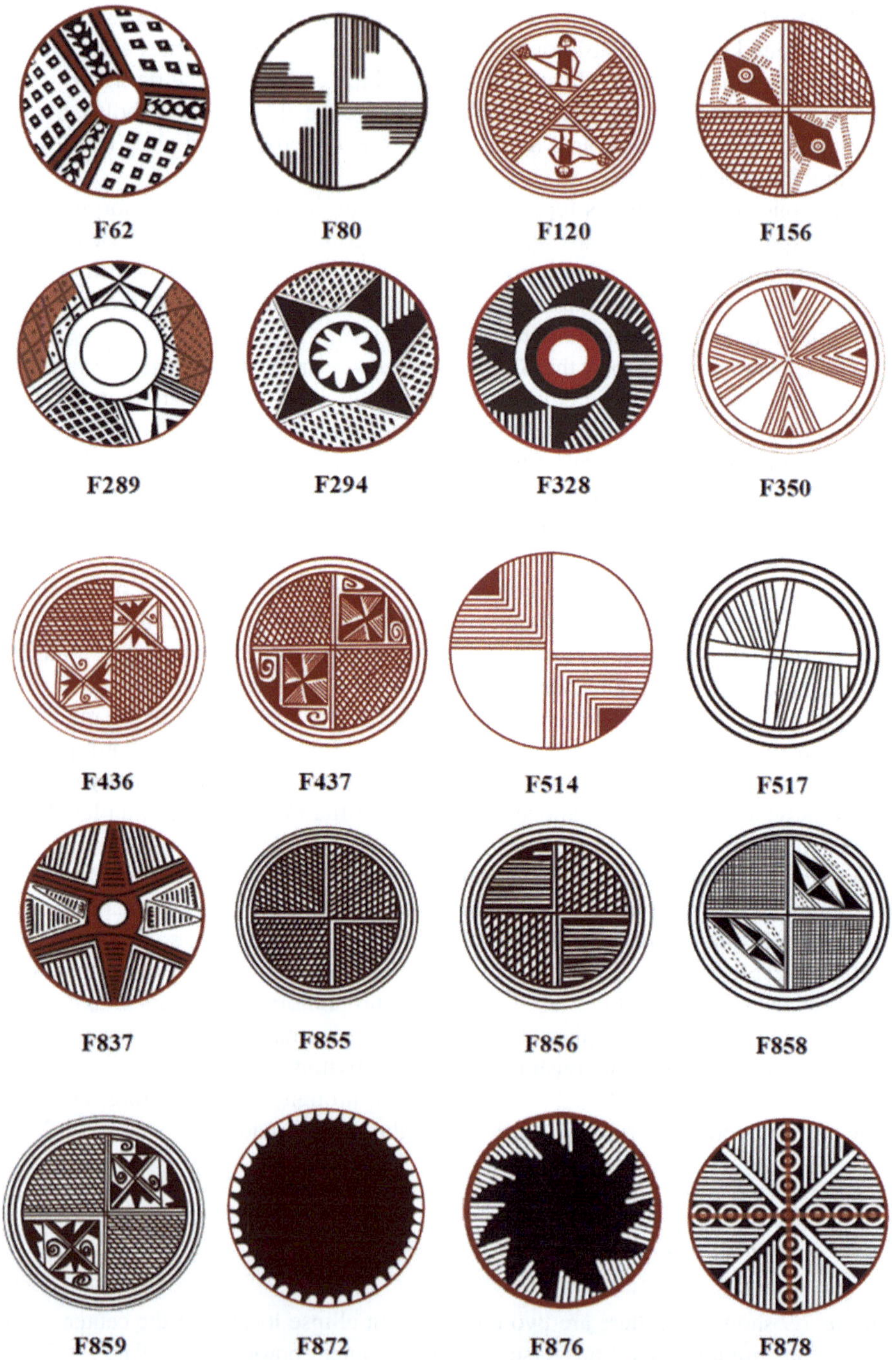

Fig. 7.4 Mural on a wall of the Congress Building of Ecuador. (Source: Frame by Oswaldo Guayasamín. https://live.staticflickr.com/65535/48206855656_338eabb8cd_b.jpg)

Fig. 7.5 Designs embodied in ceramics. (Source: Own design in GeoGebra)

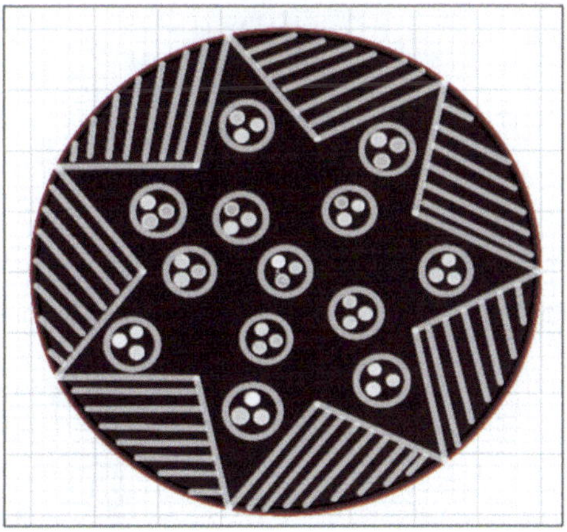

Fig. 7.6 Plate F436 Coiled snakes. (Source: Foundation Zaldumbide Rosales)

7.3.2 Rotations

Rotation: Fig. 7.10 shows that inside of plate F80 is observed a rotational symmetry because a center called the center of rotation can be found so that if we rotate the complete figure by using an angle of $\pi/2$, consecutively, the rotated figure continues to coincide with the original figure.

7.3.3 Symmetries

Symmetry: Two points of a graph have central symmetry with respect to the center of coordinates when this midpoint of the segment divides the plate in two points. In the same way, two points have axial symmetry with respect to a line, if that line is the

Fig. 7.7 Design of the bottom of the F837. (Source: Plate design elaborated in GeoGebra)

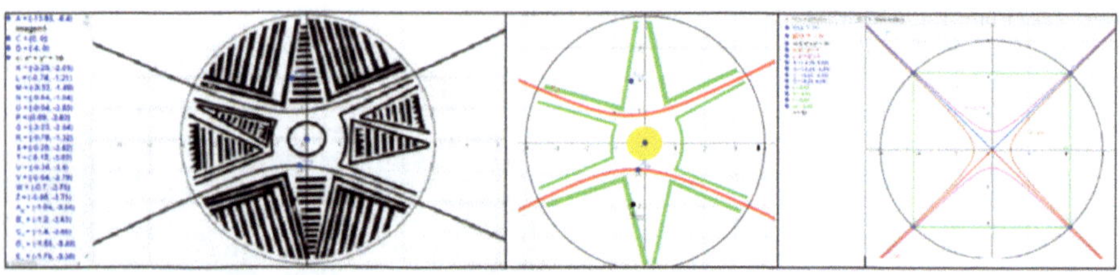

Fig. 7.8 Plate F120. (Source: Foundation Zaldumbide Rosales)

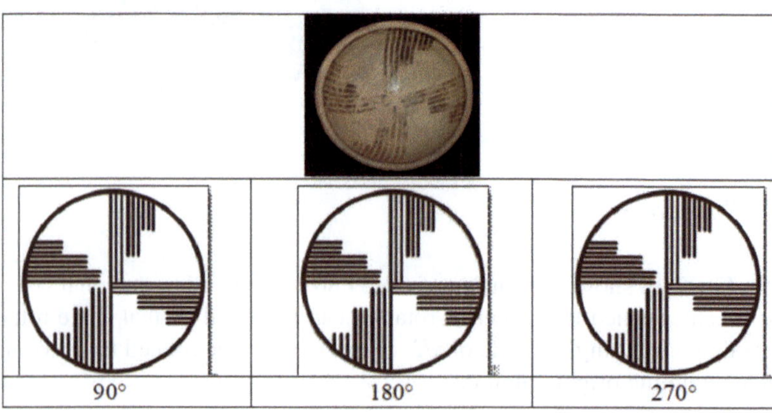

Fig. 7.9 Design of outer contour of the plate F837 plate in GeoGebra. (Source: Plate design elaborated in GeoGebra)

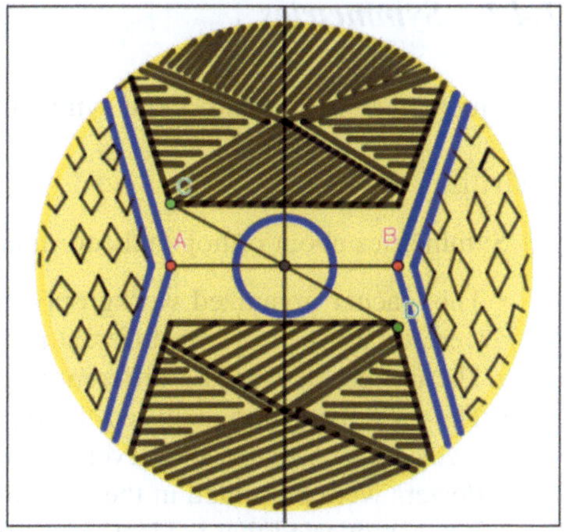

Fig. 7.10 Rotation of plate F80. (Source: Foundation Zaldumbide Rosales)

Fig. 7.11 Symmetries in the interior of plate F318. (Source: Plate design elaborated in GeoGebra)

bisector of the segment that joins the two points. Figure 7.11 shows the interior of the plate F328 in which points A and B are symmetric with respect to the y-axis and points C and D are symmetric with respect to the origin of the coordinate system.

7.4 Didactic Applications in the Classroom

A workshop was held on the different types of symmetries expressed in figures of the Cuasmal culture. This activity was developed with 30 students from 12 to 13 years old who were ninth graders, school period of 2021–2022, in the General Educational Unit Píntag School, located in Píntag, Quito Canton, Pichincha province, Ecuador.

7.4.1 Symmetries

Activities given to students about symmetries are shown below.

Activities

(a) Symmetries about orthogonal coordinate system.

- The teacher organized workgroups of 5 students and for this work 3 groups were established in two different parallels.
- The teacher delivered 5 different printed figures of different plates.
- The working group analyzed the types of axials, rotational, and central symmetry in each figure and developed a poster.
- Posters were presented in the form of an exhibition.
- It was requested that students identify the types of symmetry existing in each figure, in addition to answering the following assertations and questions:
  - Identify and trace the axis(s) of symmetry in the figure.
  - Does the graph have an axial symmetry?
  - Does the graph have a rotational symmetry?
  - Does the graph have a central symmetry?

For example, for group 1, it was proposed for students to analyze the symmetry present in the plate F120 (Fig. 7.12).

Results

Group #1 was made up of students Juan, Gustavo, Carlos, Andrea, and María. After they analyzed and graphed Fig. 7.12, plate F120, they agreed that the figure does not have an axial symmetry to the x-axis, because if it is bent by the center, both sides are the reflection, but the figure of the fisherman is reflected differently.

In the same way, they showed that the figure does not have a rotational symmetry because when it is rotated at an angle of 90°, their positions are different, because the fisherman is horizontally, different of the original vertical form. However, if the figure is rotated through an angle of 180°, it is observed that the figure returns to its original shape, which indicates that it has central symmetry. Figure 7.13 shows the symmetry with respect to the linear equation $y = -x$.

(b) Symmetry with respect to the linear function: $y = -x$.

Results

Group #2 made up of students Javier, Andrés, Camila, and Fernanda. After analyzing and graphing of plate F514, showed in Fig. 7.13, students agreed that the figure does not have an axial symmetry to the x- and y-axis (horizontal and vertical axes) because if the image is bent through the center towards both sides, the reflection does not match the original figure. However, they observed that when the figure is bent diagonally there is an axial of symmetry, and then the linear function $y = -x$ is an axis of symmetry.

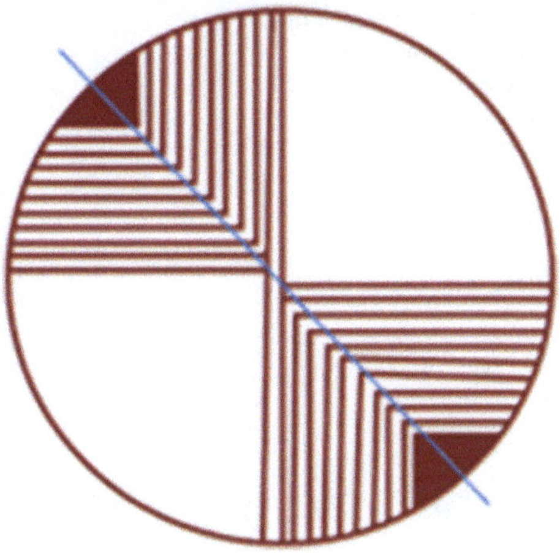

Fig. 7.12 Plate F120. (Source: Foundation Zaldumbide Rosales)

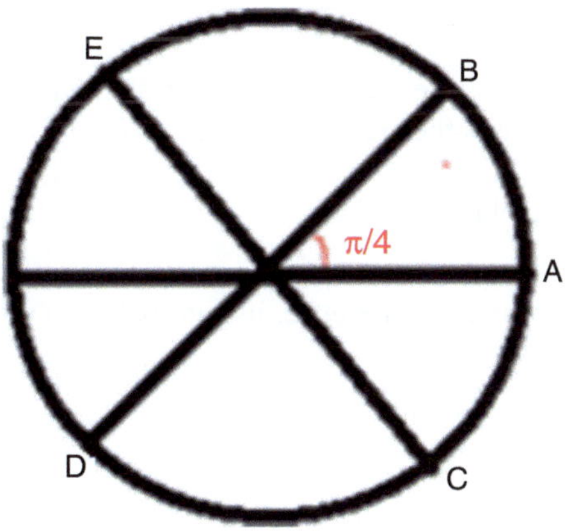

Fig. 7.13 Plate F514. (Source: Foundation Zaldumbide Rosales)

The students also stated that the figure does not have a rotational symmetry because when rotating the figure in an angle of 90°, the final position is different from the initial one. Regarding whether the figure has a central symmetry, it was established that it does since it has the same effect as a 180° turn.

7.4.2 An Agricultural Problem

A farmer of the Cuasmal culture, has planted corn in the circular sectors: $\widehat{BC}$ and $\widehat{ED}$, and the rest of the land has water. It is known that the angle formed between the arc length $\widehat{AB}$ is of $\pi/4$. It also known that $\widehat{CA}$ forms the same angle as $\widehat{AB}$ and that the diagonal $\widehat{EC}$ of the square inscribed in the circumference measures 12 m. Figure 7.14 shows a graphic representation of this problem.

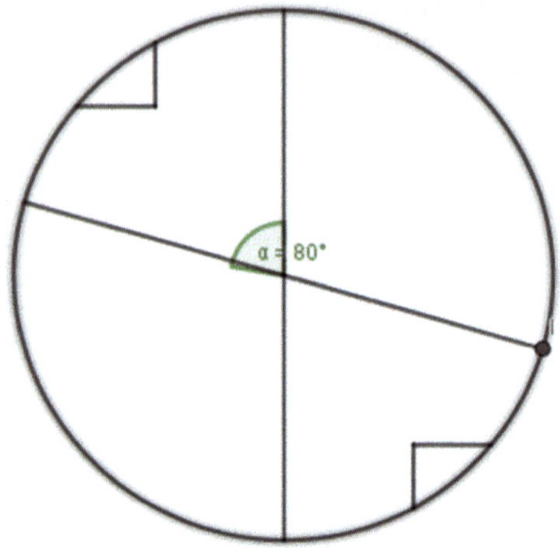

Fig. 7.14 Representation of the agricultural land. (Source: Figure elaborated in GeoGebra)

Determine:

What is the area of the part of the land that the farmer planted with corn?
What is the area of the land covered by the water?

Resolution

1. As the inscribed square shares the diagonal and at the same time the diameter of the circumference, its radius is half of this measurement.

$$d = 12\,m$$

$$r = \frac{12}{2}m$$

$$r = 6\,m$$

2. Now we calculate the enclosed circular area of $\widehat{AB}$

$$A_1 = \frac{r^2\theta}{2}$$

$$A_1 = \frac{6^2\frac{\pi}{4}}{2}$$

$$A_1 = \frac{36\pi}{8}m^2$$

3. Since the area A_1 included by the region is repeated 4 times, we multiply by 4 in such a way that we obtain the total area.
4. The area planted with corn is $A_C = 56.548$ m^2.

5. Now we calculate the total area of the land.

$$A = \pi r^2$$
$$A_L = \pi 6^2$$
$$A_L = \pi 36$$
$$A_L = 113.10\, m^2$$

6. Finally, the area covered by water is the total area of the land minus the area planted with corn.

$$A_W = A_L - A_C$$
$$A_W = 113.10\, m^2 - 56.55\, m^2$$
$$A_W = 56.55\, m^2$$

7.4.3 *A Feeding Problem*

The chef of a village of the Cuasmal culture wants to serve more meat than vegetables on a plate that has a graphic pattern representation. For this, he wants to know if the part of the plate that has the shaded pattern is smaller than the part of the plate that does not have this pattern, and in this way, he is able to serve the food in that sector.

In order to do this, it is known that the angle Θ that forms the shaded circular sector measures 80° and that the segments of the circular sector are equal to their counterparts. In addition, these segments measure the same as the radius of the plate and that the segment of the smallest circular sector formed inside of the plate has a ratio of 8:1 with respect to the largest circular sector. Figure 7.15 shows the representation of the plate in this problem.

Determine:

What percent of the plate has a shaded pattern?
In which sector should the meat be served?
Inquire about the importance of meat in the human diet?

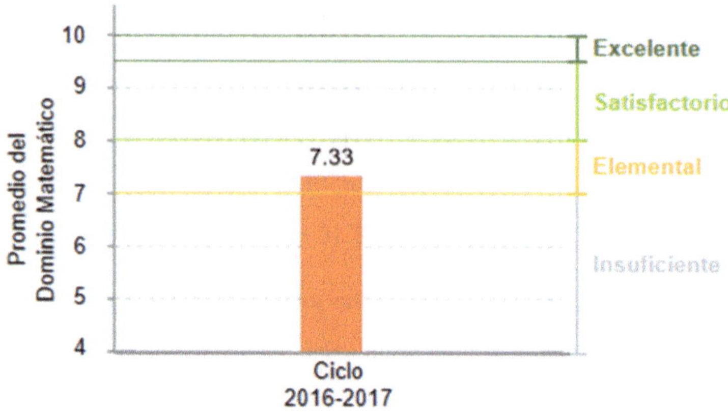

Fig. 7.15 Representation of the plate of the feeding problem. (Source: Figure elaborated in GeoGebra)

Resolution

1. In order to know the perimeter of the plate, the student measured the contour and concludes that the plate has a perimeter of 60 cm and from there we calculated its radius.

$$p = 2\pi r$$
$$r = \frac{p}{2\pi}$$
$$r = \frac{60}{2\pi}$$
$$r = \frac{30}{\pi}$$
$$r = 9.55\ cm$$

2. We calculated the area of the circular region that has the shaded pattern, for this we will use the formula. We transformed 80° to radians and obtained:

$$80^\circ * \frac{\pi rad}{180} = \frac{4\pi rad}{9}$$
$$\theta = 1.396\ rad$$
$$A = \frac{\theta . r^2}{2}$$

$$A = \frac{(1.396)(9.55)^2}{2}$$

$$A = 63.66\ cm^2$$

3. As the shaded circular sector is repeated twice, we multiplied it by 2:

$$A1 = A * 2$$

$$A1 = 63.66 * 2$$

$$A1 = 127.32\ cm^2$$

4. Now we deducted the internal part in white that is inside the shaded circular sector, as it is in a ratio of 8 to 1, then we determined that:

$$A2 = \frac{A1}{8}$$

$$A2 = \frac{127.32\ cm^2}{8}$$

$$A2 = 15.92\ cm^2$$

$$A3 = A1 - A2$$

$$A3 = 127.32\ cm^2 - 15.92\ cm^2$$

$$A3 = 111.4\ cm^2$$

5. In order to know the percentage of the surface represented by the shaded sector, we calculated the area of the plate:

$$AT = \pi.r^2$$

$$AT = \pi.9.55cm^2$$

$$AT = 286.52cm^2$$

6. Now we transformed A3 and AT into percentages:

$$x = \frac{100\% * 111.4cm}{286.52\ cm}$$

$$x = 38.88\%$$

7. Responding to question #1, about what percentage does the shaded pattern on the plate represents, we deduced that this value is 38.88%.
8. Responding to question #2, related to which part of the plate the meat should be served, we concluded that in the part that has the shaded pattern since it is the smallest.

9. In relation to the question about the importance of meat in the human diet, we concluded that meat is of great importance since it provides mostly proteins that allow for the development of our bones, muscles, cartilage, skin, and blood.

7.4.4 *Some Reflections About the Ethnomodelling Process*

Examples discussed in this chapter showed that, according to Yang (2003), glocalization becomes an expression that promotes positive dialogic relations between different cultures and worldviews. In this context, Rosa and Orey (2017) affirm that this perspective creates possibilities for generating spaces for promoting dialogue between local and global mathematical knowledge approaches.

Thus, it important to state that, in ethnomodelling research, dialogue helps to prevent the "global from overwhelming the local, while the local is still benefitting from what the global has to offer" (Fernandez, 2009, p. 46). In this regard, Rosa and Orey (2018b) highlight that:

> (. . .) glocalization may be understood as the particularization of the universal, which is the local adaptation and translation between global and local approaches. There are ways to understand mathematical ideas, procedures, and practices that are universally applicable as general templates that are modified to reflect particular cultural traits such as the development of mathematical strategies and techniques applied to solve problems members of distinct cultural groups face daily (p. 193).

The results of this investigation showed that cultural dynamism in ethnomodelling intensifies the translation between local and the global mathematical forms of knowledge through dialogue. This process captures the simultaneity of both universalizing and particularizing tendencies during the development of cultural interactions and provides an inclusive environment for addressing complementary interests during the conduction of research in mathematics education.

7.5 Conclusions

This chapter has been carried out through a historical investigation regarding the ceramics of the Cultures found in Northern Ecuador, therefore, a vision of cultural and social identity is incorporated, which implies the construction of mathematical concepts generated from the intuitive to the analytical, through the analysis of genealogical conjunctures of mathematical knowledge.

The considerations of ethnomathematics as a new look of mathematics education, with all its possibilities, intrinsic nuances, globality, locality, dialectics, and dialogicity between different cultures in time and space allows for the use of iconographies as didactic material.

An attempt has been made to systematize an ethnomodelling scheme through its stages of analysis of the ancestral elements through the emic, the etic, and the dialogic perspectives by considering the approach to mathematical concepts with a dynamic that allows for the incorporation non-traditional elements in the classroom, the use of free software to motivate the construction and semiotic interpretation of the ceramic designs of our cultures.

The arrangement of this chapter in a combination of theoretical analysis and practical application in the classroom will allow for continued dialogue in relation to further generation of innovative spaces for the management of didactic and pedagogical resources that stimulate the construction of mathematical knowledge with alternative and identity ideas. This research has been carried out on several theoretical assumptions:

- The problem of mathematics education in Ecuador based on algorithmic and decontextualized models.
- The use of identity resources such as pre-Columbian ceramics from the Northern Highlands from a semiotic and mathematical point of view.
- The implementation of ethnomodelling as a strategy for the analysis and implementation of teaching techniques and strategies.
- The analysis of learning outcomes in the classroom through a focused essay.

Based on the above, it can be concluded that interactions between archeology and the teaching of mathematics for the rescue of cultural elements such as pre-Columbian ceramics, which have not been sufficiently studied under the approach of mathematics immersed in them, will allow subsequent studies to expand the subject, and studies can be conduct that analyze the impact on mathematics education in Ecuador.

Appendix 7.1: In the Classroom: Students in Group 1 Analyzing Plate F120

Source: Personal file

Source: Personal file

Appendix 7.2: Student Answers to the Problem of Planting Corn

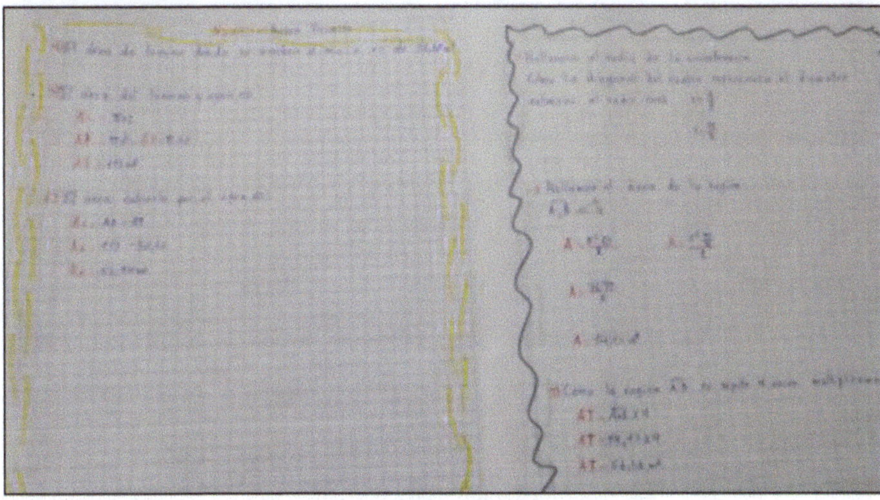

Source: Personal file

Source: Personal file

Appendix 7.3: Ceramic Plates

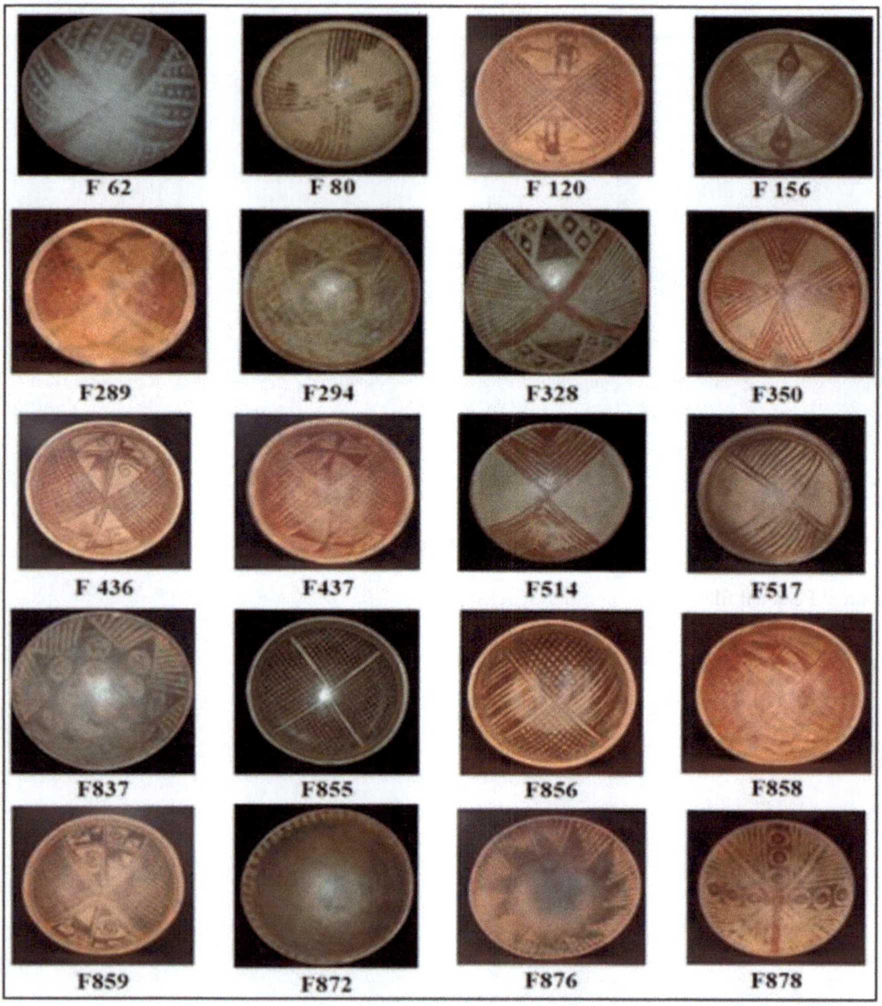

Source: Foundation Zaldumbide Rosales

Source: Foundation Zaldumbide Rosales

References

Albis, V. S. (1987). Las Proporciones del Sol de los Pastos. *Boletın de Matemáticas, XXI*(2–3), 110–134.

Bassanezi, R. (2002). *Ensino-aprendizagem com modelagem matemática. [Teaching-learning with mathematical modelling]*. Editora Contexto.

Biembengut, M. S., & Hein, N. (2004). Modelación matemática y sus desafíos para enseñar matemática. *Educación Matemática, 16*, 105–125.

Blanco-Álvarez, H., Fernández-Oliveras, A., & Oliveras, M. L. (2017). Formación de Profesores de Matemáticas desde la Etnomatemática: estado de desarrollo. *Bolema, Boletim de Educação Matemática, 31*(58), 564–589.

Blum, W., & Borromeo Ferri, R. (2009). Mathematical modelling: Can it be taught and learnt? *Journal of Mathematical Modelling and Application, 1*, 45–48.

Cadena, J., & Trujillo, J. (2016). El pensamiento mito-lógico como sistema cognitivo de las etnociencias. *Anales, 1*, 43–74.

D'Ambrosio, U. (1990). *Ethnomathematics: The art or technique of explaining or knowing*. Editora Ática.

Esterman, J. (2015). *Filosofía Andina*. Ediciones Abya-Yala.

Greefrath, G., & Vorhölter, K. (2016). Teaching and learning mathematical modelling. In *ICME-13 topical surveys* (pp. 19–21). Springer.

Fernandez, S. A. (2009). *A theory of cultural glocality. Master Thesis. College of Arts and Science. Department of Philosophy*. Jacksonville, FL: University of North Florida.

Huincahue, J., Borromeo, R., & Mena, J. (2018). El conocimiento de la modelación matemática desde la reflexión en la formación inicial de profesores de matemática. *Enseñanzas de las Ciencias, 36*, 99–115.

Ineval. (2018). *Educación en Ecuador: Resultados de PISA para el Desarrollo*. Ediciones Ineval. Retrieved from http://www.evaluacion.gob.ec/evaluaciones/pisa-documentacion/

Kragh Sørensen, H. (2014). "The end of proof"? The integration of different mathematical cultures. In *Mathematical cultures* (pp. 141–143). Birkhäuser Verlag.

Lévi-Strauss, C. (1977). *Mito y significado*. Alianza Editorial.

Lizcano, E. (2006). *Metáforas que nos piensan Sobre ciencia, democracia y otras poderosas ficciones*. SKP.

Llinares, S. (2012). Formación de profesores de matemáticas. Caracterización y desarrollo de competencias docentes. In V. Autores (Ed.), *Cuadernos de Investigación y Formación en Educación Matemática* (pp. 53–62).

Molestina Zaldumbide, M. D. (2020). La montaña: morada de los dioses: discurso de incorporación. *Boletín Academia Nacional de Historia, 98*(203), 334–357.

Orey, D. C. & Rosa, M. (2021). Ethnomodelling as a glocalization process of mathematical practices through cultural dynamism. *The Mathematics Enthusiast, 18*(3), 439–468.

Rosa, M., & Orey, D. C. (2016). Etnomodelagem, uma relación dialógica entre Etnomatemática y Modelaje. In *Etnomatemáticas pelo Brasil* (pp. 55–74). Editora CRV.

Rosa, M., & Orey, D. C. (2017). *Etnomodelagem: a arte de traduzir práticas matemáticas locais*. Editora Livraria da Física.

Rosa, M., & Orey, D. C. (2018a). The anthropological dimension on ethnomodelling. *Open Access Journal of Archaeology and Anthropology, 1*, 1–8.

Rosa, M., & Orey, D. C. (2018b). Reflecting on glocalization in the contexts of local and global approaches through ethnomodelling. *Educação Matemática Pesquisa, 20*(2), 171–201.

Saxena, R., Shrivastava, K., & Bhardwaj, R. (2016). Teaching mathematical modelling in mathematics education. *Journal of Education and Practice, 7*, 37.

Sobrevilla, D. (2008). La filosofía andina del P. Josef Estermann. *Solar, 4*(4), 231–247.

Yang, E. (2003). The circulation of East Asian trendy dramas grounded on cultural proximity. *Bansong Yeongu*, 69, 197–220.

Part III
Interdisciplinary Ecosystems: Empirical Work, Theoretical-Methodological Approaches, and Research Questions

The main objective of this section is to share some experiences and reflections related to a type of didactic pedagogical proposal that the authors conducted in the Argentinean Patagonian context. These investigations are characterized by being developed in real-world contexts and respond to the intention of facilitating the attribution of meaning to learning by students. In this section of the book, the authors summarize the main characteristics of the projects developed with students in mathematics classrooms in which mathematical modelling enables them to build rational arguments for the development of decision-making process that these projects demand.

Part III
Interdisciplinary Ecosystems, Empirical Work, Theoretical-Methodological Approaches, and Research Questions

Chapter 8
Analyzing the Availability of Renewable Energy Resources in a Project in a Real-World Context: A Framework for Making Sense of Learning

Pablo Carranza and Fabio Miguel

8.1 Introduction

When we think of mathematics teaching/learning experiences, the first step is usually selecting a disciplinary concept to teach (Chevallard, 1985) and then finding a context that gives it meaning. In this chapter, we will share a proposal that alters that dynamic while, of course, still preserving concepts to teach.

In our case, the starting point is reflecting on the conditions that could facilitate students' attribution of meaning to disciplinary concepts. Why do we focus on meaning? For several reasons. One of them is our belief that the learning process becomes more significant if students can attribute meaning or sense to it. Another reason responds to an ethical principle: students (as well as educators) have the right to carry out activities that have a meaningful impact on their lives.

The activity we present in this chapter addresses a real-life problem; more specifically, a problem for the inhabitants of the region where the students and teachers involved live. From a didactic perspective, this activity was chosen due to its potentialities to address disciplinary concepts; mainly mathematical.

From a pedagogical perspective, the activity (consisting of the calculation, construction and installation of Savonius windmills) constitutes a possibility for the students to find meaning in mathematics both from a professional and personal viewpoint.

P. Carranza (✉) · F. Miguel
Universidad Nacional de Río Negro, General Roca, Río Negro, Argentina
e-mail: pcarranza@unrn.edu.ar

M. Rosa et al. (eds.), *Mathematical Modelling Programs in Latin America*,
https://doi.org/10.1007/978-3-031-04271-3_8

8.2 Problem

As we stated in the introduction, one of the initial interests that motivated this project was searching for proposals where students could attribute meaning to learning. Nonetheless, we understand that meaning is not a quality of a didactic proposal (Develay, 1993, 2004) and knowledge itself does not carry meaning (de Vecchi & Carmona-Magnaldi, 1996). We claim that meaning is a construction made by a student according to his or her life experiences, expectations, emotions, etc.; even to their life projects (Martínez Licona & Palacios Ramírez, 2012; Pinzón, 2016).

However, even if we consider meaning as a personal construction, we also acknowledge that there are certain characteristics of the didactic proposal that help students grasp it and consequently incorporate it into their life experiences, expectations and emotions. Among those characteristics is what we call the "temporal dimension" and it is developed in two directions: present and future. Simply put, we expect knowledge to become useful for the students' present and future.

Another dimension considered when designing this proposal (and strongly related to the previous one) is *knowledge functionality*. We believe knowledge should help students understand the world around them while also allowing them to intervene in it. A third dimension considered is *knowledge significance*. We want students to gain knowledge considering it useful for them and the community.

We argue that these characteristics are enhanced if they are part of a proposal being implemented in a real-world context rather than in one staged by, for instance, a professor. By "real-world context", we mean a context that exists in the real world and students can experience it without the influence of any didactic fiction.

Nevertheless, a situation in a real-world context where students can *immerse* themselves in the problem can rarely be discussed from a monodisciplinary approach. In general, a real-world context is complex (and quite rich) and usually requires knowledge from various disciplines to reach a solution. Another characteristic of these contexts is that they frequently contain various teaching/learning situations. Therefore, we will refer to *project* as the characteristics of the general problem to be tackled in the real-world context. Likewise, we will refer to *problems to solve* as the project's internal problematic instances.

However, it is also relevant that the real-world context authentically demands disciplinary knowledge. That is, it has to be a context where concepts (Bednarz et al., 2018; Buxton, 2006; Spandaw, 2009), abilities and competencies related to, at least, the mathematics disciplinary field genuinely emerge.

Additionally, the mathematical concepts to be discussed in the proposal's framework have to be related to the possibilities and interests of the formal education in question. To meet this authenticity requirement, we searched for a context that demanded rationality in the decision-making process, precision in the decisions, and that, to a certain extent, required quantifications.

The context needs to meet a set of physical and chronological viability requirements as well. This means the proposal should be able to be carried out within the

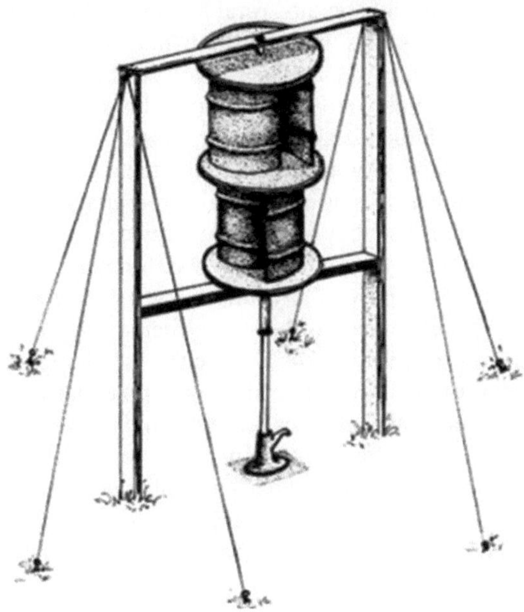

Fig. 8.1 Schematic diagram of the Savonius windmill. (Source: Authors' own work)

duration of the mathematics course. Besides, the requirements in terms of physical space, tools and technology should be met with the available resources.

So far, we presented a set of objectives which are, among others: (a) to find proposals where students can attribute meaning to learning, and (b) carry out those proposals in real-world contexts, where (c) students can understand and intervene in that reality. However, there is a didactic-pedagogical matter to address: is it possible to discuss disciplinary concepts (particularly, mathematical) in this type of proposal? In this and the following chapters, we will try to contribute some thoughts to this (in our criteria) fundamental matter.

Within the range of possibilities, the professors selected a context that tackled an environmental issue connected to the mathematics course's approach. The project involved the calculation, construction and installation of Savonius windmills (Savonius, 1922)for rural communities relying on a subsistence economy in Argentinian Patagonia.

In this chapter, we will present a summary of two "problems to solve". They appeared in the project's framework where concepts from mathematics and other disciplines were discussed. It is worth mentioning that interdisciplinarity was not a condition for selecting the project. It was an almost natural consequence of choosing a real-world context that presented the three characteristics mentioned above (temporal dimension, knowledge functionality and knowledge significance). Figure 8.1 shows the type of windmill installed. With an intentionally simple design, its rotor is built with recycled 200 L barrels.

In this project, the windmills were designed for groundwater pumping. The objective was to provide water for animals and facilitate irrigation for the trees and plants at the inhabitants' rural facilities. Throughout the project, a significant number of problems to solve appeared; however, due to space limitations, we will only discuss two of them. They involved the availability of renewable energy sources. The problems were analyzed to address a family's concern since a Savonius windmill would be installed at their house.

The family (with scarce economic resources as all of the other recipients of the windmills) was starting a small agrotourism business. Apart from pumping water for their animals, they needed to provide electricity to a refrigerator to keep fresh products for the tourists. It is worth mentioning that this property did not have an electricity supply of any kind (power grid, generators, or photovoltaic panels).

This issue had not been considered or anticipated when designing the proposal. However, it was discussed in class for two reasons: it was related to the characteristics of the context, and it was potentially didactic. In fact, the question that encouraged the analyses presented in this article was about the possible convenience of replacing the diaphragm pump originally planned for the windmill (to pump water) with an alternator. This alternator would generate electricity to charge the refrigerator's batteries. That is, we tried to analyze the possibility of adapting the windmill by replacing the pump for water pumping with an alternator for electricity generation.

Before analyzing this adaptation from a mechanical perspective (parts replacement, mechanical or electrical revolutions per minute multiplier, etc.), it was necessary to study the long-term availability of the wind resource in the area throughout the year.

Still, the study of wind resource availability was anticipated beforehand. In this case, the novelty was assessing the situation from a different perspective—considering the possibility of supplying energy to a refrigerator instead of pumping water from a well.

8.3 Method

The project involving the calculation, construction and installation of Savonius windmills for rural communities has been implemented since 2015 at the National University of Río Negro (UNRN), Argentina. It has been carried out within the first-year mathematics course in the first semester of the Industrial Maintenance Technician program (Carranza, 2015, 2017). It involves around 50 first-year students aged between 18 and 50. The students are mostly middle and mid-low socioeconomic status men.

The National University of Río Negro is an institution recently established. The Industrial Maintenance Technician program (TSMI, by its Spanish initials) is a 3-year degree. As of the date of the activities discussed here, the lessons take place in a rented facility. The available resources, in terms of infrastructure, are traditional classrooms (with desks facing a blackboard), a blackboard, a projector and a socket.

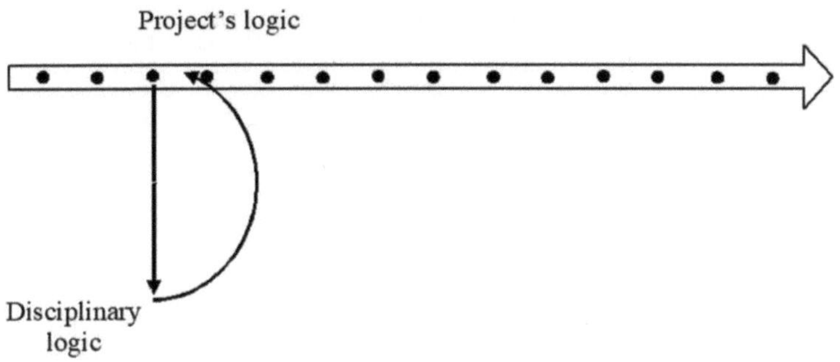

Fig. 8.2 Development of the project in two logical dimensions. (Source: Authors' own work)

Throughout its first years, the project was funded by the UNRN to buy the necessary tools and materials. Later, for the continuity of the project, we were granted funding by the National Ministry of Education. The trips to the locations were funded by the corresponding townships.

When designing the proposal, the disciplinary concepts, particularly mathematical, were thought to emerge in class as conceptual tools. Tools that allowed students to construct rational arguments for the decision-making process the project demands. In fact, due to the project's nature, this matter cannot be discussed from an intuitive perspective or following a trial-and-error strategy until finding an adequate solution. If the windmill were poorly constructed, people could get hurt or the rural property could be damaged; we could even disappoint our recipients and fail to meet their expectations. Therefore, the disciplinary concepts become essential to provide a safety framework for the students' construction/installation actions in the project.

Figure 8.2 shows a graphic representation of the project's dynamic in its two logical directions: the project's logic direction and the disciplinary logic direction. The development of the project requires undertaking actions and making decisions (project's logic). These requirements represent problems to solve (shown as the black circles) through rational arguments.

To address these problems rationally, it is necessary to introduce concepts whose use requires knowing the disciplinary logic, its objects and methods (Crombie, 1980; Hacking, 1965). One of the first problems to solve was the availability of the wind resource in the region. Wind is a windmill's source of energy and it also causes stress on its structure.

Therefore, it is crucial to know what the maximum wind speed is to design the windmill's structure. For this purpose, it was necessary to answer questions such as the following:

(a) Is the available wind resource enough to ensure water pumping throughout the year?
(b) What is the wind speed frequency distribution within the usable speed range? (5–20 m/s).
(c) What is the duration of the usable and non-usable time frames?
(d) Is the usable wind resource available sufficient in summer?

Date	Time	Temp Out	Hi Temp	Low Temp	Out Hum	Dew	Wind Speed	Dir	Wind Run	Hi Spee
01/01/10	0:10	19.9	20.2	19.9	42	6.6	4.8	NW	0.80	17.7
01/01/10	0:20	19.8	19.9	19.8	43	6.8	4.8	NW	0.80	12.9
01/01/10	0:30	19.3	19.7	19.3	45	7.1	3.2	NW	0.54	11.3
01/01/10	0:40	18.9	19.3	18.9	46	7.1	1.6	NW	0.27	6.4
01/01/10	0:50	18.6	18.9	18.6	48	7.4	1.6	NW	0.27	8.0
01/01/10	1:00	18.4	18.6	18.4	48	7.2	1.6	NW	0.27	6.4
01/01/10	1:10	18.3	18.4	18.3	49	7.4	3.2	WNW	0.54	16.1
01/01/10	1:20	18.4	18.4	18.2	48	7.2	4.8	W	0.80	11.3
01/01/10	1:30	18.8	18.8	18.4	47	7.2	9.7	W	1.61	25.7

Fig. 8.3 Partial view of the weather station database. (Source: Authors' own work)

(e) What is wind behaviour at high speeds? Is it stable or are there gusts?
(f) What is the behaviour of nocturnal winds?

These and other questions became the core of the project's sustainability. For instance, in (f) we question the behaviour of nocturnal winds. Since the windmill has a mechanical handbrake, it was necessary to know whether the windmill could operate during nighttime or not because of the strong gusts of wind. If the windmill functioning at nighttime was not a possibility, it would be advisable to always leave the handbrake off during the inhabitants' rest time.

To carry out a precise analysis of wind resource availability, we accessed a database of a local weather station of the National Institute of Agricultural Technology (INTA, by its Spanish initials). Figure 8.3 shows the first lines shown in this database. It is in .csv format.

A set of data that registered changes every 10 min between 2010 and 2015 could be accessed. It considered several variables, such as wind speed (m/s), wind direction, gusts, and even solar radiation (W/m^2), among others. In total, the weather station's database contained over 311,000 registries including 38 variables (approximately, 12,000,000 data). In order to illustrate the analysis carried out with the students, the following Table 8.1 shows the absolute frequency distribution of the wind speeds (m/s) for each month in 2013.

As regards the project's logic, this table showed the students essential information to identify the available wind resource and to calculate the stress caused on the windmill. It should be noted that during August (column A), wind speed ranged between 41.3 and 43.5 m/s 12 times. Graph 8.1 shows a representation of the information in Table 8.1.

Some students proposed visualizing wind in real-time using specialized applications or websites (Fig. 8.4). In those visualizations, they could observe the reason behind the predominance of the dry winds from the West in the region. A stable high-pressure centre is located in the Pacific Ocean while a low-pressure centre is in the Atlantic Ocean by resulting in strong winds towards the East. These winds release their humidity in the high peaks of the Andes Mountains. Then, they advance towards the region as strong dry winds.

This first analysis led to others where students carried out a seasonal wind resource assessment. That is, a study of the periods in which the windmill could be functional (wind speed ranging from 5 to 20 m/s). Night wind speed was also

Table 8.1 Distribution of wind speed frequencies during the year 2013

Intervals	J	F	M	A	M	J	X	A	S	O	N	D
2.2	2399	1129	3112	3078	2713	4320	2802	2322	2433	446	2171	1763
4.4	324	432	376	254	219	0	267	204	289	723	334	523
6.5	601	857	123	318	397	0	412	351	186	1456	559	740
8.7	253	687	132	96	202	0	167	178	257	880	230	243
10.9	214	433	176	91	176	0	159	186	166	62	178	228
13.1	331	354	245	180	237	0	190	252	302	891	261	351
15.2	113	140	130	77	90	0	70	116	129	0	121	136
17.4	67	0	80	64	67	0	62	98	114	0	106	104
19.6	87	0	41	79	135	0	107	179	99	0	147	150
21.8	24	0	28	22	37	0	40	95	29	0	39	69
23.9	23	0	10	21	31	0	44	72	24	0	39	37
26.1	21	0	0	23	57	0	61	145	54	0	48	56
28.3	3	0	10	7	23	0	10	62	12	0	23	16
30.5	2	0	0	4	17	0	14	39	0	0	15	5
32.6	1	0	0	0	35	0	17	61	0	0	24	24
34.8	0	0	0	0	11	0	13	21	0	0	10	10
37.0	0	0	0	0	8	0	8	19	0	0	7	3
39.2	0	0	0	0	8	0	8	38	0	0	7	5
41.3	0	0	0	0	0	0	3	13	0	0	1	0
43.5	0	0	0	0	0	0	2	12	0	0	0	0

Source: Authors' own work

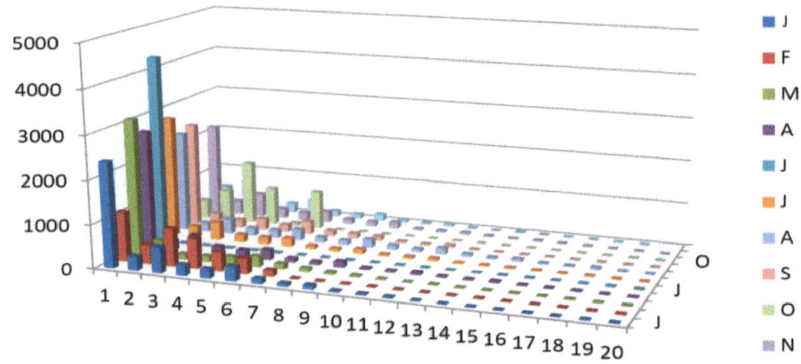

Graph 8.1 Distribution of wind speed frequencies during the year 2013. (Source: Authors' own work)

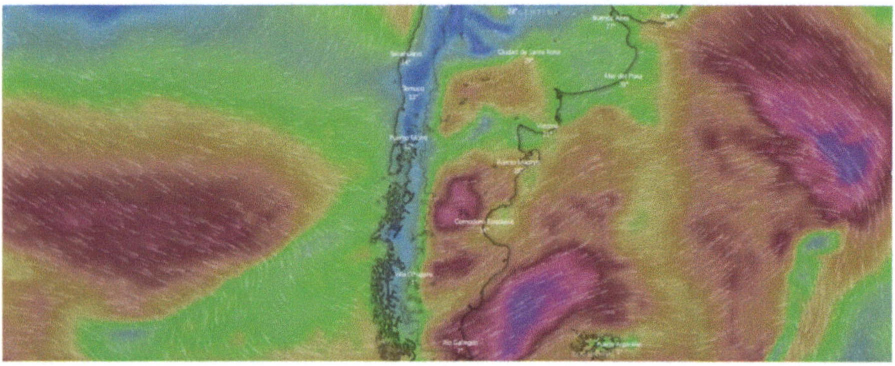

Fig. 8.4 Screenshot of Windy.com website. (Source: www.windy.com)

examined every month. Moreover, knowing the predominant direction of the wind (West) allowed students to determine the windmills' location by considering the obstacles trees and possible constructions nearby could represent.

These studies based on the database of the region's wind were incredibly useful to establish the availability of this resource and to measure the stress that would be caused on the windmill. As regards the disciplinary logic, in this case, mathematical and statistical, we discussed an important number of concepts. For instance, Table 8.1 was done using the matrix function *Frequency*. The limits of each interval were calculated through formulas considering the maximum and minimum values as well as the parameter resulting from the desired number of intervals.

Both Table 8.1 and Graph 8.1 evidenced the asymmetry in wind distribution (Weibull model). As a result, we considered the importance of both the average values and the median. It was also necessary to include position and dispersion measurements to analyze wind variability in each season. Besides, the database recorded missing or incorrect values; for example, in Table 8.1, the value of all the

Table 8.2 Windmill power values for January 2013

avg-intv	J Relative	Power	Weighted power
1.1	0.5	0.5	0.3
3.3	0.1	12.5	0.9
5.4	0.1	57.9	7.8
7.6	0.1	158.8	9.0
9.8	0.0	337.5	16.2
12.0	0.1	616.3	45.7
14.1	0.0	1017.2	25.8
16.3	0.0	1562.7	23.5
18.5	0.0	2274.8	44.3
20.7	0.0	3175.8	17.1
22.8	0.0	4287.9	22.1
25.0	0.0	5633.4	26.5
27.2	0.0	7234.5	4.9
29.4	0.0	9113.4	4.1
31.5	0.0	11,292.3	2.5
33.7	0.0	13,793.5	0.0
35.9	0.0	16,639.2	0.0
38.1	0.0	19,851.6	0.0
40.2	0.0	23,452.8	0.0
42.4	0.0	27,465.3	0.0
Total power			250.6

Source: Authors' own work

registries in June, column J, is 0. Therefore, it was necessary to address this issue as well.

Thanks to Table 8.1, it was easier to estimate the windmill's power. In fact, the analysis of power was discussed a few weeks later to estimate the possible volume of water we could lift, approximately, 7 m. The following formula was used (Avila, 2017; Menet, 2004): $Pmax = 0.18HDv^3$, where H represents *rotor's height*, D represents *rotor's diameter*, and v represents *wind speed*.

This calculation was done considering the average speed of each interval (avg-intv column). Table 8.2 shows the relative frequency for each speed interval in Table 8.1 corresponding to January 2013 (J Relative column). It also shows the power generated at that speed (Power column). Lastly, it was possible to estimate the windmill's average power each month. The table shows the estimated values for January weighting the power values by their relative frequencies (weighted average).

The power formula, in turn, evidenced the importance of wind speed (cubic participation) in relation to other elements that are related to the rotor's size (lineal participation). This led to discussing some interpretations related to tangents of polynomials. This is an example of the bidirectional project's dynamics: the project's logic requires problem-solving whereas the disciplines' logic allows us to construct arguments for our decisions.

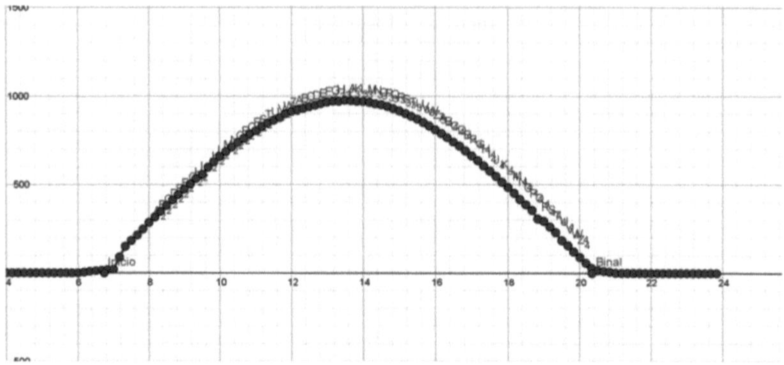

Fig. 8.5 Evolution of solar radiation for a day in January 2013. (Source: Authors' own work)

In this case, the mathematical and statistical tools allowed us to calculate wind behaviour leading to the conclusion that there are innumerable long series of time where wind speed is inferior to the windmill's start-up speed (approximately, 5 m/s). This means it would not be possible to guarantee a daily, continuous power supply to a refrigerator. At least, not relying on the wind energy resource. Some students then suggested installing one or more photovoltaic panels to provide a daily energy supply to the refrigerator.

Following the same principle of supporting our decisions with arguments, we encouraged students to use the INTA's weather station's database. We asked them to examine solar radiation as an energy source; a topic that is also related to the TSMI curricula. Among wind speed and other variables, INTA's weather station recorded solar radiation on surface. This variable (which was also registered every 10 min from 2010 to 2015) measures solar radiation in watts per square meter (W/m^2).

At an epistemic level (Hacking, 1990, 2002) the wind speed and the solar radiation on surface variables are different in nature; partly, due to the random component of each of these phenomena. Not knowing the existence of a possible explanatory model for the wind speed variable led to an exploratory data analysis (Behrens, 1997; de Mast & Kemper, 2009). At first, the students' epistemic relations as regards the solar radiation phenomenon were similar.

However, upon observing the data on the database's numerical semiotic register and the graphic illustrating evolution throughout the day, the possibility of an underlying mathematical model was suggested. Figure 8.5 shows the evolution of solar radiation on January 3, 2014 in GeoGebra (y-axis: immediate solar radiation at W/m^2, x-axis: time of the day).

Among other points, a peak of solar radiation reaching, approximately, 100 W/m^2 at around 2 p.m. can be observed. Due to the available mental models (Blomhøj, 2019), some students proposed a quadratic modelling of the daily evolution of solar radiation. It is worth mentioning that, as in many other cases, this type of modelling was not chosen because of its precision but because of the phenomenon's dynamic, which, in this case, was cyclical. Furthermore, a few weeks before, the students had

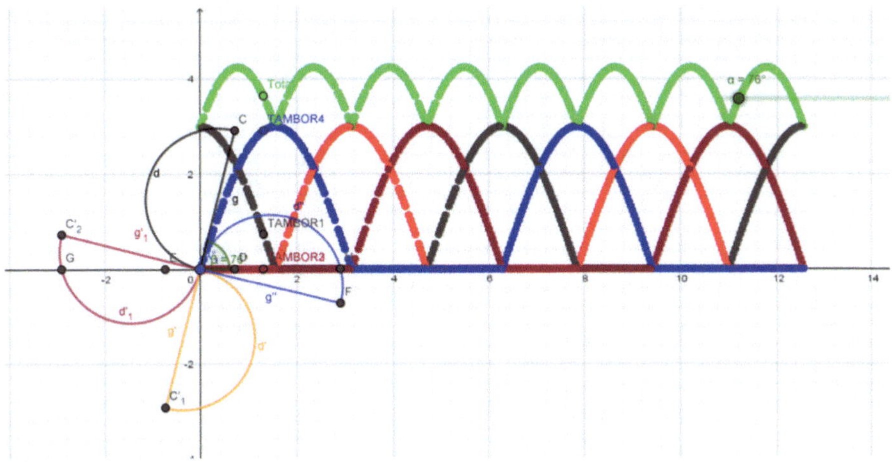

Fig. 8.6 Variation of the area to the wind in a Savonius rotor. (Source: Authors' own work)

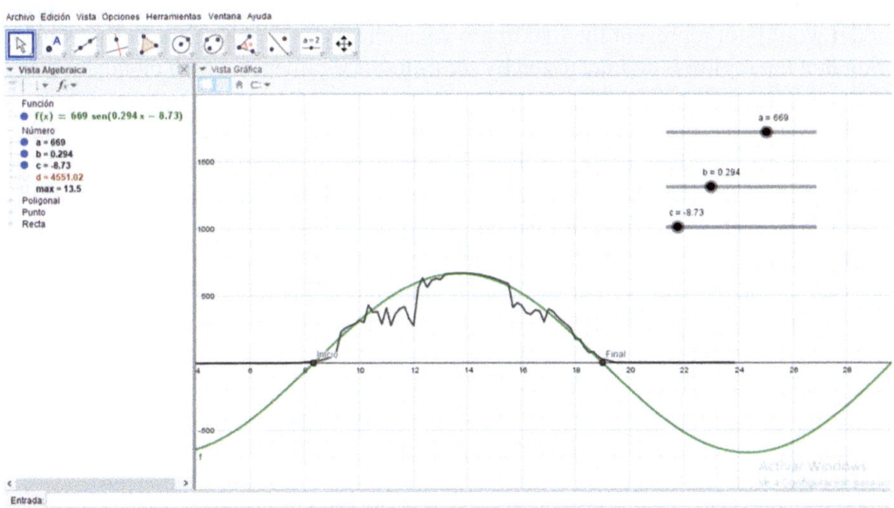

Fig. 8.7 Sinusoidal modelling of solar radiation for a cloudy day in April 2014. (Source: Authors' own work)

modelled the rotor's swept area (Fig. 8.6) and some of them commented on the similarities between these two graphics.

Eventually, the modelling of the evolution of solar radiation was carried out following a sinusoidal model and adjusting its parameters visually. This strategy was sufficient in terms of precision and didactic richness for the objectives of the problem to discuss. Figure 8.7 shows the polygonal line regarding the data on solar radiation on April 14, 2014 and the sinusoidal model carried out with the students. It should be noted that it was cloudy in the morning and the afternoon. This constitutes an

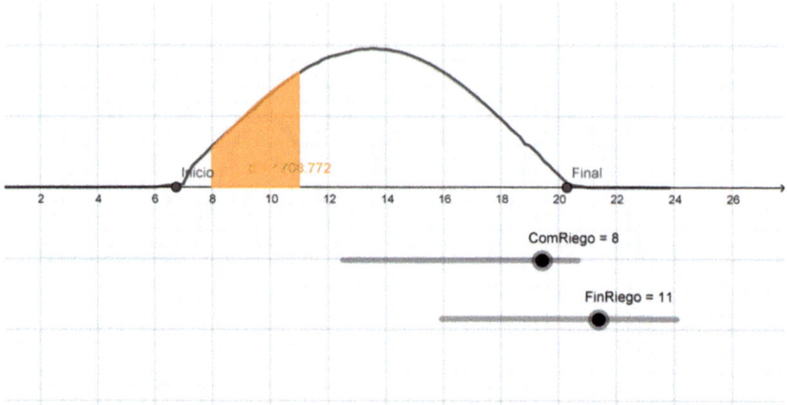

Fig. 8.8 Available power between 8 and 11 a.m. for a day in January 2013. (Source: Authors' own work)

example of returning to the context in the modelling. A least-squares adjustment model would not represent the maximum solar energy available. Especially, considering that the region is characterized by semi-desert weather in which cloudy or rainy days are rare.

The algebraic representation of the modelling of solar radiation's evolution allowed us to calculate, quite precisely, the immediate maximum energy available at any time of the day; however, this was not enough to predict power availability (Blum & Borromeo Ferri, 2009; Brown & Ikeda, 2019).

The daily power value (W/h) was crucial to determine how many hours, per m^2 of the solar panel, the refrigerator could be functioning. This matter led us to introduce concepts related to definite integrals. With the available algebraic model, we could determine the maximum amount of solar power available in the area.

On top of that, the definite integral with variable limits allowed us to estimate the power supply to different electrical appliances at different times during the day. This was possible because it could report the maximum daily power available in terms of the chosen time frames.

Figure 8.8 shows a definite integral with the value of the maximum power available per m^2 on a day of January if, for example, the energy were to be used on watering a vegetable garden between 8 a.m. and 11 p.m.

The interpretation of the obtained value according to the context was very interesting: the power equals to having a 170-W light bulb on for 10 h. When arguing about the different interpretations of the obtained power values, a student with a background in the installation of solar panels remarked:

> But, professor, that is not the power we are going to use, it is the power available on the surface. A solar panel does not convert all of the energy. A bit, actually a lot, is lost. In the course I took, I learnt that the input is between 20 and 25 percent.

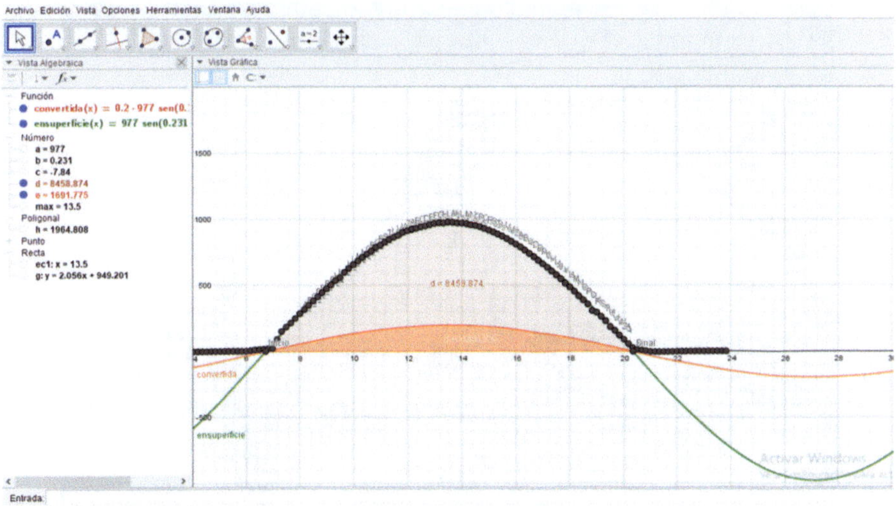

Fig. 8.9 From solar radiation to panel's generated power for a day in January 2013. (Source: Authors' own work)

The strategy to determine the real power generated by a m^2 of the photovoltaic panel resulted in interesting debates. One of them dealt with determining the 20% of the total solar power available on the Earth's surface (numerical semiotic register). Whereas another discussed the representation of the corresponding function and the calculation of its definite integral.

Figure 8.9 shows this second strategy. GeoGebra's graphic view shows the point representation of the solar radiation variable from INTA's weather station, the sinusoidal model that represents them, its definite integral, and the model of the photovoltaic panel's generated power along with its definite integral.

Observing the difference between the available power in terms of solar radiation and the power effectively generated by the photovoltaic panel allowed the students to conclude that the solar resource, in the region's latitude, is not abundant, especially in winter.

For a better grasp of the limitations in terms of power, professors and students made a table based on the algebraic models and the limits of the definite integrals. This table contained information on the energy output of a panel per m^2 on a certain day, every month. It was measured considering hours of use of some common electric appliances (computers, LED lamps, and televisions).

This analysis allowed us to understand the complexity of the variables to consider. Including the variables related to the cost of each type of solution, the maintenance, spare parts availability and trained personnel for their maintenance.

As mentioned before, this project has been carried out with first-year students during the first semester of the TSMI program since 2015. Throughout the first 2 years of the COVID-19 pandemic (2020 and 2021), the activities were limited

Image 8.1 Hoisting the first windmill built by students. Year 2015. (Source: Authors' own work)

because of the online modality implemented. However, still, all studies and analyses were carried out with the students.

In the years prior to the pandemic, the proposal was carried out in a real context where students "immersed" themselves in the problem. Moreover, their proposed solution was also developed in this context, in this case, the construction and installation of a windmill at a rural facility. Image 8.1 shows the hoisting of the first windmill built by students from the 2015 cohort. In this case, the chosen location was the rural facility of an inhabitant named Yolanda.

In the case presented in this chapter, we analyzed the availability of the wind resource and solar radiation to supply energy to a refrigerator at the rural facility "La Margarita". The video linked below shows footage of the assembling and hoisting of the windmill at this location (first installation): https://www.youtube.com/watch?v=889fvPzVK1g&t=2s.

8.4 Conclusions and Perspectives

In the experiences detailed in this chapter, we focused on presenting contexts to students that facilitate the attribution of meaning to the learning process. It is worth mentioning that more projects cover this matter. Some of them were already carried out, others are in progress and others are being planned.

These projects address several characteristics that, to our understanding, facilitate the attribution of meaning. For that purpose, we considered three interconnected dimensions: *temporal dimension*, *knowledge functionality dimension*, and *knowledge transcendence dimension*. Certain hypotheses underlie their articulation. One of them refers to students as social beings. Here, we refer to them as people who socialize outside of the classroom (Vygotsky, 1931) as well as within their community.

Another hypothesis is related to emotional and affectional matters. We see the world as something subject to improvements that can result from individual as well as collective actions. In this sense, students are encouraged to work in teams where they can design and implement an intervention to improve this world. We claim that completing this cycle of designing a solution and implementing it arises certain emotions in the students, among other matters. Hence, contributing to motivation and therefore attribution of meaning.

Considering these hypotheses, one of the first research questions we posed was about the possibility of addressing disciplinary concepts (in particular, mathematical) with this type of proposal. In other words: how can we incorporate mathematics in a proposal with such characteristics? What role could this discipline play?

Supported on the idea that students have to effectively implement the proposed solution, we contemplated a fundamental use for mathematics, intending to guarantee a genuine and authentic need to evoke knowledge. The strategy involved making mathematics appear as a set of conceptual tools that allow students to construct convincing arguments for their decisions. Those decisions would not be banal for being, precisely, a real intervention in the real world (Showalter, 2013). Therefore, mathematics would be used to construct rationality, and to perform a thorough analysis of relevant phenomena to then have supporting arguments when taking action.

The idea of constituting mathematics as a set of conceptual tools to use for rationality construction led us to the problem with mathematical modelling. The modelling cycle was repeatedly introduced in all the projects we carried out. This, in fact, becomes quite evident when considering that there are three components involved: context, the need to act rationally, and mathematics.

In this chapter, we discussed one of those mathematics models: the one referred to the analysis of solar radiation and its possible use to supply electricity to a refrigerator at *La Margarita*. This and the other models made with the students in this particular project were surprisingly rich for both students and professors. In the case of professors, it was not only because of their semantic richness but also because, along with the students' interventions, they allowed us to significantly improve our understanding of the analyzed phenomena. In this sense, the learning process was mutual.

Another learning stage for the professors' team, this time of a didactic nature, was constituted by what could be defined as meta-context permeability. On the one hand, we have the real context which, in this case, is characterized by the rural inhabitants and their difficulties in accessing underground water. On the other hand, when students approach the problem in the real-world context, they create situations with didactic potentialities that result from their interactions with the *real-world context*.

As a result, there appears what we deem the meta-context: a context constituted by the students' approach to the real-world context. This meta-context, which is dynamic and presents instances not planned in the professors' analysis carried out beforehand, creates situations with didactic potentialities.

One such example is the study of solar radiation. In the analysis made before the project, the professors had only considered the study of the wind resource since it is the source of energy for the windmill. The conversations among students upon the first visit to the rural location and their subsequent insistence to tackle the issue encouraged us to reconsider our planning and incorporate the study of solar radiation; more precisely, due to its didactic potentialities.

In addition, the teaching team has learnt (and it is continuing to do it) within what could be denominated an additional margin for uncertainty to the development of the project's logic and, consequently, to the disciplinary logic. By accepting a margin for uncertainty or permeability for the phenomena in the meta-context, we could analyze solar radiation and we could introduce the related knowledge, such as definite integrals, and parameters in trigonometric functions, among others.

This meta-context permeability also allowed us to address another problem, which was presented by the students and was related to hoisting the windmill. It will be discussed in detail in the next chapter along with a student's observation that resulted in the modelling of stress and torque.

Another instance of this permeability was observed with the students from the 2017 cohort. That group proposed a design for the rotor different from the one originally planned. This new design resembled a turbine with deflectors and its purpose was to optimize performance. Image 8.2 shows the rotor built by said student cohort.

In order to have arguments that could support the construction of this new design, it was necessary to carry out an analysis to find the optimal point among a set of variables (quantity, inclination and blade rotation) that could maximize the available power for a determined wind speed range. The complexity of that analysis exceeded the department's capabilities. Thus, to carry it out we had to request the collaboration of a research centre of the National University of Comahue. Figure 8.10 shows the simulation produced by our collaborators. To understand this simulation, students had to be introduced to concepts related to partial derivatives, among others.

As was anticipated in Sect. 8.3, the chosen real-world contexts have naturally evoked extra-mathematical knowledge that was indispensable to continue with the project's logic. One such example is the case presented in this chapter, where we studied the availability of the wind energy resource by consulting the database from a weather station.

We could confirm the almost unavoidable interdisciplinarity that this type of proposal entails. Such unavoidable communication among disciplines to construct rational arguments questions traditional education in watertight compartments. In fact, it creates tension in several directions within the traditional education ecosystem.

Image 8.2 Another rotor design for Savonius windmill. (Source: Authors' own work)

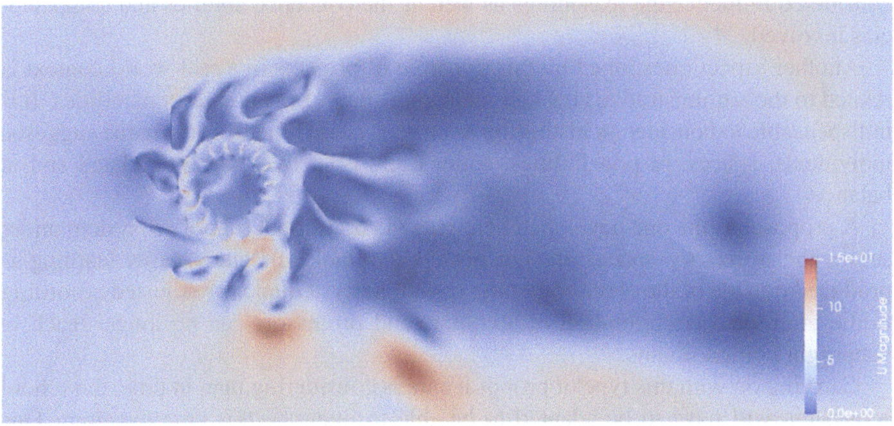

Fig. 8.10 Simulation of new Sanonius rotor design with optimized variables. (Source: Authors' own work)

One of those directions refers to the professors' concepts of teaching, learning and even planning. As regards this last aspect, the traditional didactic transposition is questioned: in a proposal constituted by projects in real-world contexts, it is necessary to not only consider the disciplinary logic but also the project's logic.

Within the project's logic, the disciplinary logic (mathematics, in our case) could likely establish essential connections with other disciplines. This could occur due to the semantics carried by the mathematical objects as a result of being formed within the project's logic. As opposed to being required by the disciplinary logic, which could be relatively autonomous within its internal development.

The different conceptions of what teaching and learning represent may also be questioned when making incursions into a proposal like this. In fact, if we agree that a concept has been learnt when the student is able to reuse it in new situations without any type of assistance or suggestion from others (Robert, 1992), then an experience like the presented in this proposal represents a sort of open laboratory where it is possible to observe if there is effective progress in the students' learning.

This projects include various instances of autonomous work in a real-world context (that is, without any help or advice from the educator) where students must consider certain concepts to solve the innumerable problems they encounter during the construction and installation stages.

Venturing into this type of proposal constitutes evidence for the pressing need to consider other elements in the teaching-learning process. Elements that go beyond disciplinary concepts: we place particular emphasis on the abilities and competencies related to modelling (Blomhøj, 2019; Maaß, 2006).

Once again, there appears a strong interrelation: the disciplinary concepts, abilities and competencies related to modelling interact and enhance each other (Blum & Borromeo Ferri, 2009; Brown & Ikeda, 2019). Moreover, as we will discuss in chapter four, the modelling tool and the kind of objects from the context to be considered influence the reasonings as well as the concepts, abilities and competencies involved.

Another aspect questioned by this dynamic of projects in a real-world context is related to the administrative area and managing space, time and responsibilities. It is indispensable to consider an institution as an ecosystem where, as the name suggests, individuals, spaces, responsibilities, actions and objectives are interrelated and at balance.

A proposal as the one developed here puts the traditional school ecosystem under strain: the necessary spaces are not always available, in many cases learning is produced outside of the classroom, time cannot be systematically adjusted according to the traditional configuration of a course, and professors can no longer teach in watertight compartments.

Carrying on with this type of proposal entails considering that, in time, the school ecosystem will have to be adapted to be able to integrate this new dynamic. This could lead to a new balanced state for the ecosystem that allows us to not only develop but also replicate these experiences. The problem then surpasses the didactic view and positions itself in the pedagogical and administrative field.

Finally, we could observe an important improvement in the competencies related to teamwork, complex objective achievement, and certain matters referring to gender equality and respect; both within the team and towards the rural inhabitants.

References

Avila, C. (2017). *Creación de una aplicación para el cálculo y diseño de turbinas savonius, para su implementación en el territorio colombiano*. (Ingeniería), Universidad Distrital Franncisco José de Caldas.

Bednarz, N., et al. (2018). Contextualiser pour enseigner les mathematiques: un enjeu de formation. *Annales de didactique et de sciences cognitives, 23*, 69–105.

Behrens, J. T. (1997). Principles and procedures of exploratory data analysis. *Psychological Methods, 2*(2), 131–160.

Blomhøj, M. (2019). Towards integration of modelling in secondary mathematics teaching. In G. A. Stillman & J. P. Brown (Eds.), *Lines of inquiry in mathematical modelling research in education. ICME-13 monographs*. Springer.

Blum, W., & Borromeo Ferri, R. (2009). Mathematical modelling: Can it be taught and learnt? *Journal of Mathematical Modelling and Application, 1*(1), 45–58.

Brown, J., & Ikeda, T. (2019). Conclusions and future lines of inquiry in mathematical modelling research in education. In G. Stillman & J. Brown (Eds.), *Lines of inquiry in mathematical modelling research in education* (pp. 233–253). Springer.

Buxton, C. (2006). Creating contextually authentic science in a "low-performing" urban elementary school. *Journal of Research in Science Teaching, 43*(7), 695–721.

Carranza, P. (2015). Molino Savonius. Proyecto de extensión y marco didáctico en clases de matemática. *Extensionismo, innovación y transferencia tecnológica. Claves para el desarrollo, 2*, 55–61.

Carranza, P. (2017). Proyectos interdisciplinarios con la comunidad. Posibilidades y dificultades. *Kimun, III*(5), 5–32.

Chevallard, Y. (1985). *La transposition didactique* (1991st ed.). La Pensée Sauvage.

Crombie, A. C. (1980). *Styles of thinking and historiography of science*. Paper presented at the Sociedad Española de Historia de las Ciencias, Madrid, Espagne.

de Mast, J., & Kemper, B. P. H. (2009). Principles of exploratory data analysis in problem solving: What can we learn from a well-known case? *Quality Engineering, 21*, 366–375.

de Vecchi, G., & Carmona-Magnaldi, N. (1996). *Faire construire des savoirs*. Hachette Éducation.

Develay, M. (1993). *Le sens dans les apprentissages: du désir au passage a l'acte*. Paper presented at the Pedagogie collégiale, Quebec.

Develay, M. (2004). *Donner du sens à l'école*. ESF.

Hacking, I. (1965). *The logic of statistical inference*. Cambridge at the University Press.

Hacking, I. (1990). *The taming of chance*. Cambridge University Press.

Hacking, I. (2002). *L'émergence de la probabilité*. Seuil.

Maaß, K. (2006). What are modelling competencies? *The International Journal on Mathematics Education, 38*, 113–142.

Martínez Licona, J. F., & Palacios Ramírez, A. (2012). Los enfoques de aprendizaje y el proyecto de vida en los jóvenes en los colegios de bachilleres de SLP. *Revista Iberoamericana de las Ciencias Sociales y Humanísticas, 1*(2), 25.

Menet, J. L. (2004). A double-step Savonius rotor for local production of electricity: A design study. *Renewable Energy, 29*(11), 1843–1862.

Pinzón, M. (2016). *Importancia del proyecto de vida, como eje motivacional para el ingreso a estudios superiores, en los estudiates de los grados 10° y 11°, de la institución educativa técnica La Chamba, del Guamo-Tolimba*. Master, Universidad de Tolima.

Robert, A. (1992). Projets longs et ingénieriès pour l'enseignement universitaire. *Recherche en didactiques des mathématiques, 12*(2.3), 181–220.

Savonius, S. (1922). *The wind rotor in theory and practice*. Savonius & Co.
Showalter, D. (2013). Place-based mathematics education: A conflated pedagogy? *Journal of Research in Rural Education, 8*(26), 13.
Spandaw, J. (2009). *Modelling mathematics 'teachers' professional development*. Paper presented at the Sixth Congress of the European Society for Research in Mathematics Education, Lyon, France.
Vygotsky, L. (1931). The development of thinking and concept formation in adolescence.

Chapter 9
Descriptive and Prescriptive Modelling in a Math Class Project: Disciplinary Concepts Participating in the Construction of Arguments for Decision Making

Pablo Carranza and Jaime Moreno

9.1 Introduction

Overall, in this chapter, we discuss the potentialities of projects in a real-world context as pedagogical-didactic proposals to approach disciplinary concepts, particularly, mathematical. We will share details on some of the stages of a project consisting of the calculation, construction and installation of Savonius windmills for rural communities in Argentinian Patagonia. We have been working on this project for several years at the *National University of Río Negro* (UNRN[1]).

To design these windmills, we deemed necessary their easy, low-cost construction as well as their simple maintenance. This led, for instance, to using recycled 200 L barrels for the rotor. Although the barrels do not optimize wind resource management, they are cheap and easy to obtain and, eventually, to replace.

The simplicity of the design is intended and not improvised. In fact, for the design to work, thorough analyses and foundations are needed to ensure the objective (in this case, to pump water) will be accomplished and long-lasting. These analyses are part of the work university students at UNRN carried out. They are also the conceptual framework that allows the students to construct and install the windmills in rural communities.

Thus, realism is a defining characteristic of this proposal and it is used from a didactic perspective as a framework for argumentation for decision making. Accordingly, this required rationality has led both students and professors to evoke disciplinary knowledge of mathematics, among other fields, which allowed them to create

[1] Universidad Nacional de Río Negro—UNRN.

P. Carranza (✉) · J. Moreno
Universidad Nacional de Río Negro, General Roca, Río Negro, Argentina
e-mail: pcarranza@unrn.edu.ar; jmoreno@unrn.edu.ar

M. Rosa et al. (eds.), *Mathematical Modelling Programs in Latin America*,
https://doi.org/10.1007/978-3-031-04271-3_9

a safety framework for their actions during the construction and installation stages. In both cases, such knowledge was evoked during the modelling process.

In this chapter, we will present a few models created with university students in the project involving the calculation, construction and installation of windmills. The models will be categorized into descriptive and prescriptive.

9.2 Conceptual Framework

As in the previous chapter, we address our interest to present didactic situations to students. Situations where they can attribute meaning to the learning process. Once again, we claim that meaning is not a characteristic or attribute of the proposal, but a construction (Develay, 1993). This construction is formulated by a student according to several elements in which some of them being their own environment, others being of a more personal nature. Nonetheless, we also understand that certain characteristics of the proposals may facilitate the students' attribution of meaning.

Among such characteristics is what we call a *temporal dimension* and it is developed in two directions: present and future. Simply put, we expect students to perceive their gained knowledge as something useful for both their present and future. Another dimension considered, and strongly related to the previous one, is *knowledge functionality*. We believe knowledge should help students understand the world around them while also allowing them to intervene in it. Finally, a third dimension taken into account is defined as *knowledge significance*. Our purpose is that students see this knowledge as something useful for themselves and the community.

In this chapter, we will focus on the second dimension, centered on the functionality of disciplinary knowledge. More specifically, we will address the disciplines' potential to construct arguments that allow students to make rational decisions in their interventions. This proposal is carried out in a real-world context to facilitate meaning attribution. By real-world context, we understand an existing context that students can experience. A context where the decision-making process is significant to them (third dimension). In other words, instead of being a context presented by the professor, it is taken from their community so students can fully experience it and observe how their decision-making process is effectively applied.

The issue of contexts and their realism is also discussed by other authors who refer to their potentialities and difficulties (Bednarz, 2018). Among the potentialities, we can find motivation (Brown, 2019), which is associated with the idea of meaning mentioned above; and what some call place-based awareness (Showalter, 2013). As regards difficulties, this last author mentions that finding mathematically rigorous cases also poses a challenge. That is, finding contexts that genuinely demand disciplinary knowledge (Buxton, 2006; Spandaw, 2009); in our case, mathematical.

In this regard, we adhere to the idea that knowledge should be genuinely required by the context and not fictitiously introduced by professors (Boaler, 2001; Brown, 2019). Because that leads to a traditional didactic contract (Brousseau, 1988) in which students focus on meeting their educators' expectations rather than on finding the rational solution the context demands.

One of the proposals we carried out is the calculation, construction and installation of Savonius windmills (Savonius, 1922) for rural communities in Argentinian Patagonia. We consider this proposal incorporates the three dimensions presented above. A more detailed description can be found in the previous chapter.

Here, we will discuss the potentialities this type of proposal presents for mathematical modelling. On the one hand, we have a real-world context that is marked by an imbalance or conflict: there is a problem or a particular situation that requires a solution. On the other hand, we cannot discuss that solution based on intuition or following a trial-and-error strategy. It is necessary to carry out a rational analysis of both the characteristics of the problem and its possible solutions.

Modelling then constitutes a space that synthesizes the real-world context. In this space, it is feasible to introduce disciplinary knowledge that allows us to carry out a precise analysis of the problem and to develop a solution supported by arguments. Hence, mathematics along with other disciplinary fields function as a set of conceptual tools for the construction of rational arguments for the decision-making process demanded by the context.

This type of proposal (particularly, the one involving the calculation, construction and installation of Savonius windmills) is referred to as a project due to its richness and complexity. Here, the project in question will be named: *Savonius Windmills Project*. Due to space limitations, we will only discuss two of the situations where mathematical modelling was necessary. One of them refers to the analysis of guy wires, whereas the other focuses on the stress involved in hoisting the windmill into its vertical position.

Both situations fit in a typical modelling cycle (Blomhøj, 2019; Caron, 2019; Czocher, 2019). Briefly described, this cycle starts with a real-world situation and a problem, then the reality's characteristical elements linked to the problem are considered. Next, those characteristical elements are associated with disciplinary methods and objects (Crombie, 1980; Hacking, 2002). Finally, at the abstraction level, reasonings and analyses are produced establishing relations previously unknown or inexistent to then carry out an argument-based intervention in the context.

This characterization of the modelling cycle is not exhaustive. Some authors discuss a validation stage (Maaß, 2006) and others the itineration of the cycle (Brown & Ikeda, 2019), among other points. While we agree on such ideas, we wish to highlight the importance of what we could call the *ending* of the cycle: the return to the context that encouraged the modelling. Such return is carried out with conceptual tools for comprehension and action; tools that were not possessed before the modelling.

Within the field of research on didactics of mathematics, the literature proposes a characterization for models: descriptive modelling and prescriptive modeling (Blum & Borromeo Ferri, 2009; Brown & Ikeda, 2019). In a few words, a descriptive modelling may characterize and show a better understanding of a phenomenon in context. Prescriptive modelling may indicate how to intervene in the context.

In the following sections, we present two examples of what we consider to be a dialectic between those two types of models. Both examples occur within the Savonius Windmills Project; however, we want to add that the same dialectic has been observed in other projects. They will not be discussed here due to space limitations.

9.3 Methods

The Savonius Windmills Project has been carried out since 2015. Nevertheless, due to the COVID-19 pandemic, some adjustments have been made through 2020 and 2021. Its authors are the professors in charge of the first-year mathematics course in the Industrial Maintenance Technician program (a 3-year degree) at the National University of Río Negro, Argentina. The project is therefore conducted with first-year university students enrolled in the program; approximately, 50 students aged between 18 and 50.

This project is the core of the mathematics course and it is developed throughout a semester, which is also the course's formal length (first-year students). The installation process usually occurs after that period because of the extremely cold Patagonian winter. The didactic strategy of this proposal is to promote as much rationality as possible in each of the project's stages (calculation, construction and installation of the windmills). Precisely, that rationality genuinely justifies the need for disciplinary concepts and methods.

In other words, for the concepts' emergence, we seek to make the most of the certainty that the project has to be successfully carried out. That certainty is important. If a windmill falls, the results can be tragic for the rural family using it; hence its fundamental aspects are thoroughly analyzed.

9.3.1 Case 1: Modelling of the Guy Wires

One of the fundamental aspects of the project is supporting the windmill in a vertical position. It is worth mentioning that the Patagonian region is distinguished by its strong winds that can sometimes reach 100 km/h in spring. Thus, even if the windmill is locked during extreme winds, it must be able to resist the wind stress on its structure. Figure 9.1 illustrates the support system used in the project for each of the eight guy wires.

The windmill stands vertically because of a set of supporting guy wires made of steel. The wires are attached to a concrete block buried in the ground. As a result, the concrete block and the ground above absorb the impact of the region's strong winds. The modelling discussed here addresses a problem that typically occurs when installing the windmill at a rural family's location. One of the tasks entrusted to

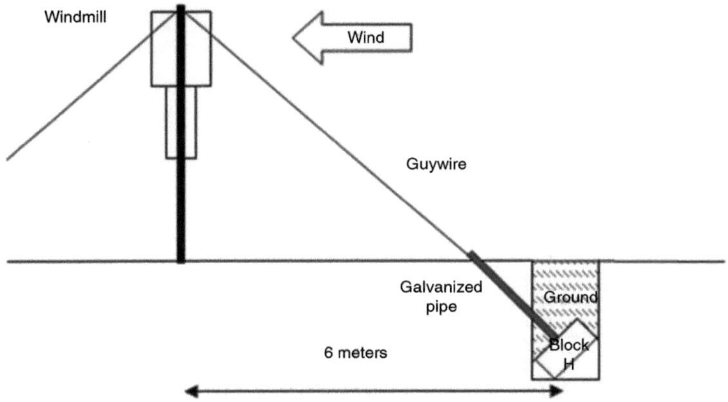

Fig. 9.1 Savonius windmill and its support system. (Source: Authors' own work)

the inhabitants is to dig four pits for the concrete blocks. A detailed plan and the necessary explanations are given to perform this task.

In some cases, the inhabitants cannot maintain the designated distance between the pits and the foot of the mill (6 m) because of blocks of stone or large roots. Since these families live roughly 400 km away from the university and have no means of communication whatsoever, in those cases, they decide to relocate the pits according to their own criteria.

This possibility is discussed beforehand with the students. In this regard, one of the most relevant questions professors frequently ask is: "If the wire is attached to the ground at a higher (or lower) distance than expected, do you think the stress on the wire and the concrete block varies due to the anchor being nearer or farther from the ground?". We will summarize the students' usual answers as follows: "Stress will not change regardless of the distance of the guy wires because it is caused by the wind".

To study this issue, the professors proposed creating a dynamic representation of the stress in terms of the distance between the wire and the windmill (Fig. 9.2). This representation was made using GeoGebra.

The curve shown in Fig. 9.2 is traced from the extreme of the vector named *tierra* (ground) which represents the vertical stress on the concrete block and the ground above it. This figure evidences how the stress module varies when the anchoring point is nearer or further; this is controlled by the *distancia* (distance) slider. This modelling allowed students to deepen their understanding of the stress dynamics on the guy wires and to identify other aspects previously unknown. For example, the vertical component of the *enrienda* vector, named *vertical*. Identifying that component led to recalculating the stress on other concrete blocks not shown in the figure, which are the blocks representing the foot of the windmill.

We believe this is a descriptive modelling since it shows the stress interactions on the support of the windmill. Its dynamism lets us observe how and why the stress varies depending on a location. Thanks to this description, iterated as many times as

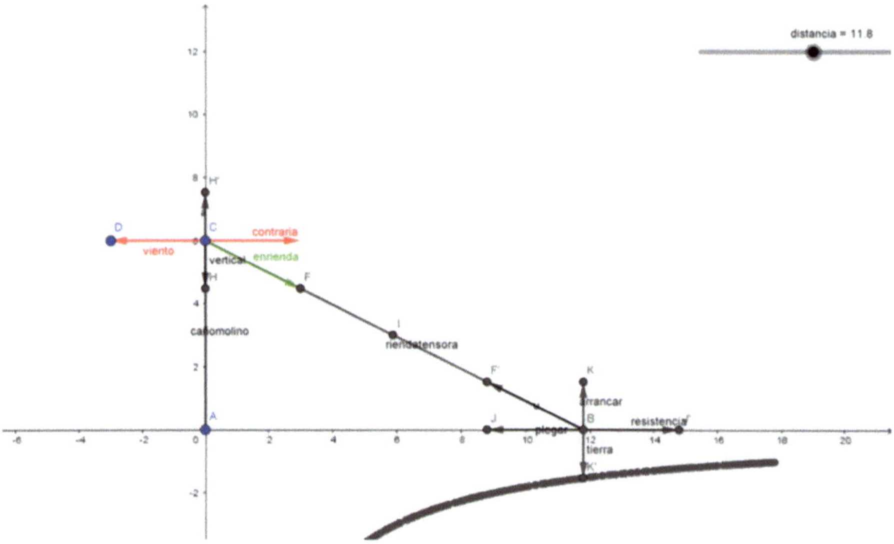

Fig. 9.2 Dynamic representation of the stress on the anchor in terms of distance. (Source: Authors' own work)

necessary, students could understand that the depth of the pit should be determined according to its distance to the foot of the windmill and the wind. In addition, they would need to inspect the depth of the pits dug by the community members at the construction site.

The project's dynamic makes the installation of a windmill an efficient yet complex task. In general, the process lasts between 3 and 4 h. Thus, work is intense and coordinated. Students are divided into groups and each group is entrusted with certain responsibilities during the installation process. In particular, the team responsible for the footing and guy wires must check the pits' depth according to the analysis done in class (Fig. 9.2).

The environmental working conditions and the challenges of the installation tasks impeded having a laptop at the location. Therefore, the team assigned to the pits cannot use one to determine depth according to the distance between the inhabitant's pits. Calculations on paper are difficult to do as well due to the same reasons. However, this difficulty could be anticipated and discussed in class thanks to the professors' gained experience working on other windmill installations. As a result, we could create a new model that allowed us to quantify the phenomenon.

We created a spreadsheet in class containing the relations and variables involved in determining the ground vector (Fig. 9.3). This work was based on the descriptive modelling in Fig. 9.2 and done before the installation at the location.

This second model was carried out once the first one was understood. The objective was to quantify the variables by algebraically expressing the relations discovered in the previous modelling. Since this model was designed on a spreadsheet, students could view it from their smartphones during the installation stage.

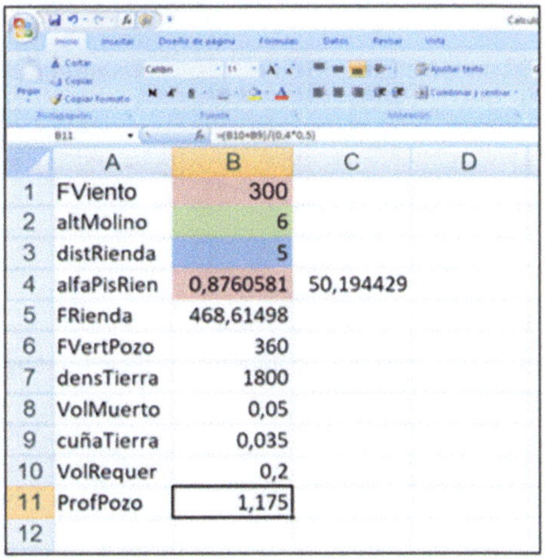

	A	B	C	D
1	FViento	300		
2	altMolino	6		
3	distRienda	5		
4	alfaPisRien	0,8760581	50,194429	
5	FRienda	468,61498		
6	FVertPozo	360		
7	densTierra	1800		
8	VolMuerto	0,05		
9	cuñaTierra	0,035		
10	VolRequer	0,2		
11	ProfPozo	1,175		
12				

Fig. 9.3 A spreadsheet on the quantification of the variables affecting the guy wires. (Source: Authors' own work)

They would only need to measure the distance of the pits dug by the community members. Then, they input those values on the spreadsheet to determine if it was necessary to deepen the pits or not; all of this was carried out quickly and safely.

9.3.2 Case 2: Lifting the Windmill at the Location

Roughly 50 students divided into teams participated in the construction of the windmill. Each team was in charge of a certain task: building the structure, building the rotor, preparing the concrete blocks and the guy wires, and building the braking system. Thus, each task was performed by a different team.

All these components are built at the students' houses or at specific places assigned for that purpose. After that, the main parts of the windmill are taken to the location. Once there, the teams work together to assemble it and hoist it next to the water well. Lastly, they install the hoses and the windmill starts working (if the wind power is sufficient at that time).

One of the critical moments in the installation at the assigned location is lifting the windmill from the horizontal assembling position to a vertical working position. The modelling presented in this section emerged from a student's question. Upon watching footage from previous years showing how the windmills were hoisted, the student asked whether it would be better to lift it differently (https://www.youtube.com/watch?v=6o0LdHIVdFE). Figure 9.4 shows a schematic diagram of the windmill's hoisting process: a vehicle with an electric winch is placed at a certain distance from the windmill. The synthetic cable of the winch is attached to the superior part of the windmill (Point A).

The student asked: "Isn't it better to attach the cable to the center of the structure (Fig. 9.4, Point B) instead of how it's being shown in the video (Fig. 9.4, Point A)?".

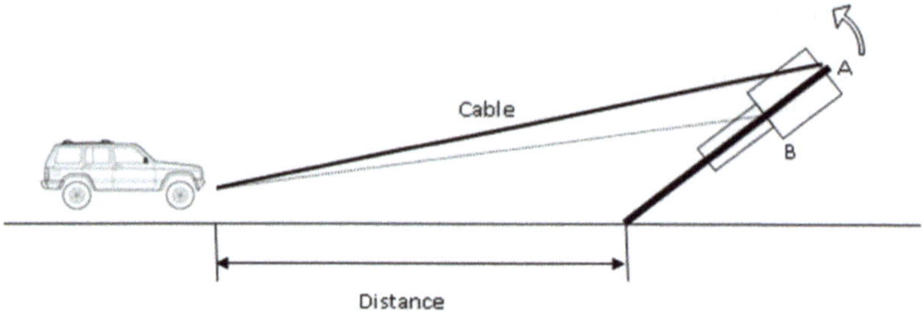

Fig. 9.4 Schematic diagram of a vehicle hoisting the Savonius windmill. (Source: Authors' own work)

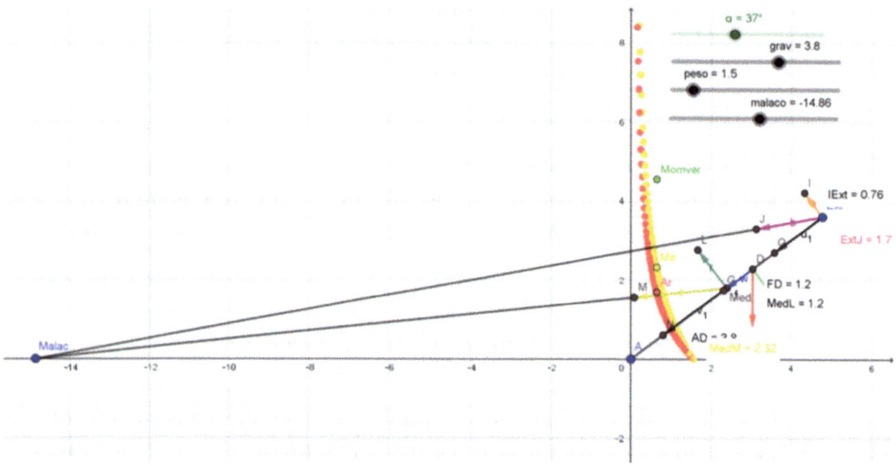

Fig. 9.5 Modelling of the stress on hoisting the Savonius windmill. (Source: Authors' own work)

Moreover, considering the debate for the modelling of the guy wires (Fig. 9.2), another student asked: "Does the distance between the vehicle and the structure affect stress?" These and other questions encouraged a new dynamic modelling in GeoGebra that quantified the stress (vectors) using sliders. This time, the modelling was done in scale.

This dynamic modelling of the stress on hoisting the windmill into its vertical position (Fig. 9.5) required different concepts than the ones needed for the analysis of the stress on the guy wires (Fig. 9.2). In that case, one of the main non-mathematical concepts discussed was the equilibrium of forces and Newton's laws. Whereas in this case, we applied an even more abstract concept: torque.

Since this modelling was done in scale, it was descriptive as well as prescriptive. We consider it was descriptive because it showed how vectors, their modules and their relations are interwoven in the hoisting process. Evidencing these objects allowed us to gain a better understanding of the stress dynamics, leading to a detailed characterization of the stress on each of the proposed alternatives.

In addition, we consider it prescriptive because as a result of understanding the stress dynamics students could produce arguments for the decisions for several matters, and for instance:

1. It is convenient to attach the winch cable to the upper end of the windmill (Fig. 9.4, Point A) because less stress is produced compared to when it is attached to the center (Fig. 9.4, Point B).
2. The farther the vehicle is from the foot of the windmill, the lesser the stress produced on the winch cable.
3. Higher stress is produced in the initial stage when the windmill is being pulled to the vertical position. Therefore, more precautions are necessary during said stage of the hoisting process. Cable resistance needs to be taken into account during this high-stress stage.
4. Initially, the windmill should be at a certain angle (not horizontal). Otherwise, the stress would solely be on (horizontal) towing and not on hoisting (vertical component).
5. There is a horizontal component at the foot of the windmill whose module is maximum at the beginning of the process. This tends to cause the foot to move towards the vehicle (these vectors are hidden in Fig. 9.5 due to space limitations).
6. The quantification of modules in scale allowed us to learn about the resistance the vehicle's winch cable has to withstand; this information was later compared with the winch's specifications.

9.4 Results

The results obtained in these studies are as follow.

9.4.1 On Semantic Richness

One of the first results we could observe from the models carried out with the students was their semantic richness. Indeed, mathematical objects are no longer abstract entities. In the models, many of them carry meaning associated with the context of origin that remains throughout their eventual transformations. This semantics establishes the relations between the mathematical object and the real-world object; thus, establishing a connection.

Still, there is one difficulty proven in class and connected to this matter. Students have to understand the objects at a disciplinary-logic level and retain their semantic. Therefore, their attention is required at two levels: disciplinary logic and semantic. However, the outcome is positive since the semantic dimension of the object, their connection to reality, helps the students gain a better understanding of them. In this regard, during the windmill's installation and construction stages, it has been observed that students integrate the arguments produced in class throughout each modelling.

This can be seen in the actions they perform autonomously. One such example is the installation stage of the windmills. In 3–4 h, the teams, in complete autonomy, recall their class debates to act diligently and rationally during the decision-making process at the location, which may demand foreseen or unexpected decisions. Another clear example can be found in the students' explanations to the community. After completing the installation, they explain to the inhabitants what precautions to take in the use of the windmill and how to perform preventive or corrective maintenance tasks.

9.4.2 On Interdisciplinarity

Another result observed is the inevitable interdisciplinarity. None of the cases presented in this chapter could have been solved without knowledge from other disciplines; specifically, from physics.

In the first case, it was necessary to include concepts related to Newton's laws. More precisely, the concepts of equilibrium of forces for bodies at rest. Without these concepts, it would have been impossible to carry on with the dynamic representation of the entwinement of vectors that showed us how the distance between the pits and the foot of the windmill affects the stress module.

For instance, in the modelling of the guy wires in GeoGebra, the mathematical concepts provided technical assistance that helped us represent and relate the vectors, whereas the physics concepts assisted us in the explanation of the phenomenon. Figure 9.6 shows the modelling of the stress on the upper corner of the windmill (Point B).

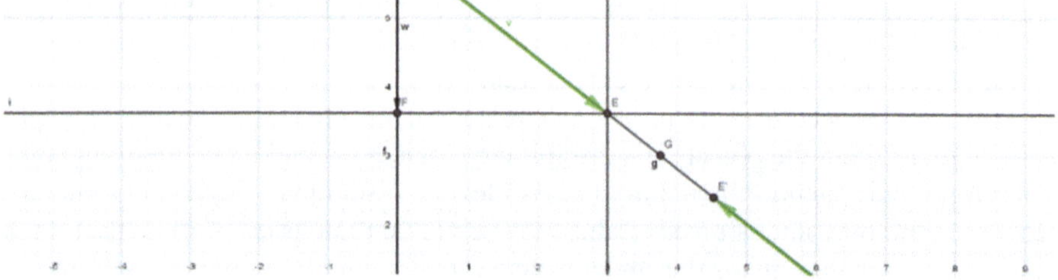

Fig. 9.6 Stress on the upper corner of the windmill. (Source: Authors' own work)

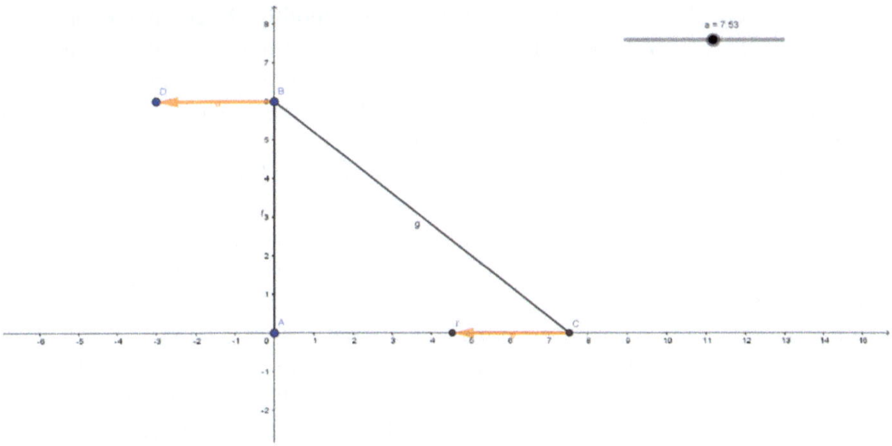

Fig. 9.7 Two related efforts. (Source: Authors' own work)

The $\overrightarrow{BD}$ vector represents, in a simplified manner, the force of the wind on the windmill. Its reaction, $\overrightarrow{BD'}$, is explained using physical arguments (bodies at rest, Newton's laws); whereas its construction was carried out using mathematical concepts (D and D′ are centrally symmetric with respect to B).

On the other hand, $\overrightarrow{BD'}$ should be considered a component of another vector since there is no object applying a force in that direction (argument from physics). The construction of the $\overrightarrow{BE}$ vector—of which $\overrightarrow{BD'}$ is a component, is done using, once again, mathematics: a line is perpendicular to the X-axis containing D′. Point E is the result of the intersection of this line and the line containing the guy wire.

In turn, there is a vertical component of the $\overrightarrow{BE}$ vector that tends to compress the structure of the pipe that transmits the stress to the foot of the windmill (argument from physics). Its construction involves mathematics as well: a line is parallel to the X-axis containing point E and intersects the Y-axis in point F, and etc.

Therefore, the interaction between the two disciplinary fields in this modelling was constant throughout its construction. The same occurs with the interpretation and comprehension of the phenomena. For instance, in Fig. 9.7 we can observe two vectors, the guy wire $\overline{BC}$ located at 7.53 m of the foot of the windmill and the main tube represented by $\overline{AB}$ (6 m tall).

When repeating the modelling in full representation (Fig. 9.2), some students observed possible module equality between the $\overrightarrow{BD}$ and $\overrightarrow{CI'}$ vectors (Fig. 9.7). This possible equality identified on the graphic semiotic register (Duval, 1993) was then observed on the numeric register (coordinates of both vectors on algebra view in GeoGebra). The confirmation of this supposition was supported by the physical and mathematical framework that emerged from the constructive and conceptual relations between the vectors.

The importance of the students' discovery should be noted: the $\overrightarrow{CI'}$ vector is the horizontal component of the stress the guy wire causes on the anchoring system. To remain in place, the ground has to exert a same-module force in the opposite direction. Students could conclude that installing a windmill on sandy ground was not recommended (return to context). It is worth mentioning that, in this case, the modelling was prescriptive since it allowed us to determine how to act in the context (opting for find firm instead of sandy ground).

This indispensable interdisciplinarity was also observed in the case 2 modelling where stress was studied concerning the hoisting process of the windmill. Once again, physics and mathematics took part in the construction of the model. As regards physics, the key concept, in this case, was torque, although Newton's laws and equilibrium of forces were also introduced. As regards mathematics, similar to the descriptive modelling in case 1, rigid transformations were crucial.

One difference with the modelling of the guy wires in GeoGebra (Fig. 9.2) was switching to semiotic, numerical and algebraic registers (Duval, 1993, 2006) since it was an analysis based on torque equilibrium. The same intrinsic interdisciplinarity was observed in the case 1 modelling done on a spreadsheet as represented in Fig. 9.3.

Let us recall that that modelling had a particular objective: once the vector relations in the guy wires were understood, it was necessary to quantify (at the installation stage) pit depth according to the distance between the foot of the windmill and the pits dug by the inhabitants. That depth can be seen on cell B11 (Fig. 9.3) and, even though the value results from algebraic and pre-algebraic calculations (Haspekian, 2005), other concepts were also introduced. For instance, estimated ground density (cell B7), and various volumes (cells B8 and B10), among others.

This intrinsic interdisciplinarity likely occurs because the project is centered on a real-world problem of the context. It is not influenced by the traditional didactic transposition (Chevallard, 1985) in which the concept to be taught is chosen first (monodisciplinary) and then the appropriate context to discuss it is selected. In this case, a real-world problem from the context was chosen first and it was then analyzed to determine if it was appropriate to address a set of disciplinary concepts. Since such problems are usually interdisciplinary, they involved knowledge from other disciplinary fields.

9.4.3 On the Construction of Rational Arguments for the Decision-Making Process

Another aspect to highlight is the models' potential to help students grasp a situation, analyze it, adopt reasonings and produce arguments for the decisions the project demands.

These models were created in spaces for debate between students and professors where even the professors discovered new characteristics to be considered; one of them being the relation between wind force and horizontal force on the ground.

9.4.4 On the Didactic Contract

The models constructed rationality that impacted the didactic contract of the mathematics class in at least two aspects: knowledge retention and mutual expectations. As regards the first aspect, the models were first introduced by the professors due to some difficulties the students had. However, the students themselves progressively acquired them and managed to propose their own analysis. Thus, the models were created in spaces for debate and discussion, not upon the professors' exposition of knowledge.

The second aspect was another interesting phenomenon; however, it was more related to tackling a real-world problem in context. Usually, in a traditional didactic design, students are interested not only in solving a given problem but also in what their professors expect from them. In some cases, their interest is in how their professor considers the problem should be solved. In this regard, their worries and their need to know about the professors' expectations are only natural since they have to pass the exams graded by them.

In this type of proposal, we can observe some changes since students are no longer exclusively focused on their professors' expectations. Now, they are also focused on what the context requires. Since this proposal demands that students produce a solution to a particular problem, their interest shifts from those expectations to other matters. For instance, meeting the community's expectations, being able to follow the team dynamics in terms of commitment and responsibilities.

In Argentina, students can usually retake mid-term exams in case they failed. This is called a "make-up exam." In this proposal, the reality of some situations does not leave room for a "make-up exam". For instance, a windmill falling would be unacceptable. This forces both students and professors alike to pay extra attention so their decisions are supported by strong arguments. Here, the disciplinary methods and concepts are the founding principles for the construction of rationality. The models become spaces for precise analysis for that construction.

It should be noted that case 1 was discussed using two models, whereas for case 2 only one was needed. This is due to, at least, two reasons: the students' knowledge and the initial objectives of each model.

In case 1, the issue with the anchor distance and stress was addressed progressively. The modelling in GeoGebra, where it is evidenced that distance affects stress, was carried out at the beginning of the semester. It also involved a set of knowledge more or less already acquired (Robert, 1998) by the students; such as Newton's laws and rigid transformations. After discovering the relationship between stress and distance, it was necessary to quantify stress to determine the depth of the pits. To do so, it was important to calculate wind stress on the windmill.

Moreover, identifying the module of that vector led to more analyses. One of them was the modelling of the rotor's swept area and the other was a statistical analysis of the region's wind. To carry out said statistical analysis, a database was accessed. It contained the registries of a government's weather station that showed 5 years of data on wind speed (m/s). This data recorded changes every 5 min.

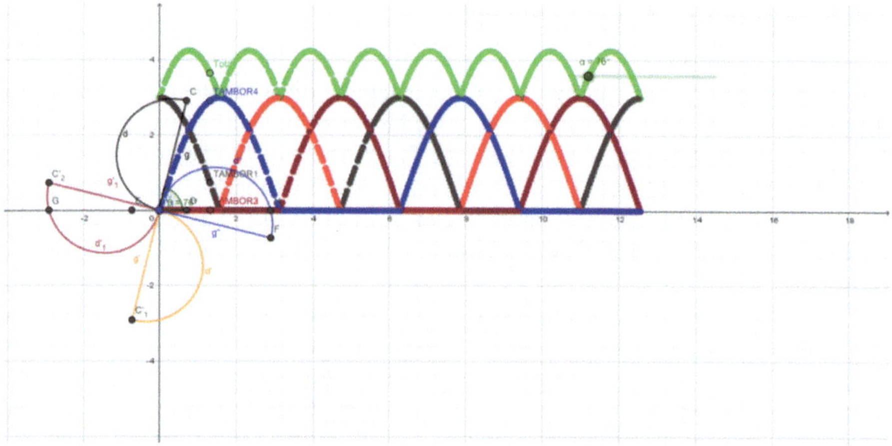

Fig. 9.8 A simplified analysis of the contribution of each half barrel in the swept area. (Source: Authors' own work)

Figure 9.8 is presented for illustrative purposes only. It shows the steps followed to quantify the "wind" vector. This figure illustrates a modelling carried out with the students where they performed a simplified analysis of the contribution of each half barrel in the swept area.

Then, once the *wind* vector was quantified, it was possible to study the modelling on a spreadsheet. A few weeks passed between these two events. As regards the modelling of the lifting of the windmill, the professors considered the limited time availability and the complexity the quantification posed on the students by considering it dealt with torque.

Therefore, the modelling was done in scale. As a result, the student's question about the convenient point to attach the wire could be answered, and an integral analysis of the moment of hoisting could be performed. Thanks to this analysis it was even possible to define the vehicle's winch cable specifications.

9.5 Perspectives

It is important to discuss some perspective on Mathematical Modelling.

9.5.1 Modelling as a Tool to Discuss Disciplinary Concepts and as Curricular Content

The models presented in this chapter, as well as others we carried out with students, prove their dual functionality. They become potential elicitors of knowledge and disciplinary methods while also becoming an object of study.

As regards knowledge acquisition, it is worth mentioning that the disciplinary concepts that emerged during the modelling were later thoroughly discussed. This discussion was done considering the disciplinary field to analyze their properties and functions. Therefore, the models contained other activities that were not strictly related to the project but that focused on discussing certain aspects of the concepts being learnt; even with instances of institutionalization (Brousseau, 1998).

As regards modelling as a learning object, the models were presented by the professors, primarily, because of the approach in which the students previously learned disciplinary concepts. They were designed to be taught as exercisable not as usable. In fact, those exercises were usually done on paper and without context or with a context assigned by their professors. The project's proposal constituted an important qualitative leap in these and other dimensions for the students. For instance, when working in teams towards different objectives.

Nonetheless, one question remains since we still need to identify the individual and collective processes that allow students to link elements from the context to the appropriate models, without the influence of professors or (in more general terms) of a didactic contract that provides a framework for those processes. We believe this is relevant considering that students will not have such framework or the aid of a professor in their personal and professional interactions.

9.5.2 Replication or Development of the Proposals

As of the date of publication of this article, the team has carried out various projects similar to the one presented here. Some of them include:

- Parabolic solar cookers.
- Savonius windmills.
- Water purification systems for families.
- A prototype for an application for phones in tracks in protected areas.
- Environmental pollution testing in urban areas.
- Mobile solar-powered water pumping systems.
- Thermal conditioning for a school library.

These projects have been carried out at urban secondary schools, rural secondary schools and universities. In terms of future perspectives, we still need to find a possible architecture or shared quality that can result in a systematization that allows us to give tools to other educators intending to work on this type of proposal.

In this sense, the models perform a key role in this possible architecture. They become a sort of factory that produces reasonings and arguments based on disciplinary concepts and enables the development of the project. The characterization of this type of proposal in a real-world context and the models become a foundational basis to determine how these factories work.

9.5.3 Modelling Instruments

We observed that modelling is not merely the result of the relation between the problem of a context and the representations of its author. The instrument or support chosen to create the modelling affects that relation; thus affecting the type of productions that can result from it.

In other words, the modelling is affected by the chosen instrument and this instrument is an influence upon what can be modeled and upon the results in terms of reasonings and conclusions. Therefore, the modelling instrument becomes a sort of active lens through which the relationship between the problem and the person analyzing it is influenced.

For instance, in case 1, the analysis of how the guy wires' anchoring distance affect the stress on them was carried out using two instruments: GeoGebra and Spreadsheets. These programs, now considered modelling instruments, conditioned the discussed aspects of the problem and what could be analyzed as well as concluded. In this regard, GeoGebra shows great potentiality because of its views and dynamic tools, such as the sliders.

However, as we will discuss in Chap. 4 when we introduce two new types of modelling, those same potentialities can facilitate a certain type of productions. We will refer to them as analogical to the detriment of others we will name analytical. This is done considering an *economics of knowledge*.

Given the facilities GeoGebra offers for representing objects, it may be more adequate to emulate a reality of the context in GeoGebra (analogical) instead of focusing on the more abstract relations between the objects (analytical), which is less cost-effective in terms of reasoning.

References

Bednarz, N. E. A. (2018). Contextualiser pour enseigner les mathematiques: Un enjeu de formation. *Annales de Didactique et de Sciences Cognitives, 23*, 69–105.

Blomhøj, M. (2019). Towards integration of modelling in secondary mathematics teaching. In G. A. Stillman & J. P. Brown (Eds.), *Lines of inquiry in mathematical modelling research in education* (ICME-13 monographs). Springer.

Blum, W., & Borromeo Ferri, R. (2009). Mathematical modelling: Can it be taught and learnt? *Journal of Mathematical Modelling and Application, 1*(1), 45–58.

Boaler, J. (2001). Mathematical modelling and new theories of learning. *Teaching Mathematics and Its Applications, 20*(3), 7.

Brousseau, G. (1988). Le Contrat Didactique. *Recherche en Didactiques des Mathématiques, 9*(3), 309–336.

Brousseau, G. (1998). *Théorie des situations didactiques*. La Pensée Sauvage.

Brown, J. (2019). Real-world task context: Meanings and roles. In *Lines of inquiry in mathematical modelling research in education* (ICME-13 monographs). Springer.

Brown, J., & Ikeda, T. (2019). Conclusions and future lines of inquiry in mathematical modelling research in education. In G. Stillman & J. Brown (Eds.), *Lines of inquiry in mathematical modelling research in education* (pp. 233–253). Springer.

Buxton, C. (2006). Creating contextually authentic science in a 'low-performing' urban elementary school. *Journal of Research in Science Teaching, 43*(7), 695–721.

Caron, F. (2019). Approaches to investigating complex dynamical systems. In G. A. Stillman & J. P. Brown (Eds.), *Lines of inquiry in mathematical modelling research in education* (ICME-13 monographs). Springer.

Chevallard, Y. (1985). *La transposition didactique* (1991st ed.). La Pensée Sauvage.

Crombie, A. C. (1980). *Styles of thinking and historiography of science*. Paper presented at the Sociedad Española de Historia de las Ciencias, Madrid, Espagne.

Czocher, J. (2019). Precision, priority, and proxies in mathematical modelling. In G. A. Stillman & J. P. Brown (Eds.), *Lines of inquiry in mathematical modelling research in education* (ICME-13 monographs). Springer.

Develay, M. (1993). *Le sens dans les apprentissages: Du désir au passage a l'acte*. Paper presented at the Pedagogie Collégiale, Quebec.

Duval, R. (1993). Registres de représentation Sémiotique et fonctionnement cognitif de la pensée. *Annales de Didactique et de Sciences Cognitives, 5*, 37–65.

Duval, R. (2006). Quelle sémiotique pour l'analyse de l'activité et les productions mathématiques. *Relime, Numéro Especial*, 45–81.

Hacking, I. (2002). *L'émergence de la probabilité*. Seuil.

Haspekian, M. (2005). *Intégration d'outils informatiques dans l'enseignement des mathématiques*. Doctorat, Paris 7 Denis Diderot, Paris.

Maaß, K. (2006). What are modelling competencies? *The International Journal on Mathematics Education, 38*(2), 113–142.

Robert, A. (1998). Outils d'analyse des contenus mathématiques à enseigner au lycée et à l'université. *Recherche en Didactiques des Mathématiques, 18*(2), 139–190.

Savonius, S. (1922). The wind rotor. In theory and practice. : Savonius & Co.

Showalter, D. (2013). Place-based mathematics education: A conflated pedagogy? *Journal of Research in Rural Education, 8*(26), 13.

Spandaw, J. (2009). *Modelling mathematics' teachers' professional development*. Paper presented at the Sixth Congress of the European Society for Research in Mathematics Education, Lyon, France.

Chapter 10
Designing and Building a Mobile Support for Solar Panels: A Project for 12-Year-Old Students that Requires Concepts, Among Others, of Mathematics

Pablo Carranza and Ailen Morales

10.1 Introduction

As in the previous chapters, here we focus on proposing situations where learning is seen as meaningful. In this case, we worked with first-year students at a secondary school in a town quite singular due to its rapid growth. In this town with significant migration flows, we found that students' ages varied considerably within the same grade level, and they also came from different locations. For the school system, these students presented significant learning difficulties which resulted in high rates of grade repetition or even dropouts. In this situation, it became crucial to consider what we deem a fundamental matter: how can we help students find meaning in learning?

To address this matter, we considered a set of hypotheses that we named dimensions and that tend to characterize situations that facilitate the attribution of meaning to learning. This led us to proposing the students creating a solution for a real problem that affected a substantial number of inhabitants in the area (even their parents and grandparents): groundwater pumping for irrigation and animals. The proposed solution for this problem consisted of building a support for photovoltaic panels that would supply electricity to a small water pump. Given the lack of the energy source (solar radiation), the students had to consider the possibility of regulating the angles of the support to maximize energy usage.

The didactic objective was addressing mathematical disciplinary concepts that would appear as a set of rational tools to construct arguments for their decisions. However, this discipline alone proved to be insufficient to understand the problems of the project and to propose the necessary solutions. In this chapter, we focus on our reflection on the introduction in unison of different disciplines in order to understand

P. Carranza (✉) · A. Morales
Universidad Nacional de Río Negro, General Roca, Río Negro, Argentina
e-mail: pcarranza@unrn.edu.ar

M. Rosa et al. (eds.), *Mathematical Modelling Programs in Latin America*,
https://doi.org/10.1007/978-3-031-04271-3_10

and solve problems in real-world contexts. Moreover, in many cases, this introduction involved teaching certain concepts because they were not previously known.

10.2 Problem

The problem we will address in this chapter establishes a relation between learning contexts, modelling and interdisciplinarity. In this case, the term *learning context* refers to the didactic context; that is, to the space designed by the teacher to promote learning concepts and developing abilities and competencies. Some refer to it as a didactic situation (Brousseau, 1998); however, due to certain possible particularities, we will use the term learning context. Precisely, one of those particularities may be the type of didactic context presented: a real-world context.

It is necessary for us to explain that by "real-world context" we mean one that exists or it is likely to exist in the real-world (Bednarz, 2018; Brown, 2019) but also allows students to be embedded in it and fully experience it. On the contrary, if the learning context exists or it is likely to exist, but students cannot experience it, we consider it as an evoked or mediated context by, generally, the educator. There are other possibilities for learning contexts, such as simulations (Bush et al., 2018; Chance & Rossman, 2006; Heuvel-Panhuizen, 2018; Rider & Stohl Lee, 2006).

Nonetheless, we understand that these and other options present their potentialities and difficulties. In this chapter, we discuss real-world contexts due to the students' possibilities to immerse themselves in a given problem. Discussing real-world contexts and mathematics learning leads, almost naturally, to consider modelling. In fact, modelling allows us to connect these contexts to the mathematical world. Precisely, we argue that modelling is a mental construction that establishes a connection between an individual's representation of context and a structured field of knowledge such as mathematics.

Why relate a real-world context to mathematics through modelling? One answer seems evident: the analyses and conclusions of a real-world context that can be obtained using mathematics are qualitative and quantitatively far superior to what could be obtained through direct observation of reality. And why relate these contexts to mathematics learning through modelling? In this regard, interesting evidence can be found: modelling contributes to the students' attribution of meaning to mathematics; besides, when modelling a problem from the context, it facilitates discussing concepts and their resignifications (Boaler, 1993, 2001; Spandaw, 2009; Urrieta Jr., 2007).

Therefore, both real-world contexts and mathematical modelling are topics of didactic interest. Several studies (along with their corresponding experiments) appear to question this relationship between real-world contexts and mathematical modelling. More precisely, our questions on this matter are: is mathematics sufficient to create a mathematical model of a real-context problem with students? Or is it necessary to introduce other disciplinary fields? We acknowledge that this is no trivial matter due to the possible impact on the whole school organizational system that having to integrate a connection of disciplines when addressing mathematical activities in a real-world context could have.

Moreover, if what we could call mathematics' lack of self-sufficiency to discuss mathematical modelling in real-world contexts is proven, we should probably consider an interdisciplinary didactic. In this chapter, we will share some details on an activity related to this matter as well as some considerations towards this direction. The results, conclusions and perspectives presented here are related to the previous chapters. However, we detailed some changes that precisely evidence certain phenomena we consider relevant.

One of those changes is that we, the authors, were not the educators in charge of the class. Nonetheless, we did conduct the activity presented in this chapter. In this particular case, an organization aspiring to strengthen secondary school students' mathematics learning invited us to conduct a workshop. This workshop was held in the second semester of 2019 with first-year students (aged between 11 and 13) during out-of-school hours. The location was the library of a school in Añelo, Neuquén province, Argentina. In fact, the students' age is another change in comparison to previous projects where adult university students participated (as mentioned above, we worked with children and pre-teens). Another change is related to the location and institution where this activity took place. Añelo is quite a fast-growing town because it is located in the epicenter of a formation called Vaca Muerta.

This formation is a significant oil and gas reservoir where resources are extracted through fracking. The intense drilling activities during the last years have impacted the town's dynamics. What once was a calm rural town is now affected by exponential growth. As a result of this exponential growth, no homogeneous profiles could be observed within the students at the mentioned school. Some students came from families relying on a subsistence economy in which their main activity was livestock (sheep) farming, whereas others came from different countries or regions due to their parents' jobs in the petroleum industry. What is more, their family dynamics also varied.

In this industry, and particularly in this area, work shifts usually consist of 14 days working at the drilling site followed by 7 days off. Therefore, apart from the mentioned heterogeneity, most of the students did not have the same family dynamics. Lastly, another aspect to consider was the high rates of grade repetition of first-year students: almost 50% of first-year students are held back a year; this is their first step into the secondary education system.

Considering all these aspects, we contemplated the idea of facilitating students' attribution of meaning to learning and proposed a project related to the region's context. This activity would be conducted with the students. For this purpose, we considered the school's first-year curricula, its available facilities, and the characteristics of the students, the school, and the region's inhabitants. It is worth mentioning that an important number of students come from two rural areas (Sauzal Bonito and Los Chihuidos). Each of these villages has an approximate population of 500 people. They are about 100 km away from Añelo and can only be accessed through gravel roads.

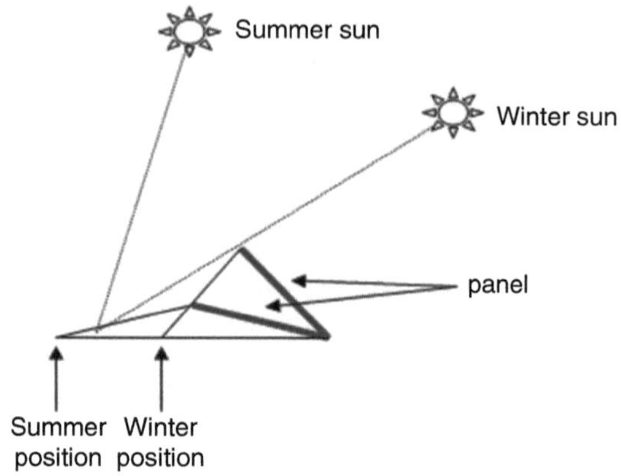

Fig. 10.1 Supporting structure for photovoltaic panel and two possible positions. (Source: Authors' own work)

The project consisted of the development of a mobile supporting structure for a photovoltaic panel. This panel would be used to supply energy to a water pump that would allow people to extract water for irrigation and animals from a small unconfined aquifer. The panel and its structure were given to an inhabitant named Ms. Zenaida (we will refer to her as Doña Zenaida) (38°30′20.53″S, 69°10′19.69″W). Personnel of the National Institute of Agricultural Technology referred her, and she did not have a water pumping system for this small aquifer.

We opted for a mobile supporting structure in an attempt to optimize the use of the solar radiation available in the area. The purpose was to try to set the photovoltaic panel in a perpendicular position to the solar rays throughout the year. This is especially relevant since, at the region's latitude, the variation in the angles of the rays is significant. For instance, if the solar panel is oriented towards a summer-based position, its performance during winter will be insignificant; and vice versa. Moreover, the students could easily replicate this optimal design at their families and neighbours' houses. Therefore, other people apart from Doña Zenaida could be benefited from this.

As regards the development of this project done with students in the second semester of 2019, in this chapter we will discuss aspects related to the articulation between different disciplines for the construction of arguments for the students' decisions. Particularly, we will focus on what could be called the unavoidable interdisciplinarity resulting from proposals based on real-world contexts.

The project's objective was to build a supporting structure for the photovoltaic panel. Said structure had holes on its inferior part to allow Doña Zenaida to move the structure's supporting bar each month in order to tilt the panel to a perpendicular position to the sun. Figure 10.1 shows the supporting structure in question and its two possible positions (for summer and winter).

In terms of didactics, one of the main objectives of this project was to help students grasp the meaning of learning. For this purpose, the concepts would appear as conceptual tools to construct the rational arguments for the decisions the project demanded. One of these decisions was determining the location of the holes on the horizontal structure. These holes would allow the panel to be perpendicular to the sun. Nonetheless, this was one of the last decisions to make. Before that, it was necessary to understand the phenomena and perform several tasks.

In this sense, we argue that comprehension is a necessary condition to make decisions and act upon them. From our view, the process is not about having students follow a set of instructions or indications to build the supporting structure. On the contrary, we intend to develop creativity and autonomy habits in them (Beauvais, 2006; Beauvais & Haudiquet, 2010) that lead to their own production of a solution. This creativity should be supported by arguments that allow them to have a degree of confidence in the solution's viability. As in the projects presented in previous chapters, students will not only have to produce a solution but also implement it.

In the following section, we will describe some of the project's initial stages where the dynamic of the introduction of mathematical concepts is evidenced. It is also shown how this knowledge indispensably had to interact with other disciplines; mainly, Physics.

10.3 Method

After introducing the project to the students, the first stage involved understanding what a photovoltaic panel does. The presentation of this and other stages will be schematized by detailing some of the stages' elements in order to summarize each description.

10.3.1 Stage 1: What Does a Solar Panel Do?

Objective: to understand the function of a solar panel—converting solar radiation into energy.

Location: schoolyard and library. Materials: photovoltaic panel, cables, 12 V lamp.

Activities: the students learnt about the parts of a solar panel and verified its energy production by connecting it to a 12 V lamp.

Concepts and disciplines introduced: principles of electricity, incandescence, electrical conductivity, short circuits, and electrical safety (physics).

10.3.2 Stage 2: Is the Energy Production of the Solar Panel Constant?

Objective: to identify the factors affecting the panel's energy production. Location: schoolyard and library.

Materials: solar panel, cables, 12 V lamp.
Activities: the students studied the lamp's brightness intensity in relation to (a) adjusting the panel's tilt angle, (b) changing the panel's orientation towards the sun, and (c) covering the panel completely or partially.

The students made several guesses proposing a connection between the tilt angle and the voltage of the electric current generated. The concept of voltage constituted an available term (Robert, 1998) for them due to their daily interactions with electric appliances.

Concepts and disciplines introduced: angles, rotation, lines, electricity, voltage (mathematics and physics).

10.3.3 Stage 3: Does Voltage Change?

Objective: to validate the hypothesis about the connection between the tilt angle and the voltage of the electric current generated going from a visual assessment (lamp's brightness) to a quantification (measuring voltage using a multimeter).
Location: schoolyard, library, and computer lab.
Materials: photovoltaic panel, cables, 12 V lamp, multimeter, sheets of paper, computers with spreadsheets installed, protractor.
Activities: we developed a protocol in which students set the panel facing the sun while registering three factors: the panel's tilt angle (quantitative, sexagesimal system), lightening of the lamp (categorical, yes-no) and voltage measured by a multimeter (quantitative, decimal).

In the following meeting, the students used spreadsheets for the first time in the computer lab to represent these three variables and to create a line chart for the two quantitative variables (independent variable: tilt angle; dependent variable: registered voltage).

Concepts and disciplines introduced: sexagesimal system, parallax error, decimal system, tabular representation of data, ordered pairs, line charts, functional relations, principles of the scientific method, introduction to algebra (Bruillard & Haspekian, 2009; Haspekian, 2005), and voltage (physics and mathematics).

Conclusion: despite the graphics showing a voltage fluctuation connected to the panel's tilt angle, we concluded there was a contradiction in the *voltage hypothesis*. In most of the observations, the voltage level was higher than required to light up the lamp; however, the lamp was not turned on.

10.3.4 Stage 4: Another Factor? Amperage

Objective: to discuss a new hypothesis that explains the lightning of the lamp as well as the brightness intensity variation in relation to the panel's tilt angle. To measure a new variable using the multimeter: the observed amperage according to the angle's value.

Location: schoolyard, library, and computer lab.

Materials: photovoltaic panel, cables, 12 V lamp, multimeter, pencil and paper, computers with spreadsheets installed, protractor.

Activities: we repeated the tilt angle protocol to register the amperage shown in the multimeter (quantitative variables) and the lightning of the lamp (categorical variable).

At the next meeting, we entered the data registered on paper into the computers' spreadsheets and created the corresponding graphics.

Concepts and disciplines introduced: sexagesimal system, parallax error, decimal system, tabular representation of data, ordered pairs, line charts, principles of functional relations, introduction to algebra, amperage, scientific method.

Conclusion: the graphics showing the relations between angle, amperage, voltage, and lightning of the lamp allowed us to establish the causal relationship between the lightning of the lamp and the amperage produced by the panel. Simultaneously, they showed how the produced amperage was related to the panel's angle.

10.3.5 Stage 5: Is the Tilt Angle Relevant?

Objective: to rectify the reference angle for the solar panel.

Location: schoolyard, library.

Materials: photovoltaic panel, cables, multimeter, adhesive tape and wood stick.

Activities: some students proposed the idea that to obtain the panel's optimum amperage, we should not consider an angle to the ground, we should focus on an angle perpendicular to the sun.

To confirm it, we used adhesive tape to fix a wood stick in a perpendicular position to the panel. Then, we measured the amperage produced based on the stick's shadow on the panel rather than on the angle.

Concepts and disciplines introduced: perpendicularity, decimal numbers, projections, scientific method, amperage (physics and mathematics).

Conclusion: we could observe that the maximum amperage was produced when the stick did not cast any shadows on the panel. Therefore, the panel's perpendicular position to the sun was optimal.

The stages of this project succeeded each other. Figure 10.2 shows the project's dynamics where two directions of development can be observed. On the one hand, the direction of the project's logic.

On the other, the direction of the introduced disciplines particularly referred to the concepts that emerged in the successive stages.

We will now summarize one of the final stages of the project: determining the structure's measures as well as the distance between the holes for each of the panel's positions.

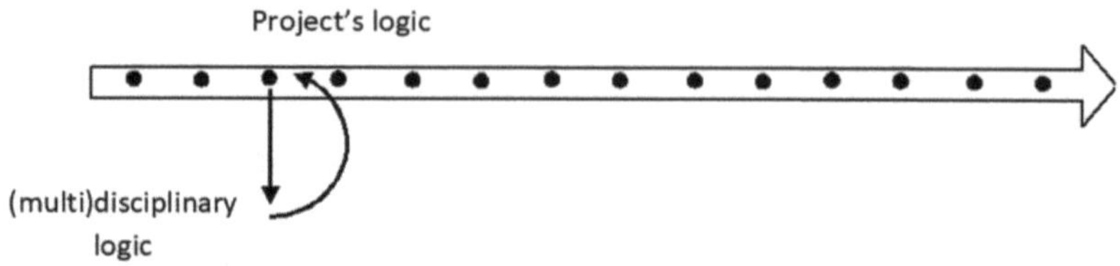

Fig. 10.2 Development of the project in two logical dimensions. (Source: Authors' own work)

10.3.6 Stage K: Measures and Distance of the Mobile Supporting Structure

Objective: to determine the support's measures, the distance between the holes and the type of materials to use.
Location: schoolyard, library, computer lab.
Materials: photovoltaic panel, paper and pencil, wood strips, measuring tape, set square and calculator.
Activities: the activities were conducted in two main phases. In the first phase, we tried to make a scale representation of the support. In the second phase, we made a dynamic modelling of the support in GeoGebra.

In this stage, we gave the students information about the angle of the solar rays at Añelo's latitude, available on the Internet. We considered the solar rays' angle to the ground at 2 pm (maximum solar radiation) on a typical day of each season.

According to this context, we considered four possible angle values: one for summer (December 21st), one for spring (September 21st), one for autumn (March 21st) and one for winter (June 21st).

10.3.6.1 First Phase: Scale Model Representation

To determine the measures and the location of the holes that would determine the distinct positions of the panel, the students suggested making a scale model. However, briefly put, this strategy was inefficient to determine measure and distance in scale. With the available tools for the workshop, students (nor teachers) could move with precision all of the scale model's articulations.

10.3.6.2 Second Phase: Representation in GeoGebra

After analyzing a set of conditions including the date, students' responsibilities with other school subjects, and the conceptual requirements of each strategy, among others; we decided to propose making a dynamic representation of the supporting structure in GeoGebra.

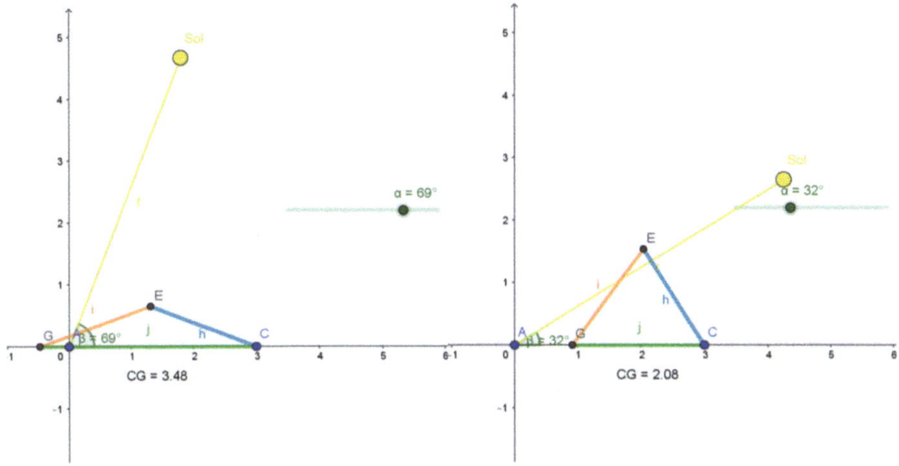

Fig. 10.3 Model for winter and summer positions. (Source: Authors' own work)

While the scale model strategy did not allow us to determine the measures, it did help students make a representation of the support's parts and movements. Therefore, despite being imprecise and jumbled, it facilitated the mental representation of the parts and articulations.

Relying on those mental representations and the disassembled scale models, we assisted the students with the modelling in GeoGebra that established the relations between the elements of the scale model, its movements, and the representation in this software. It is worth mentioning that the students were not familiar with GeoGebra. In fact, it was necessary to install it on the school's computers. Therefore, we had to teach them how to use the software's interface.

The construction of the modelling produced a sort of analogical relation between the scale model and the GeoGebra representation. In other words, the parts of the supporting structure, its articulations and proportions were represented in GeoGebra exactly as they would be built in the real support. Figure 10.3 shows GeoGebra's graphic view displaying the modelling for the winter and summer positions.

CE segment: photovoltaic panel.
ASol segment: solar rays.
EG segment: panel's supporting bar.
CG distance: distance at which the hole should be punched to introduce the bolt that will set the panel in a perpendicular position to the solar rays. Figure 10.4 shows all the objects involved in the modelling.

The panel is represented on a perpendicular line to the solar rays that contains point C. The supporting bar EG, whose longitude is determined by the ratio of the circumference d, intersects the lower part of the support in G 2.08 m away from point

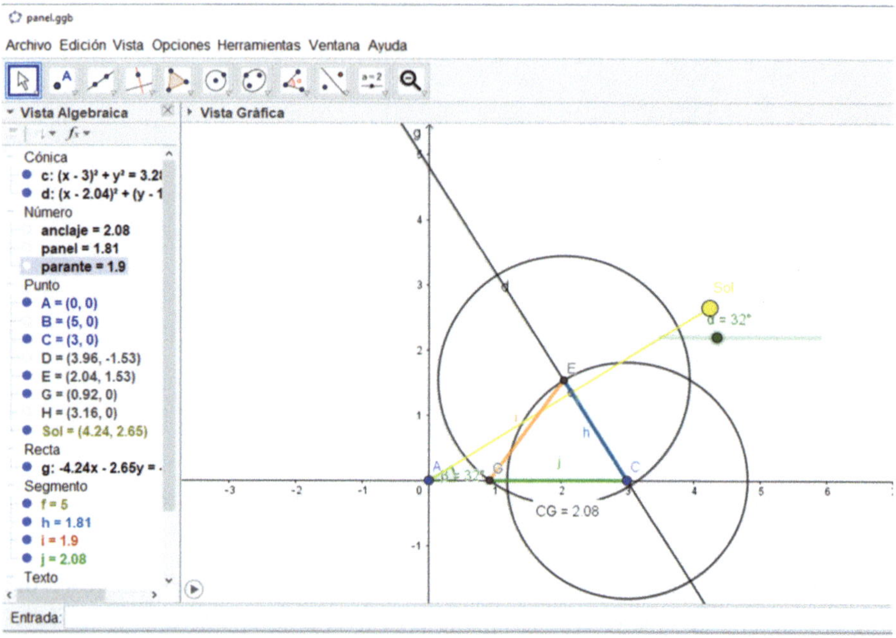

Fig. 10.4 Different elements of dynamic model. (Source: Authors' own work)

C. The CG distance is our reference point to punch the holes securing the panel's perpendicular position to the solar rays in that season (winter).

Concepts and disciplines introduced: perpendicularity, circumference, angles, decimal numbers, sexagesimal system, rotation, lines, points, and segments, ordered pairs representations, properties of triangles (mathematics).

Conclusion: the scale representation and the analogy between the model and the supporting structure to build allowed us to determine the distance (CG) at which the holes should be punched on the base of the structure. Since the sun's angles in autumn and spring were similar, we only punched one hole for those two positions to facilitate Doña Zenaida's tasks. Thus, the support had three positions: summer, winter, and autumn-spring.

10.3.7 Stage K + 1: Cutting the Materials—How Can We Save on Materials?

Objective: to optimize cutting the square-tube iron profiles to minimize the number of tubes to buy.

Location: schoolyard, library.

Materials: measuring tape, blackboard, paper, and pencil.

Activities: the modelling in GeoGebra in stage X allowed us to determine the measures of the parts of the photovoltaic panel's mobile support. Since the iron profiles sold are 6 m long, the students had to establish the optimal combination of cuts to reduce the number of profiles to buy.

Concepts and disciplines introduced: decimal numbers, basic mathematical operations, rounding numbers up (mathematics).

Conclusion: we could establish the possible series of longitude values of the parts to be cut as well as the number of profiles to buy.

10.3.8 Stage K + 2: Buying Square Tube Iron Profiles, Cutting, and Welding

For safety reasons and lack of infrastructure at the school, the cutting and welding of the rigid parts was done by the professors at their own houses. The support's parts were given to the students afterwards.

10.3.9 Stage K + 3: Will It Work?

Objective: to assemble and test the functionality of the mobile support, the photovoltaic panel, and the water pump.

Location: schoolyard.

Materials: support's parts, photovoltaic panel, water pump, cable, basic tools, electric drill, and paint.

Activities: the students assembled the mobile support and used an electric drill to punch the holes for the panel's different positions. They also painted the indications for each hole. Then, they connected the water pump to the panel and verified the system's functionality.

The students could observe and confirm two aspects: the system's functionality and how changing the panel's position from one season to another resulted in the pump lifting less water.

Concepts and disciplines introduced: integration of general concepts.

Conclusion: the solution for Doña Zenaida's rural facility could be materialized by the students. This stage can be seen in Image 10.1 (the students' faces are blurred to protect their identities).

Image 10.1 Students testing parts. (Source: Authors' own work)

10.3.10 Stage K + 4: Assembling the System at Doña Zenaida's Location

Objective: to install the mobile supporting structure, the photovoltaic panel, and the water pump at Doña Zenaida's location.

Location: school and Doña Zenaida's location.

Materials: support's parts, photovoltaic panel, water pump, cable, basic tools, and an electric drill.

Activities: when we concluded the previous stage (Stage K + 4), the school calendar was about to end. The administrative formalities to request permits from the local authorities of the Ministry of Education could not be presented on time. Therefore, the students were not allowed to travel to Doña Zenaida's location.

To avoid the possible damage caused by long-term storage and upon the uncertainty caused by the spreading pandemic (December 2019), we decided to install the supporting structure without delay. The transportation and assembling tasks at Doña Zenaida's rural facility were conducted in mid-December 2019 by the professors and personnel from the organization that funded the project.

We installed the system and checked its proper operation. Then, we explained to Doña Zenaida and her collaborator the maintenance tasks and the different panel's positions according to each season. This stage can be seen in the Images 10.2 and 10.3.

Image 10.2 Final installation of panel. (Source: Authors' own work)

Image 10.3 Final installation of support. (Source: Authors' own work)

10.4 Results

This project was conducted within a semestral workshop with students with "significant learning difficulties in mathematics" from a school in Añelo, Neuquén, Argentina. It was developed in what we could define as the periphery of the *school organization*: in parallel to the school classes and their corresponding schedules, teachers, and curricula. That is, we worked with the students during out-of-school hours.

We could argue that this type of proposal could only be carried out within the framework in which it was effectively developed: a workshop during out-of-school hours, without the restrictions and demands of the school ecosystem. Nevertheless, such affirmation can have at least two objections. First, an innumerable number of experiences like this can be observed within formal education systems; indeed, some examples of this can be found in other chapters from this book. Second, this type of proposal is precisely promoted in the objectives of the curricula.

A possible explanation for the observable gap between the curricula's objectives and what effectively occurs at school is that the materialization of those objectives requires a school ecosystem that grants a space to develop them. One of the ecosystem's aspects that could facilitate this type of proposal is the expansion of the physical workspace, which is usually the classroom.

Especially in mathematics classes, since the generalized epistemological reference shows that the mathematical objects and methods are abstract entities that can be discussed using paper, computers, and a board. This type of proposal questions the idea of the classroom as the sufficient workspace to develop mathematical concepts.

In this project, it was indispensable to use other spaces like the schoolyard (and even Doña Zenaida's rural facility). It is worth mentioning that this aspect was not part of the project's objectives. It was a consequence of intending to facilitate the students' attribution of meaning to the learning process through an activity that relates mathematical concepts to a problem in context.

Apart from workspace expansion, time extension should also be considered in this type of proposal. We could observe that to carry out certain activities, more than the time assigned by the formal system was necessary. In this school, lessons are organized into 80-min modules. We could corroborate that to work on the stages that could not be interrupted, we needed to extend the time of those modules.

Another expansion to be considered, and one we wanted to explicitly show here, is related to the interaction between disciplines. The purpose of describing the stages in the previous section was to evidence the unavoidable introduction of, at least, two disciplinary fields: mathematics and physics.

The inseparable interaction between these fields maintained an argumentative coherence in the development of the project. For instance, stages 1 to 5 were characterized by a constant introduction of physical and mathematical concepts that were necessary to understand the phenomenon: the influence of the panel's angle in the electric power generated.

Without discussing or at least introducing the concepts of electricity, voltage, and amperage it would have been impossible to develop the argumentative processes that led to studying the panel's angle as an optimizable variable.

In these cases, the introduction of physical knowledge was not an *application* or *re-signification* (Robert, 1998, 2003). In other words, the students could not recall those concepts with or without their teachers' assistance because they did not know them. They simply had no point of reference for them; especially, for the concept of amperage.

In this regard, we could argue there are two different types of interactions between the disciplines: concept application or resignification, and concept introduction. We will refer to concept application or resignification as the interaction in which previously discussed knowledge is recalled.

In some cases, this knowledge may already be known by the students, and they recall it on their own or with the teacher's assistance. The concept of amperage cannot be linked to this type of interaction. The students did not have a point of reference to consult or a conceptual construction to rely on to understand this concept. Therefore, it was necessary to spend time explaining the basics of amperage to them.

These two types of interactions determine distinct roles for the teacher as well. In the first case (application or resignification of concepts from another discipline) students can introduce the concepts and integrate them into their argumentative discourse without major deviations from the monodisciplinary line. In general, in this type of interaction, the teacher looks for indications or points of reference that allow the students to recall those concepts and apply or re-significate them.

However, if it is a concept introduction interaction, the situation is quite different. The teacher must stop the activity in question and create a learning opportunity for what they consider the fundamental aspects of the concept from another discipline. Once done, they can resume the activity.

As a result, the teachers enter into what they may consider an unexplored epistemological and didactic territory. Therefore, their teaching work is affected by a new stressful phenomenon that can be studied in terms of their comfort zone (Carranza et al., 2017; Miravalles et al., 2014) and work ergonomics (Roditi, 2000).

As regards teachers exploring extra-disciplinary areas, we could observe stress associated with the difficulties of controlling the uncertainty produced by such explorations. This stress could influence the teachers into not conducting or continuing with this type of proposal, especially, if the school ecosystem shows indifference or reluctance towards the initiative.

As mentioned above, certain expansions need to be considered and this type of proposal should be introduced into the traditional school ecosystem. They are necessary conditions to support and facilitate the development and consolidation of a proposal like the one we present. If not present, tensions may arise in the school ecosystem. Unless all the parts involved in the ecosystem act with conviction, it is highly likely that the traditional ecosystem itself will stifle the initiative and nothing will change.

With this, we wish to emphasize the importance of contemplating the institution as a whole when approaching a proposal like this. It is then indispensable not to think of teaching as a set of watertight compartments in which each teacher focuses on their discipline. Instead, it should be reconsidered as a progressive interrelation of communicating vessels and shared spaces.

Thus, these proposals are not only didactic but also pedagogical. This occurs because of the interrelations between the proposal and the school administration, teachers, non-teaching staff, and parents. Such interrelations occur to ensure the community can efficiently use the obtained solutions.

As regards the modelling in GeoGebra, it was interesting to verify its didactic potential to construct reasonings and determine the actions to take. Then, we wish to highlight something we consider fundamental: modelling constitutes an abstraction space that, supported by a mathematical conceptual framework (in this case), allowed us to understand the system and reach conclusions that would have been very difficult to draw through direct observation of reality.

Therefore, modelling presents significant potentialities to construct rational arguments to support the actions demanded by the project. In a way, it becomes a space for the construction of rationality for teachers and students. This allows them to move forward with the project on a firm rational footing.

In this sense, interesting phenomena have been observed in relation to the didactic contract (Brousseau, 1988). The argumentative construction achieved with the modelling allows the teachers to switch positions. They are no longer in the knowledge holder position; instead, their position is now marked by rationality. In this rationality, different agents are involved and the results' validity is supported on the constructed arguments and not on the teacher's status.

The modelling in GeoGebra has allowed us to corroborate relevant phenomena that had already been discussed in the literature: modelling is a vehicle for learning disciplinary concepts as well as a learning object (Blomhøj, 2019; Blum & Borromeo Ferri, 2009; Boaler, 2001; Brown & Ikeda, 2019; Spandaw, 2009). As regards future perspectives, we underscore three major relevant aspects:

- Is it possible to systematize this type of proposal to facilitate the teachers' appropriation? In other words, what are the common denominators shared by these proposals? Is it possible to conceive an architecture of proposals based on real-world contexts?
- How can the interdisciplinary instances be tackled in this type of proposal as regards didactics and management? As regards didactics, is there an interdisciplinary didactic or is it sufficient with expanding the teacher's epistemological field? As regards management, is it possible to design a plan or progressive adaptation program to adapt the traditional school ecosystem to this type of proposal?
- As regards modelling and the constant evolution of technological tools, is there a classification of modelling by the type of elements considered in the context? Related to this matter, in the next chapter we will present our views on another modelling analysis. We will study models in terms of being centred on the objects from the context (analogical modelling) or on the relations between the context (analytical modelling).

Finally, we understand that proposals based on real-world contexts have a great potential for learning disciplinary concepts; especially considering the concepts' possibilities to construct rational arguments. We also understand as well that these dynamics are not affected by the singularities of a region or country. Each community has spaces that can be improved and where students and teachers alike can intervene and feel active members of their community.

References

Beauvais, M. (2006). *Des postures de l'accompagnateur â la posture de l'accompagnant: Projet, autonomie et responsabilité*. Paper presented at the 7e Colloque Européen sur l'Auto-formation, Faciliter les apprentissagesautonomes, Auzeville.

Beauvais, M., & Haudiquet, A. (2010). *Accompagner des acteurs éducatifs: Des contextes en changement(s), des valeurs en question(s)*. Paper presented at the l'Actualité de la Recherche Enéducation et en Formation, Geneve.

Bednarz, N. E. A. (2018). Contextualiser pour enseigner les mathematiques: Un enjeu de formation. *Annales de Didactique et de Sciences Cognitives, 23*, 69–105.

Blomhøj, M. (2019). Towards integration of modelling in secondary mathematics teaching. In G. A. Stillman & J. P. Brown (Eds.), *Lines of inquiry in mathematical modelling research in education* (ICME-13 monographs). Springer.

Blum, W., & Borromeo Ferri, R. (2009). Mathematical modelling: Can it be taught and learnt? *Journal of Mathematical Modelling and Application, 1*(1), 45–58.

Boaler, J. (1993). The role of contexts in the mathematics classroom: Do they make mathematics more "real"? *For the Learning of Mathematics, 13*(2), 12–17.

Boaler, J. (2001). Mathematical modelling and new theories of learning. *Teaching Mathematics and Its Applications, 20*(3), 7.

Brousseau, G. (1988). Le Contrat Didactique. *Recherche Endidactiques des Mathématiques, 9*(3), 309–336.

Brousseau, G. (1998). *Théorie des situations didactiques*. La Pensée Sauvage.

Brown, J. (2019). Real-world task context: Meanings and roles. In *Lines of inquiry in mathematical modelling research in education* (ICME-13 monographs). Springer.

Brown, J., & Ikeda, T. (2019). Conclusions and future lines of inquiry in mathematical modelling research in education. In G. Stillman & J. Brown (Eds.), *Lines of inquiry in mathematical modelling research in education* (pp. 233–253). Springer.

Bruillard, E., & Haspekian, M. (2009). *Behind students's spreadsheet competencies: Their achievement in Algebra? A case study in a French vocational school*. Paper presented at the CERME 6, Lyon, France.

Bush, J., Webb, D., Kress, N., Yang, W., & Perkins, K. (2018). *Classroom activities for digital interactive simulations to support realistic mathematics education*. Paper presented at the 6th International Realistic mathematics Education Conference, Georgetown.

Carranza, P., Chrestia, M., Goin, M., Quijano, T., & Sgreccia, N. (2017). Ambientes de aprendizaje y proyectosescolares con la comunidad. *Revista Latinoamericana de Etnomatemática, 10*(1).

Chance, B., & Rossman, A. (2006). *Using simulation to teach and learn statistics*. Paper presented at the ICOTS7, Bahia, Brasil.

Haspekian, M. (2005). *Intégration d'outils informatiques dans l'enseignement des mathématiques*. Doctorat, Paris 7 Denis Diderot, Paris.

Heuvel-Panhuizen, M. v. d. (2018). *International reflections on the Netherlands didactics of mathematics. Visions on and Experiences with Realistic Mathematics Education*. Springer Open.

Miravalles, A. F., Sánchez i Valero, J.-A., & Sancho Gil, J. M. (2014). Salir de la zona de confort. Dilemas y desafíos en el EEES. *Tendencias Pedagógicas, 23*.

Rider, R., & Stohl Lee, H. (2006). *Differences in students' use of computer simulation tolls and reasoning about empirical data and theoretical distributions*. Paper presented at the ICOTS 7, Bahia, Brasil.

Robert, A. (1998). Outils d'analyse des contenus mathématiques à enseigner au lycée et à l'université. *Recherche Endidactiques des Mathématiques, 18*(2), 139–190.

Robert, A. (2003). Analyse de vidéo de séances de classe. *Documents pour la formation des enseignants, 2*.
Roditi, E. (2000). *L'enseignement de la multiplication des décimauxensixième. Etude des pratiques ordinaires*. Doctorat, Université Paris 7.
Spandaw, J. (2009). *Modelling mathematics' teachers' professional development*. Paper presented at the Sixth Congress of the European Society for Research in Mathematics Education, Lyon, France.
Urrieta, L., Jr. (2007). Figured worlds and education: An introduction to the special issue. *The Urban Review, 39*(2).

Chapter 11
Analogical Modelling and Analytical Modelling: Different Approaches to the Same Context?

Pablo Carranza, Mónica Navarro, and Mariana Letourneau

11.1 Introduction

In this chapter, we present a set of models created with students in which we tackled problems from different projects in real-world contexts. Only three of those projects carried out between 2015 and 2021 are presented. These models occurred in complete modelling cycles (reality, modelling, reality) and they allowed both students and professors to significantly improve their understanding of the analyzed phenomena. Moreover, several of those models were fundamental for the decisions taken in the projects.

The types of models are different as well. Some are characterized for retaining elements or objects from the context, whereas others favour more abstract relations. They result from how students establish connections between the problem's context and the mathematical world.

Besides, we could observe links between these types of models that evidence the evolution in the students' mental processes. The models could also be didactically considered as strategies for progressively difficult modelling; allowing students to deepen their understanding of the phenomena and the appropriation of relatively complex disciplinary concepts.

P. Carranza (✉) · M. Navarro · M. Letourneau
Universidad Nacional de Río Negro, General Roca, Río Negro, Argentina
e-mail: pcarranza@unrn.edu.ar

M. Rosa et al. (eds.), *Mathematical Modelling Programs in Latin America*,
https://doi.org/10.1007/978-3-031-04271-3_11

11.2 Problem

The literature on the role of modelling in mathematics teaching highlights two major uses: modelling as a vehicle for learning disciplinary concepts, and modelling as a teaching object (Blum & Borromeo Ferri, 2009; Blomhøj, 2019; Brown & Ikeda, 2019; Caron, 2019; Czocher, 2019).

As regards the learning vehicle role, modelling has been acknowledged to show potential for the discussion of mathematical concepts. Indeed, the models, as an abstraction space for a given problem, become the appropriate place for the learning and resignification of mathematical concepts.

However, modelling a problem with mathematical objects and methods constitutes a non-evident cognitive act. Complex mental processes appear in the interaction between the elements of the problem, in the person's perception of them, in the representations of the mathematical concepts, and the available modelling tools and creative abilities, among others.

The relationship between these matters does not appear to be a linear and determined sequence. Constructing a mathematical modelling of a problem does not constitute a bijective relationship between the problem and the mathematical tools. A problem can be modelled in different ways and vice versa. This relationship between the problem and the mathematical tools requires certain abilities and competencies that are nowadays considered necessary. Therefore, learning how to create models becomes a teaching-learning object.

By modelling, we understand a cycle that begins with a problem, continues in a space of mathematical objects, and then returns to the problem. What is the purpose of carrying out this cycle? Going through the mathematical objects allows us to analyze and construct arguments supported by a rich mathematical scaffolding. Besides, these arguments are many times impossible to obtain through direct observation of reality. To sum up, using mathematical modelling elevates the level of comprehension of the problem and allows us to approach it with solid arguments.

In this sense, the literature distinguishes two different types of modelling: descriptive modelling and prescriptive modelling (Blum & Borromeo Ferri, 2009; Brown & Ikeda, 2019). Descriptive models allow us to understand how a problem works, whereas prescriptive models help us determine how to act in the problem in question.

In this chapter, we will present a set of models created with students for projects in real-world contexts. In the examples we detail, the two types of models as well as their constant interactions can be observed. In fact, it is possible to observe how the models constitute vehicles for learning mathematical concepts and how they can become learning objects.

However, our focus is on proposing a categorization for modelling based on the type of problem's elements considered. Thus, we will discuss the concepts of analogical modelling and analytical modelling. In the former, physical elements of the problem as well as its relations, connections and articulations are taken into consideration. In the latter, abstract relations non-observable through our senses are considered.

In this chapter we present these and other differences with illustrative examples of the models created with students for projects in real-world contexts. We will refer to projects rather than problems due to several reasons. We claim that a project, due to its complexity, contains a set of different problems and it is developed within a period bigger than a problem. Moreover, these projects are claimed to be in a real-world context because said context exists and students are embedded in it. Besides, their solutions are implemented in it; that is, the context is not created, it is experienced.

In general, the models presented here are prescriptive since they facilitated the construction of arguments to intervene in the projects' problems. However, they are also descriptive because they helped to understand the analyzed phenomena. To present the models in terms of being analogical or analytical, we will first introduce a summary of the projects in which they were created.

11.3 Method

The models discussed here were created in three projects carried out in real-world contexts: the "Savonius Windmills" project, the "Mobile Panel" project, and the "Water Purification Systems" project.

11.3.1 Savonius Windmills

Course: mathematics (semestral).

Level: tertiary education. First semester of the first year.

Students: aged between 18 and 50.

Project's summary: the students designed, calculated, and constructed windmills to pump underground water for irrigation and animals. The windmills were given to rural communities relying on a subsistence economy in Argentinian Patagonia.

Year: the project has been carried out since 2015. One or two windmills are built each year.

Funding: National University of Río Negro (2015–2017). Argentine National Ministry of Education (2018–2022).

Illustrative media: Fig. 11.1 shows the type of windmill built. The following hyperlink redirects to a video summarizing the installation stages of two windmills at two different rural facilities: https://www.youtube.com/watch?v=889fvPzVK1g.

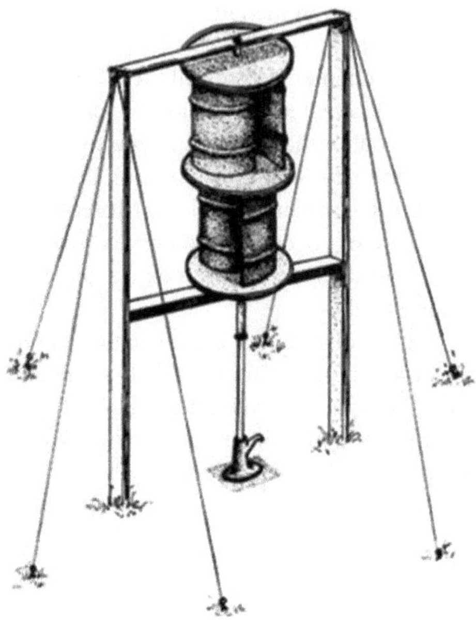

Fig. 11.1 Schematic representation of Savonius Windmill. (Source: http://cepechile.blogspot.com/)

11.3.2 Mobile Photovoltaic Panel

Course: extracurricular mathematics workshop (semestral).
Level: secondary education. Second semester of the first year.
Students: aged between 11 and 13.
Project's summary: the students designed, calculated, and built a mobile supporting structure for a photovoltaic panel. The purpose was to optimize the use of the available solar radiation. The panel and a water pump were given to a rural inhabitant in Argentinian Patagonia.
Year: 2019.
Funding: organizations and companies.
Illustrative media: Fig. 11.2 shows the supporting structure during the assembling stage. The students assembled the structure at the schoolyard.

11.3.3 Water Purification Systems

Course: mathematics (annual).
Level: tertiary education. First year.
Students: aged between 18 and 40.
Project's summary: the students designed, calculated, built, and tested prototypes for low-cost water purification systems for families. These systems are based on technology developed by researchers at the University at Buffalo, in the United

Fig. 11.2 Photovoltaic supporting structure. (Source: Authors' own work)

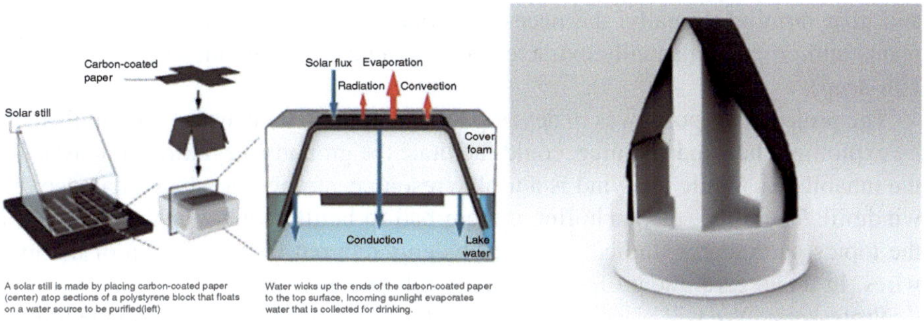

Fig. 11.3 Solar distillation. (Source: Buffalo University)

States (Gan & Zhang, 2017). The recipients are any inhabitants who are not connected to drinking water distribution systems.

Year: 2021.

Funding: organizations and university.

Illustrative media: Fig. 11.3 shows the first and second versions of the system's technological principle while Fig. 11.4 shows the culmination of the building stage of one of the five prototypes built by the students.

The set of models we will present were created in these three projects. All of them shared common ground: they allowed students to understand the phenomena in the projects and to construct rational arguments for their decisions. Although there are points in common between all the models exposed, they are differentiated by the type of elements they retain from the context.

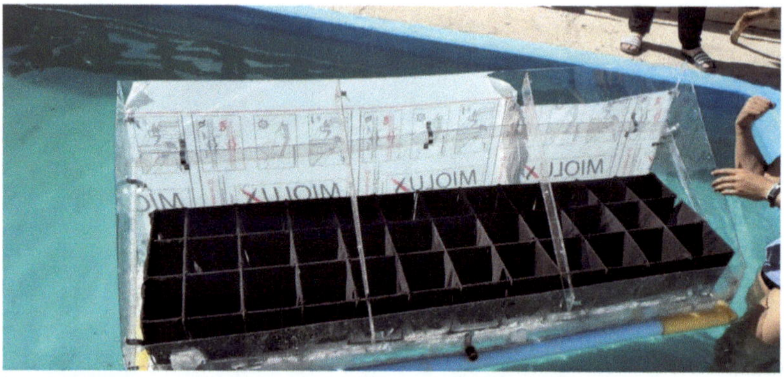

Fig. 11.4 Students' prototype of solar water distilling system. (Source: Authors' own work)

11.3.4 Modelling 1 of the Guy Wires in the Savonius Windmill: Analogical Modelling

The plateau region in Argentinian Patagonia is characterized by low precipitation levels (200 mm average per year) and strong winds. In this region, there live inhabitants relying on a subsistence economy where their main activity is sheep and goat farming. Usually, the necessary water for the animals, and for watering plants and crops is manually extracted from groundwater supplies (between 5 and 7 m deep).

The project's proposal was to develop low-cost, low-maintenance windmills that, by exploiting the wind resource, could facilitate the groundwater extraction tasks for the inhabitants. While the wind is a usable resource, strong gusts could damage the windmill. Therefore, the anchoring system had to be thoroughly analyzed. One of the topics particularly studied with the students referred to the location of the guy wires. In this regard, the professors usually ask one question:

Professors: "We will place the anchors of the guy wires 6 m away from the foot of the windmill. As you saw during your visit to the field, we may not be able to place the anchors where we planned. There were big shrubs and blocks of stone that may force us to relocate the anchors at a different place. The question is: if the anchor is placed nearer or farther from the original location, will the stress on the anchor and the guy wire change? Or will it remain the same?"

The students usually answer: "The force depends on the wind. If the wind does not vary, there is no reason for the stress to change". Figure 11.5 illustrates the windmill and one of the guy wires holding the mill to a concrete block buried on the ground 6 m away from its foot.

The purpose of the professors' questions was to analyze whether the stress on the anchor (concrete block and ground) changed according to the distance between the anchor and the windmill (6 m in the figure). Being one of the first considerations analyzed, professors usually propose a dynamic representation of stress in GeoGebra aiming to understand the relations between the vectors.

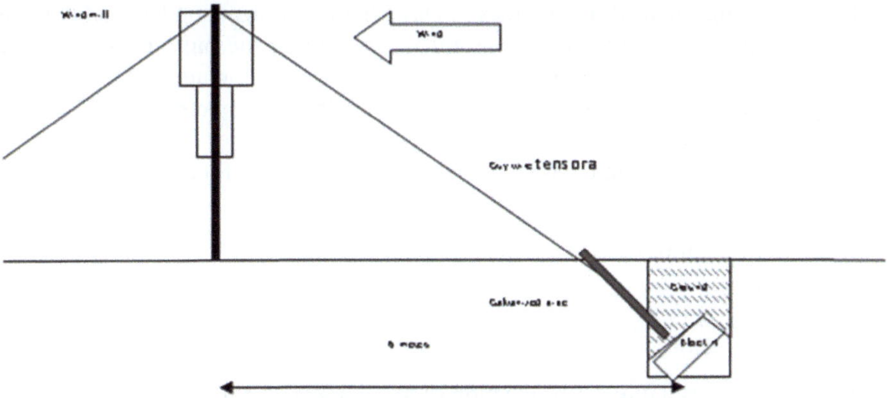

Fig. 11.5 The Savonius windmill and its support system. (Source: Authors' own work)

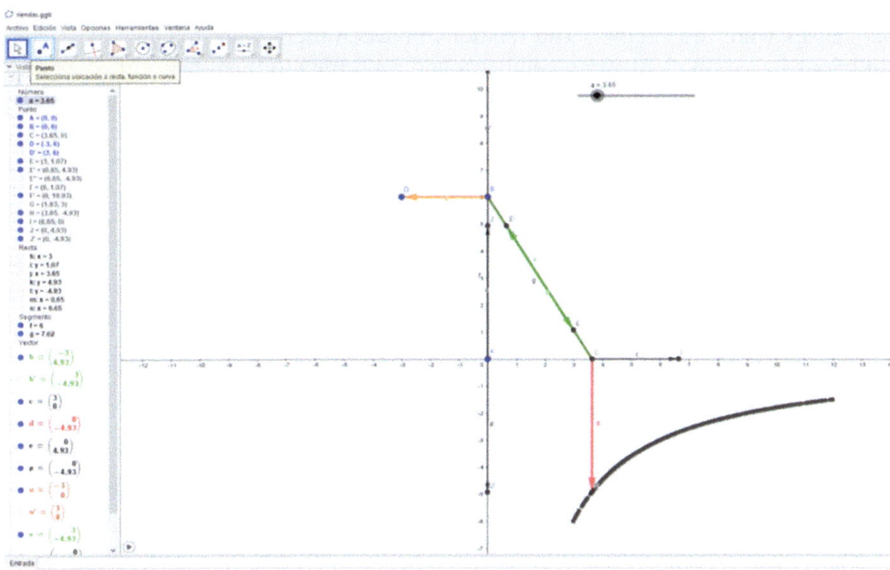

Fig. 11.6 Dynamic representation of the stress on the anchor in terms of distance. (Source: Authors' own work)

Briefly put, they started the modelling process and students progressively acquired the autonomy to complete it on their own. In this case, the modelling was descriptive because it allowed us to understand the relation between the stresses. Figure 11.6 shows the model produced. The slider controls the distance at which the guy wire is attached to the concrete block.

Due to space limitations, we will not share the details of the creation of the modelling. The curve observed shows the evolution of the rear end of the d vector. This vector represents the stress caused on the concrete block and the ground above it when holding the windmill in its vertical position.

The modelling, iterated many times in class, helped students understand that the anchor's distance is a variable to consider when digging the pits for the concrete blocks. To assist their comprehension, some of the iterations included a visualization of the geometrical elements that structure those relations (symmetries could only be visualized through the objects).

This model was created based on the reproduction of fundamental physical elements of the context and the relations between them:

- Windmill's main pipe.
- Guy wire.
- Pit location.
- Associated vectors.

Therefore, it was a direct reproduction of the elements from the context: the pipe was represented by a segment, the wire was represented by another segment and the anchors appeared as intersection points. Indeed, there is a replication of the elements from reality in the modelling. The only elements that could not be observed through our senses were the vectors. The relations between them were mathematical constructions based on elementary geometry and rigid transformations.

We argue this modelling was analogical since it was a replication of the objects from the context, which was done by geometrical objects. At a semantic level, the students could establish a direct, bijective relation between the elements from the context seen through their senses and their representations in the model. We will present another model referred to the problem of the guy wires; it will be a continuation of Modelling 1.

11.3.5 Modelling 2 of the Guy Wires in the Savonius Windmill: Analytical Modelling

The previous modelling done in GeoGebra helped the students understand a phenomenon relevant to the project: the stress on the guy wires was influenced not only by the wind speed but also by the distance at which the pits for the concrete blocks were dug. Of the various conclusions drawn, here we focused on one in particular: *the nearer the pits, the deeper we must dig them.*

That modelling was descriptive as well as analogical and was aimed at understanding the phenomenon. Once it was understood, the next step was its quantification. That is, it was necessary to determine with precision the pits' depth according to their distance to the foot of the windmill. It is worth mentioning that, while students were constructing the windmills, the inhabitants had to dig the pits for the concrete blocks. For this purpose, an explanatory plan with a schematic diagram of the distance between the pits and the foot of the windmill was given.

However, sometimes, they could not maintain the designated distances because of large shrubs' roots or blocks of stone. In those cases, they decided to relocate the pits. As a result, during the windmill's installation stage, students had to revise the

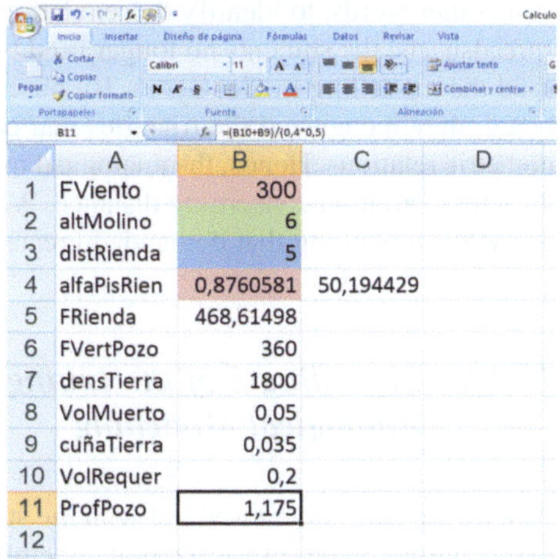

Fig. 11.7 A spreadsheet on the quantification of the variables affecting the guy wires. (Source: Authors' own work)

pits' depth according to their final location. Moreover, the context determined that this revision task had to be expeditious. There was no time to make the necessary calculations at the location.

Likewise, it was not convenient to use the previous modelling in GeoGebra due to the location's conditions: the ambient light and the risks of damage for a laptop. Since that model was not appropriate for revising the pits' depth at the location, a new modelling was created on a spreadsheet that students would be able to access from their smartphones (Fig. 11.7).

Most of the cells for this modelling already had input values or formulas, the main data to input during the installation at the location was the pits' distance (B3 cell). Once it was introduced, the B11 cell showed, in meters, the optimum pits' depth. For the purpose of this chapter, we considered some characteristics of this modelling in relation to the type of elements from the context retained. This time, there were no physical (pipe and guy wires) or geometrical objects to represent in the modelling.

In this case, we retained, mainly, the abstract relations between the objects that could not be seen through our senses. As regards mathematics, algebra (Haspekian, 2005; Bruillard & Haspekian, 2009) and trigonometry are introduced instead of geometry. Likewise, the modelling abstraction level required the students to identify the trigonometric relations using paper and pencil; then, they had to materialize them in a spreadsheet. In this case, there is no longer a direct relationship between the objects from the context and the objects from the modelling. The abstraction level is greater.

Due to its role in the context, this modelling was prescriptive since it helped the students determine how to proceed (as regards pits' depth). Besides, it was analytical because of the type of elements considered. Additionally, it is worth highlighting the transitional role of the analogical modelling regarding the analytical modelling. The geometrical representation of the objects in the modelling in GeoGebra introduced mathematical objects that allowed the students to identify the trigonometric relations of the analytical modelling.

In other words, to identify those relations that would be used in the modelling on spreadsheets, the students no longer used the windmill's context as a reference; they used the modelling in GeoGebra. By observing this model on their laptops and the board, they recognized the triangles that consequently led to considering the trigonometric relations. Hence, the analogical modelling assisted in the comprehension of the stress variation concerning distance. What is more, it also helped transition to the analytical modelling that demanded greater abstractions.

11.3.6 Modelling 3 of the Lifting of the Savonius Windmill: Analogical Modelling

Among the analyses carried out with the students, there was one referred to hoisting the windmill from its horizontal assembling position to its vertical functional position. For this purpose, they watched footage from previous years showing how windmills were lifted. The hoisting stage at the location is critical due to the risk of the mill falling and the consequent damage if one of the parts involved in the hoisting process broke. Figure 11.8 shows a schematic diagram of the windmill's hoisting process. The structure is lifted to its vertical position using the synthetic cable of a vehicle's electric winch. Figure 11.8 shows a schematic diagram of a vehicle hoisting the Savonius windmill.

The winch's synthetic cable is attached to the superior part of the windmill (A). Upon watching the footage, one student proposed changing the attachment point from point A to point B to reduce stress. Other students proposed placing the vehicle as far as possible from the windmill, while others suggested the opposite; others also questioned whether the synthetic cable would resist the strain. In the end, the proposals were analyzed through a modelling.

To facilitate the comprehension of the stress dynamics at the hoisting moment, the professors induced a modelling we consider as analogical. That is, a reproduction of the physical elements from the context in, in this case, GeoGebra. Figure 11.9 shows this modelling.

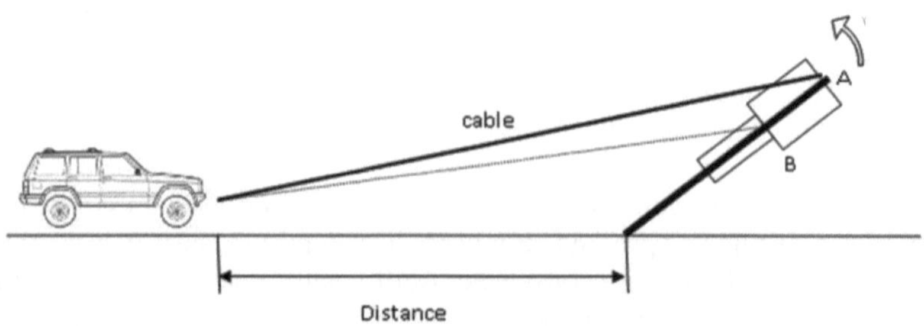

Fig. 11.8 Schematic diagram of a vehicle hoisting the Savonius windmill. (Source: Authors' own work)

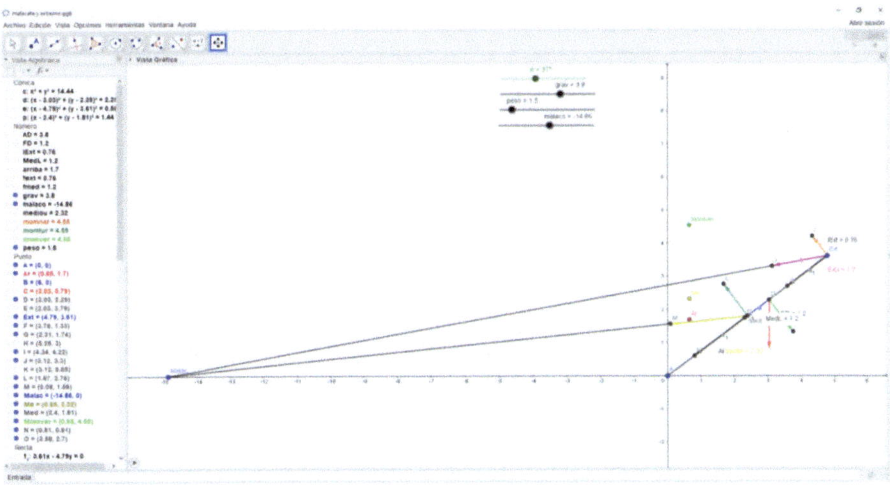

Fig. 11.9 Modelling of the stress on hoisting the Savonius windmill. (Source: Authors' own work)

This model assisted the comprehension of the stress produced during the hoisting stage and its changes as the windmill is closer to its vertical position. The lifting is controlled by the slider *alpha angle* and the location of the vehicle is controlled by the slider named *malaco*.

To quantify stress, the vectors' modules were made in scale. This allowed us to determine, for example, if the winch's cable would resist the strain during the hoisting. Unlike the modelling of the stress on the guy wires in GeoGebra, in this case, it was necessary to draw on elementary algebra since the physical concept that relates the vectors is torque (M = Fxd).

We consider this modelling as analogical because it consisted of the replication of objects from the context. This replication was achieved by using, mainly, geometrical elements. Figure 11.10 shows the graphic view with all the geometrical elements involved.

The modelling allowed us to understand the stress dynamics as well as to reject the idea of attaching the winch's cable to the windmill's point B, the stress would be greater if the cable was attached there and not in point A.

Likewise, the modelling confirmed that the winch's cable was appropriate for hoisting the windmill (according to its specifications, it resists 4 times more strain than the one caused at the hoisting stage). Besides, it allowed us to answer, among other matters, the question on the distance between the vehicle and the windmill, the farther the vehicle, the lesser the stress produced.

This modelling was descriptive because it assisted the comprehension of the stress dynamics. However, it was also prescriptive since it helped determine the decisions on how to act when hoisting the windmill. Moreover, because it was strongly based on the representation of objects from the context, it was analogical. Its creation was mainly based on geometrical elements, although elementary algebra was introduced for the physical concept discussed: torque.

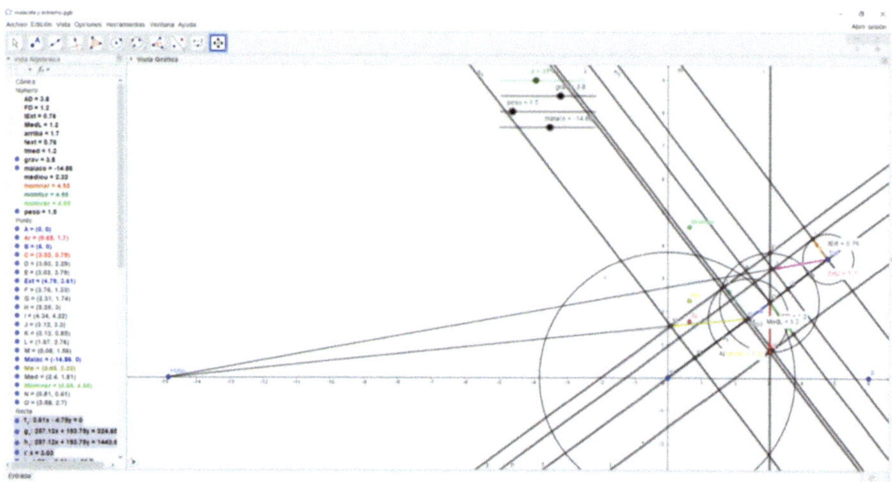

Fig. 11.10 Graphic view of all geometrical elements involved in modelling. (Source: Authors' own work)

11.3.7 Modelling 4: Analysis of the Savonius Windmill's Rotor—Analogical Modelling

One of the objectives of the Savonius windmills project is to let the inhabitants produce their solutions to the underground water extraction problem. For this purpose, we chose an easy, low-cost design. Additionally, given the distance between the rural facilities and the urban centres, we opted for a design that could be built using disused objects from the fields. That is why the windmill's rotor is built using recycled 200 L barrels.

As regards the rotor, various questions usually arise among the students: "Why are we using two rows of barrels instead of one?", "Will we produce more power if we use more rows?", and "Is it true that the rotor vibrates?". In these situations, students are usually asked to create a modelling for the rotor in GeoGebra that, even if it is simplified, allows them to understand the functionality and consequences of the design.

The modelling induced by the professors was analogical: physical elements of the rotor and their relations were taken into consideration, and then, a model was created introducing, mainly, geometrical elements and rigid transformations (rotations, translations, and symmetries). Figure 11.11 shows the simplified modelling of the rotor built using four half-barrels placed in two rows (two up and two down).

In this modelling, the rotor's spin is controlled by an angle-type slider named alpha. The modelling represents the four half-barrels as well as each half-barrel's contribution to the total area, although it is a simplified representation. The contributions of each half-barrel are represented by points; their traces allow us to understand the dynamics of the rotor's swept area. Figure 11.12 shows, in a

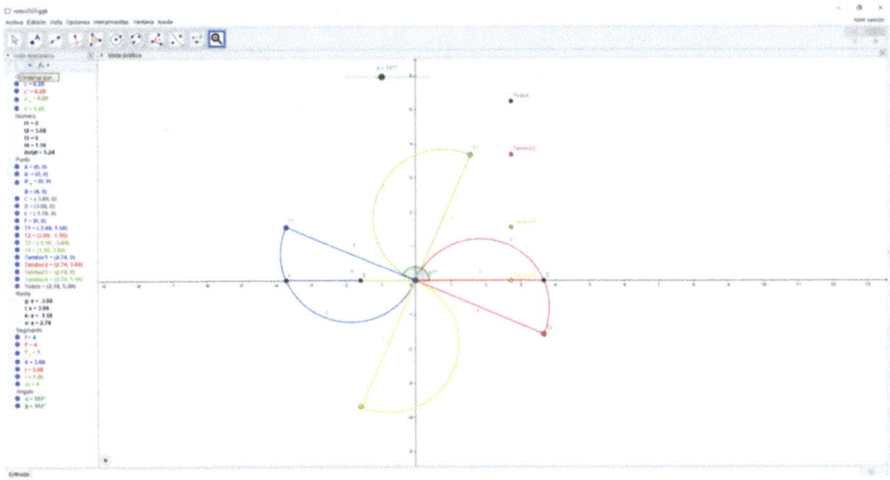

Fig. 11.11 Analysis of the contribution of each half barrel in the swept area (first steps). (Source: Authors' own work)

simplified manner, the swept area of each half-barrel and the total area of the four half-barrels when they are all spinning.

This modelling was descriptive for assisting us in understanding the rotor's functionality: in this regard, a fundamental aspect is the one referred to the variable rotor's area (superior trace). This variation in the rotor's swept area led to warning the students about the windmill's vibration. The reason for such vibration was the variability of the swept area, not the supposed irregular centres of gravity of the axis of rotation.

Thanks to the modelling, we could answer the question on the (in)convenience of using only two half barrels instead of four. Not only the total area would be minor but there would also be greater vibration as a result of the increased variability of the area. Therefore, we could determine that placing the four half-barrels in the same row would be inconvenient since they would obstruct each other.

This model was also prescriptive. Among other conclusions, the ones related to the unavoidable vibration of the rotor led the students to prioritize bolting rather than welding. From the current perspective, the model was analogical: there is a replication of objects from the real-world using geometrical objects; besides, the rigid transformations are predominant.

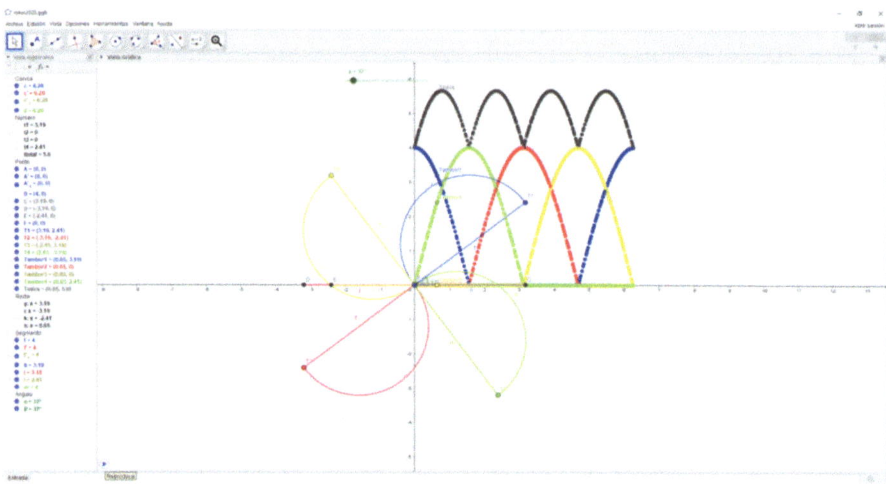

Fig. 11.12 Analysis of the contribution of each half barrel in the swept area. (Source: Authors' own work)

11.3.8 Modelling 5: Analysis of the Possible Changes to the Savonius Windmill's Rotor—Analytical Modelling

In the previous modelling, the students discovered that the windmill would vibrate due to certain matters inherent in the design: the variable swept area. A group of students suggested analyzing a possible modification to reduce or, in the best-case scenario, eliminate the vibration. The conversation ensued as follows:

Group 1: "Since there is increased variability using two barrels, and less variability using four barrels. . .is it possible that there will be even less variability using six barrels?"

Group 2: "But they shouldn't be at the same level, otherwise they would collide."

Group 1: "It could be a windmill with three rows of barrels".

Professors: "Are you sure that that is how you reduce vibration? That is, by increasing the number of barrels?"

This debate continued and encouraged performing a new analysis about increasing the number of half barrels symmetrically placed and their effect on the variability of the total swept area. Unlike the previous modelling created for four half-barrels, in this case, we tried to create a model for six, eight or more half-barrels.

The constructive method of analogy with the real world was now perceived a tedious due to the costs in terms of time and the elements involved. Meanwhile, in the construction of the modelling of the four half barrels, more precisely on the section of the fourth half-barrel, some students identified the trigonometric relation

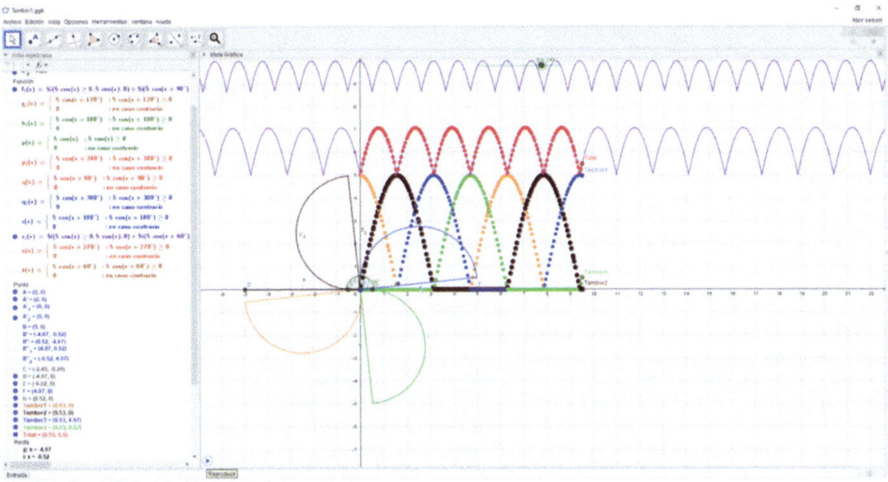

Fig. 11.13 Analysis of the contribution of each half barrel in the swept area (trigonometric functions). (Source: Authors' own work)

that described the area contribution of each half-barrel. Two elements constituted clear evidence: the shape of the curve that described the tracing and the triangular geometry of the contribution of each half-barrel.

To model the six-half-barrels rotor's swept area, the students proposed a more economical option as opposed to the analogical model: using trigonometric functions. A brief representation of the type of modelling created by the students with the professors' assistance can be seen below (Fig. 11.13).

Figure 11.13 shows the graphic and algebra view of the original modelling (analogical) of four half-barrels, the curve of the total area of the four half-barrels obtained by the analytical model and the curve of the total area of a hypothetical windmill with six half-barrels. The trigonometrical models of each half barrel are hidden in the graphic view, but they can be seen in the algebra view.

Similar to the guy wires case, we observed a switch in the type of modelling: from analogical to analytical. In the guy wires case, due to the characteristics of the context, it was convenient to carry out the modelling of the pits depth calculation on a different support (Spreadsheets). In this case, the support (GeoGebra) was appropriate, although that was not the case for the type of modelling used (analogical).

It is worth mentioning that the students found in the analogical modelling the indications to create an analytical modelling: the shape of the curve of the area of each half-barrel, and the triangular geometry of the area of each half-barrel. Figure 11.14 shows this triangular geometry.

Considering the wind comes from below, the simplified wind area of the half-barrel in the figure is determined by the projection of the barrel's diameter on the X-axis (multiplied by the height of the barrel). This geometrical representation and the shape of the curve that describes the projection based on the angle constituted

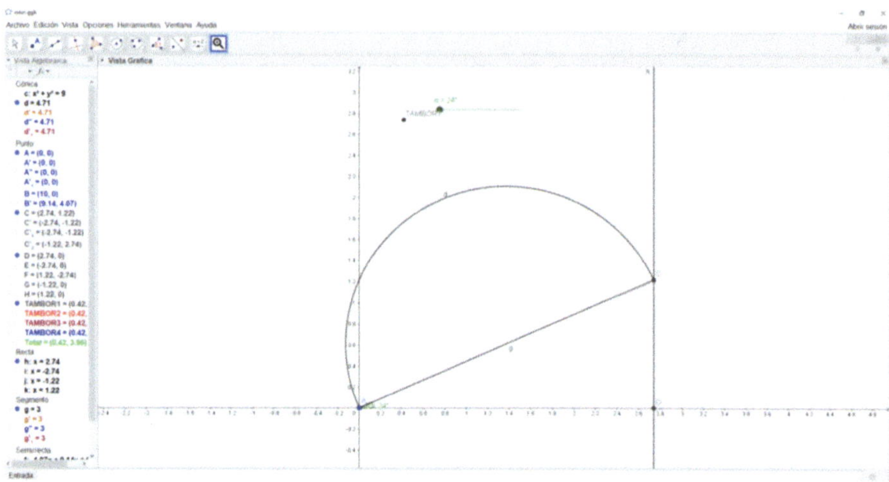

Fig. 11.14 Triangular geometry of the area of each half-barrel. (Source: Authors' own work)

enough evidence for the students to propose the trigonometric functions as more economic models to address the problem.

Therefore, due to economy reasons and, again, parting from the indications of the analogical modelling, the students created an analytical modelling of the rotor's swept area. As a result, they could confirm that as the number of half barrels increases, the variability in the swept area decreases, which results in a reduction of the vibration. Parting from this analysis, the students could provide a basis for a helical design (Fig. 11.15) that would have no vibration in the rotor.

11.3.9 Modelling 6: Building a Supporting Structure for a Solar Panel—Analogical Modelling

Within the framework of an extracurricular workshop conducted at a secondary school, we presented the students (aged between 11 and 13) a project to build a mobile supporting structure for a photovoltaic panel. The purpose was to improve the usage of solar radiation. The panel would supply energy to an electric 12 V water pump used to extract water from an unconfined aquifer at a rural facility in Argentinian Patagonia.

After weeks of exploring the causal relationships that optimize the performance of the solar panel (see Chap. 3), the students discovered that the optimal amperage was obtained when the photovoltaic panel was in a perpendicular position to the solar rays.

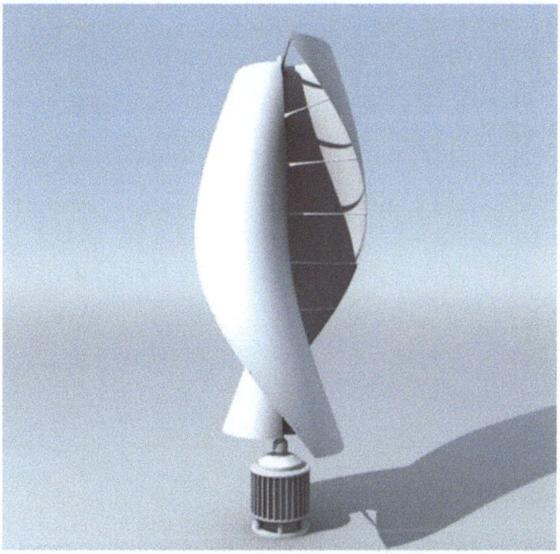

Fig. 11.15 Helical windmill. (Source: https://www.turbosquid.com/es/3d-models/3ds-max-helical-savonius-wind-turbine/804711)

Students and teachers then determined to provide four different positions for the photovoltaic panel, one for each season of the year. However, given the proximity between the angles for spring and autumn, only one position was considered for those two seasons. As a result, the supporting structure presented three positions: summer, autumn-spring, and winter.

One of the potentialities of the analogical models in GeoGebra is that the scale work leads to quantifiable results based on basic geometrical elements. Considering the students' age and their difficulties in mathematics, we opted for that strategy to determine the placement of the holes for each of the panel's positions. Figure 11.16 shows the construction made in GeoGebra that allowed them to determine the distance between the holes for the photovoltaic panel.

This modelling helped the students determine placement by simply moving the slider (sun's angle) until reaching the angle for each season and then observing the distance of the CG segment. This representation was descriptive since it allowed the students to understand how the mobile support worked. It was also prescriptive because it helped determine the actions to take.

From the perspective of the type of elements considered, it is an analogical modelling. In its construction, physical objects from the context were replicated as well as the movement and contact relations between them. Moreover, this modelling led to identifying—through elementary geometry elements—the distances at which the holes had to be punched to guarantee the panel's perpendicular position to the solar rays.

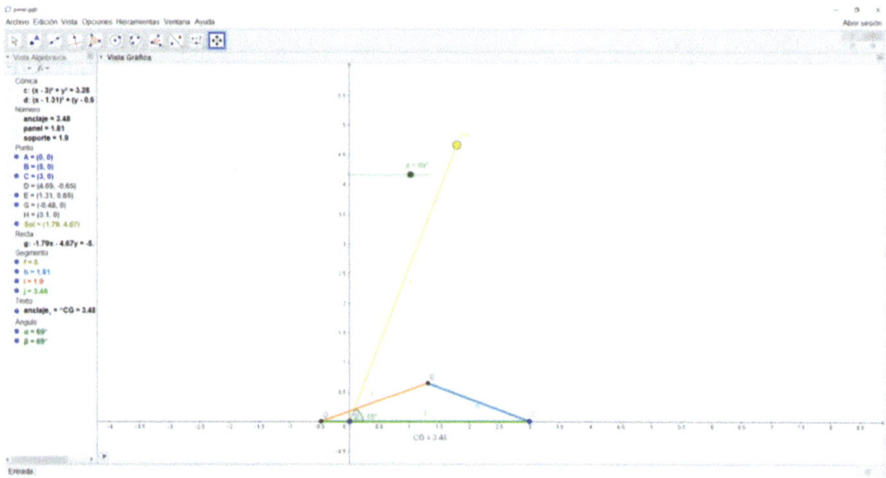

Fig. 11.16 Supporting structure for photovoltaic panel. (Source: Authors' own work)

11.3.10 Modelling 7: Optimization of the Lateral Sides of the Water Purification System—Analogical Modelling

In the framework of a project carried out in the Geology and Palaeontology programs (Bachelor's degree), first-year students (in 2021) designed, calculated and, built and tested prototypes for low-cost, solar-powered water purification systems. The systems' technological principle was based on research carried out at the University at Buffalo (the USA). The professors involved in the project maintained contact with the researchers under memorandums of understanding between the two universities. This agreement authorized the non-profit use of this technological development (Gan & Zhang, 2017).

The project with students consisted of implementing that technology into prototypes to be tested. Therefore, while the technological principle was already determined, designing the prototypes involved an important number of decisions to make. Especially considering the shortage of the needed energy source (solar radiation) in Argentinian Patagonia.

The system consisted of using solar radiation to evaporate water retained in an absorbent fabric. Then, that steam was condensed in a transparent cover and transported in gutters into a container. The water collected (distilled) was then purified using the appropriate salts. Figure 11.17 shows a picture of the first version of this technology. It was published by the researchers at the University at Buffalo.

To optimize the system's functionality, one of the variables analyzed was the surface of the lateral sides of the cover. The modelling we present here was created with students from the mathematics course. Its purpose was to determine the possible existence of a minimum area of lateral sides for a prototype with a base area of 1 m^2.

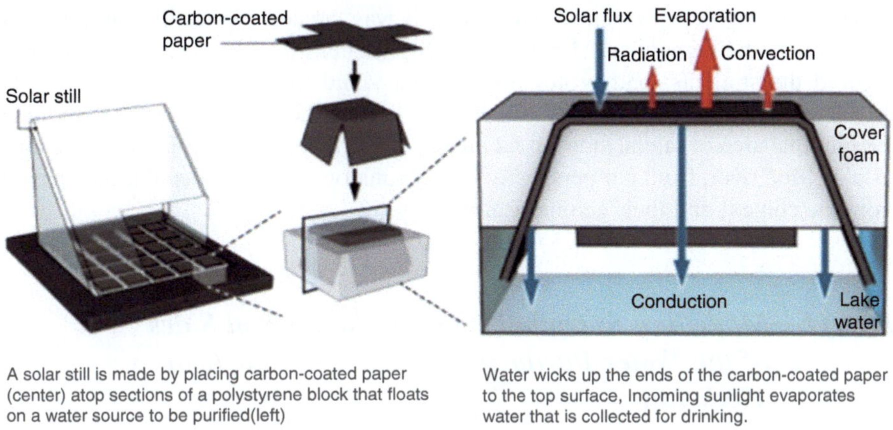

Fig. 11.17 Solar distillation. (Source: Buffalo University)

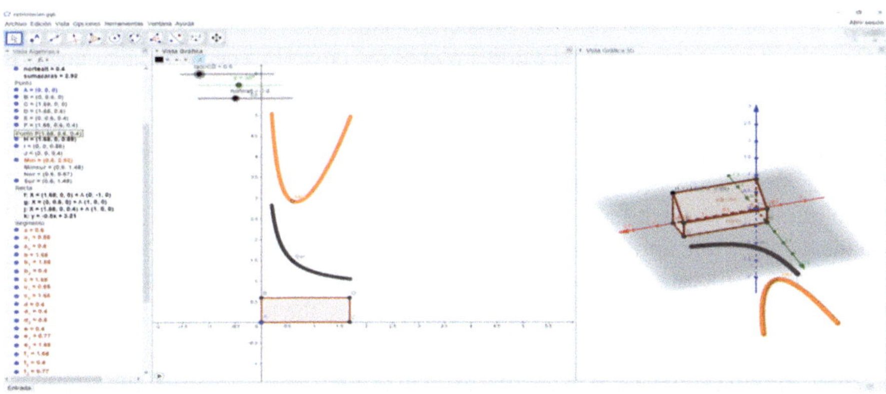

Fig. 11.18 Modelling the surface of the lateral sides of the cover. (Source: Authors' own work)

Figure 11.18 shows the students' construction in GeoGebra; they were assisted by their professors.

While performing the analysis with the students, a set of variables remained undefined. For instance, the angle of the cover's upper part and the height of the north side. Therefore, the professors induced a modelling with sliders on three variables: the measure of a side of the base (lado CD), upper part's angle (alfa) and north side's height (nortealt). This modelling was also constructed based on physical objects and their geometrical relations.

The 2D graphic view shows the rectangular base that allowed us to obtain the minimal surface of lateral sides for the 39° upper part's angle and the 40 cm height of the north side. The minimal surface of the lateral sides was obtained by placing the lado CD slider at 60 cm, approximately. The inferior curve shows the evolution of the south side's surface. This side was also relevant for not receiving direct radiation, being the coldest one and probably the one that condenses more water.

The 3D graphic view shows the cover. It was created replicating the physical objects and their geometrical relations. This modelling was descriptive since it allowed the students (and professors) to understand the problem in question. At the same time, it was prescriptive because, by finding the dimensions that determine the minimal area of lateral sides, it facilitated determining the criteria for the actions to take. Moreover, from our perspective, it was analogical due to the objects retained from the context and their geometrical relations.

11.3.11 Modelling 8: Optimization of the Lateral Sides of the Water Purification System—Analytical Modelling

For this project, the students (around 40) were divided into teams. In total, there were five teams and each of them made their own decisions, which in some cases were different. Therefore, there were five prototypes with some variations. For instance, as regards the strategy of keeping the upper part of the cover in a perpendicular position to the solar rays, some preferred to place regulable floaters based on the summer position, while others were based on the winter position. This led to certain differences in some of the variables, in particular, the angle of the upper part of the cover.

In the analysis of the optimal measures, it was necessary to work on the three variables (sliders). Each team assigned their values to the upper part's angle (alfa) and the height of the north side (nortealt) to determine the measure of the side named caraslat. For a more evident analysis of the change of the minimal surface of the lateral faces (caraslat) according to the other variables, the teachers proposed making an algebraic representation of the addition of the lateral sides' surfaces.

Thus, the independent variable would be lado CD, while the dependent variable would be the addition of the lateral sides' surfaces, and the parameters would be the upper part's angle and the height of the north side. Figure 11.19 shows the curve obtained; first, supported on paper and then in GeoGebra (p(x)).

The tangent line to the curve for a given x value became a visual indicator of the location of the minimal point. The function p(x) that models the total surface of the lateral sides was done on paper based on the analogical modelling from the representation in GeoGebra's 3D algebra view.

Again, this modelling was descriptive as well as prescriptive due to the same reasons detailed for the analogical modelling. In this case, resorting to algebra for the relation between the measure of lado CD and the total surface of the lateral sides constituted a conceptual leap different from the previous modelling. This model was then analytical. It was no longer based on the physical relations between the objects. Instead, it was based on the more abstract relations that exceeded sensorial perception. These relations had to be analyzed in the algebraic and functional plane, among others.

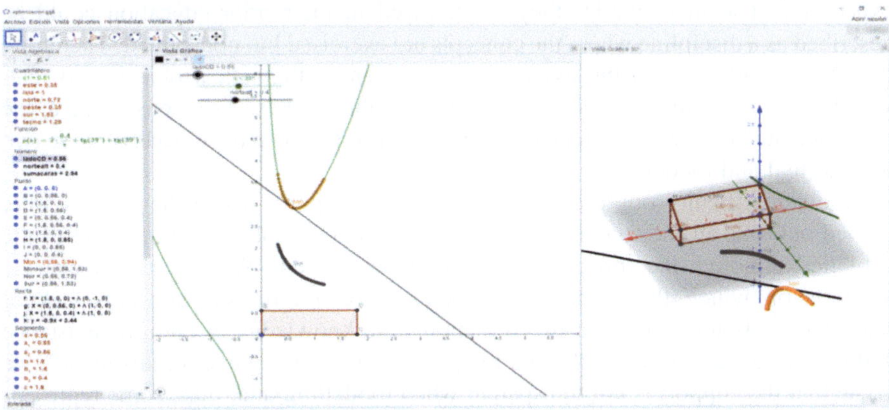

Fig. 11.19 Optimization of the lateral sides of the water purification system. (Source: Authors' own work)

11.4 Results and Perspectives

In terms of preliminary results, we claim that, as the literature states, modelling is an appropriate place for learning disciplinary concepts; in particular, mathematical. The eight models presented here have allowed us to discuss new concepts while re-significating concepts already known by the students. The list of mathematical concepts introduced in each modelling will not be listed here as it is not relevant for this chapter.

We do wish to highlight the diversity of the branches of mathematics that are interrelated in the same modelling and the possibilities for the resignification of elementary concepts that become fundamental to characterize objects as well as parts' movements. To illustrate, we could consider the circle; a geometrical object which at the same time is used as an instrument to measure distance and even to determine rotations. The modelling experiences with the students have enabled them to interpret and reinterpret concepts not only by the semiotic registers in which they appear (Duval, 1993, 2006; Hitt, 2004) but also by the phenomena that they can explain and the objects that can represent the real-world.

Another interesting dynamic observed in the models with students refers to the tool-object dialectic (Douady, 1986, 2002; Czocher, 2019). Because of the models, it was possible to deepen the learning of concepts in that dialectic. The concepts are tools for understanding and describing phenomena, but they are also considered learning objects. As such, one can be familiarized with their characteristics, rules and types of use leading to instances of institutionalization of the characteristics of the mathematical objects.

As regards modelling as a teaching object connected to abilities and competencies, we could verify a considerable number of difficulties among the students at the beginning of the modelling. Partly, this could be explained by their lack of experience. The mathematics they experienced in their prior education could be described as a discipline where the concepts are exercised but not applied. Therefore, the abilities referring to the associations between mathematical objects and real-world relations were scarce at the beginning of the models. As a consequence, the professors had to intervene and induce relationships between the objects of these two worlds (real and model).

Another difficulty observed is related to the supports where the models were carried out. For all the students, GeoGebra was a new tool at the beginning of the activities, although the spreadsheet was familiar to them. We understand that the supports on which the models are produced are not transparent. This lack of transparency is understood in two directions: the knowledge of both the existence and use of the supporting tools, and the possibilities they enable. In this regard, we argue that the support is not transparent when modelling and, thus, some representations are feasible on a certain support while others are expensive in terms of procedure, or plainly impossible to carry out. Hence, a support has a space of possibilities linked and this space conditions what can be modelled with it. Thus, the students also presented difficulties due to their initial lack of knowledge about the GeoGebra support.

There were two more phenomena linked to difficulties observed at the initial stage. On the one hand, there was confusion caused by the change in the didactic contract. On the other, there was a lack of experience working in teams in complex contexts. As regards the didactic contract, the students entered their project classes with habits quite different from the ones in project-based learning. Their prior experiences were marked by doing individual activities using pencils and paper, their educators' expository lessons and final exams at the end of the course. The dynamic of working in projects in a real-world context and, particularly, the models where they had to produce arguments caused surprises at the initial stages of the projects. The same applies to teamwork and collective commitments.

In addition to that, the proposals presented here were carried out throughout the first year of their education. It is important to highlight that dropouts characterize this period. Thus, the students who remained in the courses also had to learn how to solve complex issues in the projects related to their classmates' absence. In this sense, we highlight that the models presented were carried out in educative experiences (projects) that lasted 4 or 8 months, depending on the case. This relatively long-time frame allowed the students to improve their previously mentioned difficulties.

We also consider relevant the interdisciplinarity, unavoidable at times, to discuss modelling. In particular, although not limited to, the analogical models. However, since analogical models replicate characteristics and phenomena of the context, they appear to have an evident tendency to resort to interdisciplinarity in comparison to analytical models. This could be observed in the development of all the analogical models presented here.

Moreover, the analogical models seemed more appropriate than the analytical for the general and basic understanding of the analyzed phenomena. The analogical modelling possibility to establish direct relations between the objects from the context and the objects of the model appear to have facilitated the students' appropriation of the problems as well as their understanding.

This potentiality presented in the analogical models did not pose any difficulties in the possibilities of constructing arguments or making decisions based on them. On the contrary, precisely for establishing relations between the model and reality, they helped the students construct the arguments for their decisions. In this direction, we have observed certain difficulties in the analytical models. Their level of abstraction posed difficulties in the students' understanding and production of potential analyses relevant to the problems being discussed. This was verified in the answers to open questionnaires we gave to them.

In those questionnaires, they were asked about matters observed in an analytical modelling. In general, their answers did not contemplate elements of the model. Instead, there was a tendency of returning to the real-world context to explain it, which was inappropriate considering the impossibility to observe the phenomenon in the real-world context. Thus, the analogical models appear as an interesting intermediate instance in relation to the analytical models. In fact, the analytical models presented in this chapter were produced based on analogical models.

There appears a sort of sequence of levels of abstraction where the analogical modelling constitutes the first step of abstraction on which it is convenient to support more abstract models that enable other types of analyses. As a result, at least two potentialities can be observed. On the one hand, the possibility of creating different interpretations and arguments of each type. On the other, the staggering or sequencing where analogical modelling is relevant for two reasons: its results and its support for a more complex analytical model.

References

Blomhøj, M. (2019). Towards integration of modelling in secondary mathematics teaching. In G. A. Stillman & J. P. Brown (Eds.), *Lines of inquiry in mathematical modelling research in education* (ICME-13 monographs). Springer.

Blum, W., & Borromeo Ferri, R. (2009). Mathematical modelling: Can it be taught and learnt? *Journal of Mathematical Modelling and Application, 1*(1), 45–58.

Brown, J., & Ikeda, T. (2019). Conclusions and future lines of inquiry in mathematical modelling research in education. In G. Stillman & J. Brown (Eds.), *Lines of inquiry in mathematical modelling research in education* (pp. 233–253). Springer.

Bruillard, E., & Haspekian, M. (2009). *Behind students' spreadsheet competencies: Their achievement in Algebra? A case study in a French vocational school.* CERME 6, Lyon, France.

Caron, F. (2019). Approaches to investigating complex dynamical systems. In G. A. Stillman & J. P. Brown (Eds.), *Lines of inquiry in mathematical modelling research in education* (ICME-13 monographs). Springer.

Czocher, J. (2019). Precision, priority, and proxies in mathematical modelling. In G. A. Stillman & J. P. Brown (Eds.), *Lines of inquiry in mathematical modelling research in education* (ICME-13 monographs). Springer.

Douady, R. (1986). Jeux de cadres et dialectique outil-objet. *Recherche en Didactique des Mathématiques, 7*(2), 5–31.

Douady, R. (2002). *Rapport enseignement apprentissage: Dialectique outil-objet, jeux de cadres.* Cahier de Didactique des Mathématiques, IREM.

Duval, R. (1993). Registres de représentation Sémiotique et fonctionnement cógnita de la pensée. *Anales de Didactique et de Sciences Cognitives, 5*, 37–65.
Duval, R. (2006). Quelle sémiotique pour l'analyse de l'activité et les productions mathématiques. *Relime, Número Especial*, 45–81.
Gan, Q., & Zhang, N. (2017). Extremely cost-effective and efficient solar vapor generation under nonconcentrated illumination using thermally isolated black paper. *Global Challenges, 1*(2), 1600003.
Haspekian, M. (2005). *Intégration d'outils informatiques dans l'enseignement des mathématiques*. Ecole Doctorale Savoirs scientifiques: Épistémologie, histoire des sciences et didactique des disciplines, Paris 7 Denis Diderot, Doctorat, Paris.
Hitt, F. (2004). Les représentations sémiotiques dans l'apprentissage de concepts mathématiques et leur rôle dans une démarche heuristique. *Revue des Sciences de L' Éducation, 30*(2), 329–354.

Part IV
Mathematics and People: Empirical Work, Theoretical-Methodological Approaches, and Research Questions

The category of socioepistemological modelling is an educational program to learn about the use of mathematical knowledge of communities in their different scenarios: school, work, and life. This study reveals the emergence of people's mathematical knowledge. Within this framework, the orientation of the program is to conform epistemologies of the use of mathematical knowledge as a basis for all educational levels: basic, intermediate, and higher education. In this section, a modelling category is defined as a theoretical variety to express the functionality of mathematical knowledge in a plane formed by two axes: transversality of mathematical knowledge and institutionalization of scenarios. This category is the re-signification of uses of mathematical knowledge in *a horizontal and reciprocal relationship between mathematical knowledge*. Empirical evidence is offered for the research questions on *the relationship* through the re-signification and the educational impact in the domains on some engineering and engineering students. But also, about *the role of the relationship* in the initial training of mathematics teachers to generate disciplinary identity. Furthermore, on *the relationship* to define the contemporaneity of learning, which consists of handling technology to obtain adequate representations of data and generate symbolic processing. And finally, on *the connectivity that the relationship* provokes between contemporary natural phenomena such as Covid 19 and the design of school situations in mathematics.

Chapter 12
A Category of Modelling: The Uses of Mathematical Knowledge in Different Scenarios and the Learning of Mathematics

Francisco Cordero, E. Johanna Mendoza-Higuera, Irene Pérez-Oxté, Jaime Huincahue, and Jaime Mena-Lorca

12.1 Introduction

A sui generis *category of mathematical modelling* is formulated, due to the epistemological, ontological, and educational stance of our research approach. This does not correspond to the classical definitions of modelling in the field of mathematics that in general draw attention to the interests of the work of modelling, in which major aspects intervene: represent reality and apply a knowledge structure to a real situation, with empirical or analytical models according to disciplinary interests (Bissell & Dillon, 2012).

F. Cordero (✉)
Centro de Investigación y de Estudios Avanzados del Instituto Politécnico Nacional, Ciudad de México, Mexico
e-mail: fcordero@cinvestav.mx

E. J. Mendoza-Higuera
Unidades Tecnológicas de Santander, Bucaramanga, Santander, Colombia
e-mail: edithmendoza@correo.uts.edu.co

I. Pérez-Oxté
Universidad Abierta y a Distancia de México, Ciudad de México, Mexico
e-mail: Irene.perezoxt@nube.unadmexico.mx

J. Huincahue
Universidad Católica del Maule, Talca, Región del Maule, Chile
e-mail: jhuincahue@ucm.cl

J. Mena-Lorca
Pontificia Universidad Católica de Valparaíso, Valparaíso, Chile
e-mail: jaime.mena@pucv.cl

M. Rosa et al. (eds.), *Mathematical Modelling Programs in Latin America*,
https://doi.org/10.1007/978-3-031-04271-3_12

The *category* is a *variety*[1] of mathematical modelling, which is a process that legitimate of the mathematics in use that happen in different scenarios and to the crossings between scenarios of those uses. We will call the first one *epistemological plurality* and the second *transversality* of uses of mathematical knowledge. Both aspects define the mathematical functionality of the mathematical knowledge communities that occur in the scenarios: school, work, and city.

An invariant that has been identified in the different meanings of mathematical modelling is the *relationship with reality*. This invariant, although, in generic terms, is common in modelling, it can also be the tip of the iceberg that denotes the different mathematical modelling programs to improve mathematical teaching and learning. One aspect of our interest, in this article, is to discuss what is meant by reality in our modelling program. This meaning of reality will lead to a category of modelling that will value the uses and meanings of mathematical objects, which will be formulated and justified later.

12.1.1 That What Is Called "Reality"

Entering the subject: Mathematics in education must be accountable for reality; and this should be a fundamental basis at the different educational levels. But it is necessary to question the term *reality* in the relations between school mathematics and mathematical work to gain precision. Cordero et al. (2016) makes us see that this so-called *reality* has different meanings, but also the discussion can be extended to the various philosophical currents as mentioned by Pollak (1979).

Nevertheless, for questioning, it is necessary to direct it to empirical aspects and to keep in mind the fundamental principles of Kant's critique of practical reason (1788/1998). In this sense, aspects such as the subjective reality and others linked to the human sensations, which favour the functional sense of knowledge, where their uses and re-significations happen in the discipline work and life. For example, a reality in modelling may be the behaviour of a particle in an electromagnetic field, and the model that is generated in this respect responds to the disciplinary interests of science, as well as in some engineering.

However, *reality* in mathematics classrooms, as shown by Cordero et al. (2016), is not clear nor trivial: to bring mathematics to the student's reality and to create environments of everyday life mathematics, are slogans that have not been able to be carried out fully. School mathematics, in its tradition, has not been oriented to this end.

Therefore, *reality* should be restricted, in order to standardize it to the education of mathematics: to consider all levels of education and the diversity of disciplines, as well as the work and life of people. Perhaps, *reality* should be interpreted in what is

[1] Variety expresses the idea of creating an alternative definition for mathematical modelling as explained in the Constructs of the Modelling Category section: a variety.

usual in all these scenarios, where routine uses are expressed; that is to say, the everyday life of the disciplinary specialist, the worker, and the people (Cordero, 2016; Zaldívar et al., 2014; Mendoza & Cordero, 2018).

In general terms, with this account of reality, functional knowledge means a useful knowledge of people in situations of mundane life, work, and the profession (Arendt, 2005). This useful knowledge is composed of uses and meanings, which are re-signified in the transit of the situations. In that sense, it can be said that functional knowledge is the result of the transversality of the use of the knowledge of the people in the different situations, which are re-signified.

The definition of re-signification is articulated by the definition of use, then: the uses of mathematical knowledge **U**(CM)[2] are organic functions of situations (functioning), which are manifested by the "tasks" that make up the situation, and the form of use will be the class of those 'tasks'. The tasks can be activities, actions, executions, and alternations of domains. When tasks alternation happens, a new organic function is generated, which will be debated with the forms of the uses.

Thus "act of use" it was agreed to call it a re-signification of mathematical knowledge uses (**Res**(**U**(CM)) (Cordero & Flores, 2007). *Re-signification* is, in some way, the construction of knowledge. Later, in the approach, *transversality*, *re-signification*, and *functionality* will become constructs to observe and analyse in different communities of knowledge.

12.2 The Problems and Socioepistemology

In this Modelling Program the interpretation of reality must build the reciprocal relationship between mathematical knowledge and everyday knowledge of the disciplinary specialist, the worker, and the people. For example, a community of bionic engineers in their day-to-day knowledge of Control Systems builds a category of mathematical knowledge called Behaviour Reproduction. This category is in reciprocal relationship with the Stability of a differential equation.

The first expresses the mathematical uses and meanings of the engineering community and the second expresses the mathematical object of school mathematics. For these engineers, a differential equation is "an instruction that organizes behaviours" while for school mathematics is *finding a solution that is not known* (Mendoza & Cordero, 2018). These will be the realities that this Modelling Program will meet (Cordero, 2016; Cordero et al., 2015; Buendía & Cordero, 2005).

To carry out scientific research with the Modelling Program, it requires theoretical-methodological constructs. What is formulated here belongs to the *Socio-Epistemological Theory of Educational Mathematics* Cantoral (2013, 2019) has provided this with fundamentals, which consist of four principles: the normativity of social practice, contextualized rationality, epistemological relativism,

[2]The acronym **U**(CM) comes from Spanish, which means *Mathematical Knowledge Uses.*

and progressive signification (resignification). With these principles, socioepistemology explains the enigma of the social construction of mathematical knowledge and its institutional diffusion.

A fundamental construct is the social practice, with which, in a complex system of the social dimension processes, in relation to the cognitive, epistemological, and didactic dimensions, mathematical knowledge is problematized, considering the sage, technical and popular knowledge to synthesize them with human wisdom. Research has been conducted where the mathematics put into use by communities is problematized when successive variation is modelled in mathematics (high school students) and in science (cardiologists) (Cantoral et al., 2018).

Cordero et al. (2016) considers that the construct *social practice*, in the Socio-Epistemological Theory, has revealed that in school mathematics (basic, middle, and higher education) aspects of the social dimension, such as: reality, the uses of knowledge, and in more generic terms, people, have been *forgotten subjects*. These are necessary to recover, in order to alleviate the problem of the teaching and learning of mathematics (Cordero et al., 2015).

For example, to teach parabola, a mathematical object that appears in middle school mathematic courses, between 15 and 17 years old, it is difficult for a teacher, with their teaching resources, to have a frame of reference to incorporate the parabola in situations of variation, approximation, and transformation to generate prediction arguments, local behaviour, and trends, respectively (Morales & Cordero, 2014; Mendoza et al., 2018).

With this approach, Cordero formulates a *General Socio-Epistemological Program* called *Forgotten Subject and Transversality of Knowledge* (SOLTSA, acronym derived in Spanish), where the *Modelling Program* is immersed (Cordero, 2016). Its foremost objective is *to reveal the uses of mathematical knowledge* and its resignifications that occurs in the mathematical knowledge communities in the different settings: school, work, and city.

The Program SOLTSA is developed through two simultaneous lines of work: the Re-signification of Mathematical Knowledge, and its Educational Impact. In the first one, mathematical knowledge categories that occur in the communities between different scenarios of knowledge that obligatorily comes into play, are discussed: school mathematics, disciplinary field, and everyday life of the community.

For example, in the first one it can be a *differential equation*, in the second it can be the *heating of a focus* and in the third scenario it can be the *reproduction of a behaviour*. The three aspects mentioned are *resignifications of the stability uses*, as it will be seen later (Cordero, 2016; Mendoza & Cordero, 2018). In the second line of work, the multi-factors and stadiums that contribute to the quality alliance of mathematical teaching, to lead to the transformation and educational change of mathematics. *Identity*, *socialization*, *and inclusion*, among others, are the multi-factors for this purpose (Opazo-Arellano et al., 2018; Pérez-Oxté & Cordero, 2016; Medina-Lara et al., 2018).

After this necessary digression, the stance on mathematics in the SOLTSA program is shown below. First, it is stated that this study is concerned with the social function of mathematical knowledge.[3] This fact has led to questions about the uses of knowledge from diverse communities. For example, this reveals that these uses are different for mathematicians as compared to engineers (Cordero et al., 2015; Mendoza & Cordero, 2018; Pérez-Oxté & Cordero, 2016), but they are also different in school and in everyday life (Carraher et al., 1997).

The meaning of this evidence opens questions that require their contrasts to be precise. It is not enough to ask, in this case, what is mathematical knowledge; but also, what mathematics, which entails considering its epistemological plurality and its transversality of knowledge (Cordero, 2016; Cordero et al., 2015). Perhaps because of this, there are investigations that proposes to integrate non-mathematical entities that contribute to learning into classical mathematical classrooms, for example, Lim et al. (2010) analyze the role of prediction in learning mathematics.

The previous stance leads to a *stance on modelling*. It is a category that responds to what is useful to a human in a specific situation. It is something more robust than a mathematical application (to a real situation), it is a specific community practice, in its scenarios: school, work and city. This situation is composed of significations and re-significations with their respective procedures, regulated by an instrument: both are constructed according to the operations that the participants can perform, with the conditions that they can capture and transform and with the concepts that they are progressively building (Cordero, 2016).

This category of modelling carries out multiple realizations and makes adjustments in its structure to produce a desirable pattern; it is a medium that supports the development of reasoning and argumentation (Suárez & Cordero, 2010). It is in itself a construction of mathematical knowledge. *The category of modelling is the process where mathematical knowledge is re-signified, which values the elements, in the environment of the object to which they give meaning* (this statement will be discussed more carefully in Fig. 12.6).

12.3 Research Questions and Tasks

The theoretical approach consists of constructing a reference frame, where the teacher and the student legitimize the uses of mathematical knowledge (epistemologies of uses) for the benefit of teaching and learning. Its construction is not obvious; it requires scientific research that can endow with truthful elements of these epistemologies. The category of modelling plays the role of providing a form and functioning of mathematical knowledge uses to observe, analyse, and shape them in specific situations. These categories are inferred in the study, communities using

[3] By mathematical knowledge, it refers to the mathematics put into use by the different communities in their different scenarios: school, work, and city. The ages of the people that make up the communities can correspond to children, adolescents, young people, and adults.

empirical theoretical-methodological aspects, but what is not known (that needs to be known) a priori is the educational impact.

For that, it will be necessary to design school situations for all educational levels (basic, middle, and high) and work on the following research questions: (a) What are the extensions of learning episodes in the classroom when the reciprocal relationship between mathematics and reality, is restored? and (b) What are the fundamental factors in the new permanent program to maintain the environments of reciprocal relationships? These tasks should guide the educational transformation that favours the reciprocity between mathematics and everyday life in the classroom.

12.4 The Status of the Uses of Mathematical Knowledge and Modelling in Mathematics Education

The appearance of mathematical modelling in the mathematical education has to do with the following facts. Depending on what is meant by *mathematical education*, different pronouncements are found in relation to the *application of mathematics* in education.

Henry Pollak (1979) says that the applications of mathematics and mathematical modelling already played an important role in school mathematics in the nineteenth century in Europe and North America. Felix Klein, mathematician and mathematics educator, introduced applications to school mathematics in Germany and other parts of Europe through the development of an innovative curriculum that integrated the applications of mathematics into higher education. Moreover, Klein defended, in the teaching of mathematics, a balance between applications and modelling on the one hand, and pure mathematics on the other. Surely, these pronouncements have achieved educational impacts on the world.

For example, in the Common Core State Standards for Mathematics (CCSSM) of the United States, mathematical modelling had minimal attention in 1989 and 2000, but it is currently prioritized: On the one hand, there is the Standard for Mathematical Practices (SMP), and, on the other hand, for secondary education, it is a conceptual category (Hirsh & McDuffie, 2016). Blum and Borromeo-Ferri (2009) argue that mathematical modelling can support the learning of mathematics regarding motivation, comprehension, and retention.

Progress is undeniable; mathematical modelling in education today plays a significant role in mathematical education (Kaiser & Sriraman, 2006). However, attention to another aspect is given in this research. In this scenario, the basis for defining what mathematical modelling is, is the *ambit of science*, but not in the *ambit of people*. For example, CCSSM states that modelling means to use mathematics or statistics to describe a real-world situation and to deduce additional information from the situation by mathematical and statistical calculation and analysis (CCSWT, 2013).

This definition is feasible, surely for mathematical modelers, and also for those who believe in modelling for mathematical education. But here is the questioning (and in return the point of interest) is: (a) How do they model? and (b) How do people use modelling? These people can be scientific, but it is not necessary.

12.5 Construct of the Modelling Category: A Variety

The structure of the modelling category ζ(Mod) is composed of the uses of mathematical knowledge **U**(CM), and by the re-significations of those uses, **Res**(**U**(CM)), in specific situations (*S*). Such situations are part of that environment (mutual relations) that occur in communities of mathematical knowledge (*CCM*).

Each specific situation S_i is formed by sequential elements that construct what is deemed mathematical: signification, procedure, and instrument, to derive the argumentation of the situation (*Arg*(*CM*)). In generic terms, $Arg(CM)_i$ is a **Res**(**U**$(CM)_i$) constructed by CCM_i in S_i (see Fig. 12.6) (Cordero, 2016). It is a situational mathematical knowledge; it does not correspond to the *emulation* of the mathematical object in the situation, but to the *revelation* of the uses and meanings of the object, community-owned, regulated by the situation.

The **U**(CM) re-signify themselves in each *S*. Additionally, also when transversality (T_i) occurs between scenarios or domains of knowledge (D_i). Still, in the situations and transversality, moments happen (Mo_i), and between them, the uses are re-signified also. The Mo_i's are phases in the situational process.

12.5.1 The Variety: From the Mathematical Object to the Functionality of Mathematics

It is believed that a convenient way to face the enigma of the category is to bring it to the notion of *variety*, which consists of creating a difference without losing unity. This means that mathematical modelling has a *unit* that defines or distinguishes it as such. In that sense, a *principle* prevails both in the approaches to science and in the approaches to education, which has to do with a *cycle that connects the real world and mathematics*.

In this context, Blum's model (2011) is an example of this principle: it defines the cycle with a sequence that, in general, begins with a real-life situation, which in turn, in simplifying and idealizing, becomes a real-life model, since which, through mathematization processes, a mathematical model is derived. Then, with mathematical tools, it is interpreted and validated in the real-life situation.

The cycle, with its sequence of phases, is consistent with the principle: it connects the reality of the world with mathematics. And, that cycle is taken to different educational levels to improve the learning of mathematics. An important aspect, of this study's interest, is to point out that in the usual educational approach the cycle *pre-exists*[4] the experiences of the teacher and the student; the processes between the phases of the cycle are monitored or studied.

It is believed that from this fact the variety can formulated. For this purpose, let us consider the principle (P) of mathematical modelling (Mod). *P is the cycle that connects the real world and mathematics.* Then the variety must be based on a principle P' of P. In some sense P' is a derivative of P; it can be a sub-principle of P or, at best, something that gives rise to P. However, in this case *P' is the functional aspect of the reciprocal relationship between mathematics and everyday life*. This P' generates the category of modelling ζ(Mod), that is, the uses of people's mathematical knowledge.

Hence, on the one hand, mathematical modelling, in generic terms, is formulated as follows:

- Principle P assumes the existence of a mathematical knowledge (M) and an existence of reality (R). Given R there is a specific mathematical knowledge M' that *mathematizes R:* $M'(R) = R'$, where R' is an interpretation of R.
- $M'(R)$ is a mathematical object: it is the knowledge generated by the mathematical modelling.

On the other hand, the variety of modelling is formulated as follows:

1. Now, P' is functional; hence, neither R nor M pre-exists.
2. The use of people's mathematics is functional, **U**(CM).
3. People live in different situations, S_k.
4. In the transition between the S_k, there are epistemologies E_j (plurality) and transversality T_n (re-significations).
5. The S_k might be on domains of knowledge D_m and in the alternations between the D_m's.
6. The modelling category is the re-signification of uses, **Res**(**U**(CM)), when a transition between S_k and S_m happens, even in alternance of domains. This is the knowledge that generates ζ(Mod).

The category ζ(Mod) is composed of two axes: the institutionalization and the transversality of knowledge, where situations S_{ij}, domains D_j and alternations of scenarios occur: the school-academic scenario, the work-profession scenario, and the city-everyday life scenario. The scheme of the variety, which is called the *Mathematical knowledge framework* of ζ(Mod), is represented in Fig. 12.1.

[4]This pre-existence means that a priori the teacher and the student are not mathematical modelers.

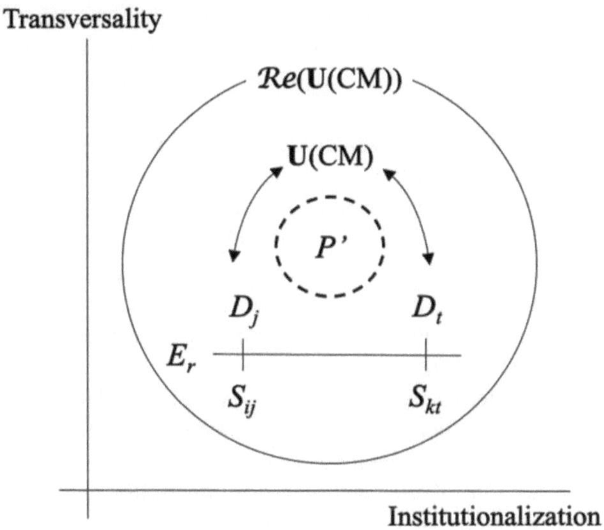

Fig. 12.1 Mathematical knowledge framework of ζ(Mod) (The interpretation of this framework of mathematical knowledge of the modelling category was widely discussed considering the synthesis of three investigation experiences: the functionality of mathematical knowledge, mathematical modeling and initial mathematics teacher training. The collaborators were F. Cordero, J. Mena-Lorca, and J. Huincahue. A version of this interpretation was agreed upon in Huincahue's doctoral thesis (2017).). (Source: Cordero (2022))

12.6 An Example: Start-Up of ζ(Mod)—From Stability to the Reproduction of Behaviour

The development of the category of modelling ζ(Mod) is outlined in two episodes, considering empirical research that has been carried out with communities of engineers in the profession and in training: *Re-signification of stability* and *transversality of knowledge*.

12.6.1 Data Collection and Definition of the Study Community

The selection of the communities consisted of the availability, of its members, to be video-recorded and interviewed, preferably, in the scenarios of professional work. The selected communities were Bionic Engineers and Industrial Chemical Engineers.

With ethnographic methods (Guber, 2001) and a case study by each engineering community, this disciplinary work was characterized. The characterization consisted of identifying the routine situations where they use mathematical knowledge and the problematizations of their mathematical knowledge. The analysis of these characterizations was carried out through the constructs of the modelling category with the Socio-epistemological Theory and considering the documentary technique and semi-structured interviews.

To problematize, the mathematical knowledge was analysed, through the resignification of those uses in the school-academic and work-profession settings. The problematization formed an epistemology of resignifications of the mathematical knowledge uses that emerges in the community when considering the specificity of the scenario. On the one hand, with the documentary technique of analysis, and on the other, with the semi-structured interview technique, patterns, and relationships among them were identified, alluding to the tendency or reproduction of a behaviour in the contexts of the situations.

The behavioural patterns and relationships were organized by an instrument (instruction that organizes behaviours) accompanied by their meanings (graphic and analytical patterns) and procedures (variation of parameters). And with the unit of analysis composed of the constructs: use, resignification and transversality, the transformation situation was conformed (resignification of uses of stability) (see Fig. 12.6). Subsequently, the transformation situation was taken as a basis to design activities and analyse their emergence in the community of engineers during training of different semesters.

12.6.2 Community of Bionic Engineers

The community of mathematical knowledge of bionic engineers *CCM* (BE) that were observed and interviewed, are assigned to the Interdisciplinary Unit of Engineering and Advanced Technologies of the National Polytechnic Institute (UPIITA-IPN), Mexico (Mendoza & Cordero, 2018). Bionic Engineering is conceived as the set of interdisciplinary knowledge between electronics and biology whose purpose is the creation of artificial systems to reproduce the characteristics and structure of living organisms.

The subject of Modelling and Control of Biological Systems is the backbone of the curricular program, for which it was observed, for 3 months, the professional practice (as a teacher) of a bionic engineer and the practices of the engineers in training (the students). One of the central concepts is the study of control systems. The artificial devices that are built are made up of processes that are required to be controlled, in such a way that the desired characteristics and structure can be reproduced.

Likewise, the stability of these systems, which are modelled by differential equations, was one of the characteristics that led to the emergence of different concepts and techniques both for Control Theory and for mathematics itself. In general, the objective of a control system is *to control the outputs in some prescribed way by means of the inputs through the elements of the control system* (Mendoza & Cordero, 2018).

Table 12.1 Core and specific situations: transformation and control systems (Source: Mendoza and Cordero (2018))

Construct of what is deemed mathematical	Core situation	Specific situation
	Transformation situation	**Design of Control Systems**
Meanings	Graphical and analytical behaviour patterns	Behaviour of the input signal and the output signal
Procedures	Variation of parameters	Feedback transfer function
Instrument	Instruction that organizes behaviours	The differential equation models the behaviour of the signals and the stability of the system
Argumentation/re-signification	Tendency behaviour/ reproduction of Behaviours	Reproduction of the behaviour from the exit to the entrance

The design of control systems is the specific situation of the *CCM* (BE) in the school setting, which gives meaning to the category of modelling (behavioural reproduction) whose structure is the core situation (transformation situation) that rules the mathematical knowledge uses of the engineer. The meanings, the procedures, the instrument, and the arguments are specific to *CCM* (BE) (Mendoza & Cordero, 2018) (see Table 12.1).

With the evidence of the ζ(Mod)'s mathematical knowledge framework, it was decided to justify it in one of the laboratory practices developed by the teaching engineer. The practice is called control the lightbulb's temperature, where the main problem is: once assigned a reference value, the temperature of the lightbulb reaches it. For this purpose, students assemble a physical model with the following elements: Arduino board, lightbulb, AC solid state Relay and temperature sensor (Fig. 12.2).

Initially the students analyse the behaviour of the output signal in the system: the temperature of the lightbulb: "The temperature must stay within a certain range (...) what we are going to see is an asymptote that corresponds to the maximum temperature that the lightbulb can reach" (Mendoza & Cordero, 2018). The teacher draws on the blackboard and comments on the following (Fig. 12.3).

The system is characterized by the system gain, the system time constant and the transfer function that relates the input and output signals. The students, in laboratory practice, must adjust the parameters of the system gain and the time constant, based on the graph provided by the Arduino software, allowing it to reach its maximum temperature. A group of students problematize the adjustments of the parameters:

A1. That's why I tell you, it's 77 degrees, I have to put it in ...
A2. And how many were there?
A1. 8000 ohh! Do not
A2. Ahhhh still needed to stabilize
A1. I told you (Class observation, cited in Mendoza & Cordero, 2018).

Fig. 12.2 Connection diagram: lightbulb, Relay and Arduino. (Source: Mendoza and Cordero (2018))

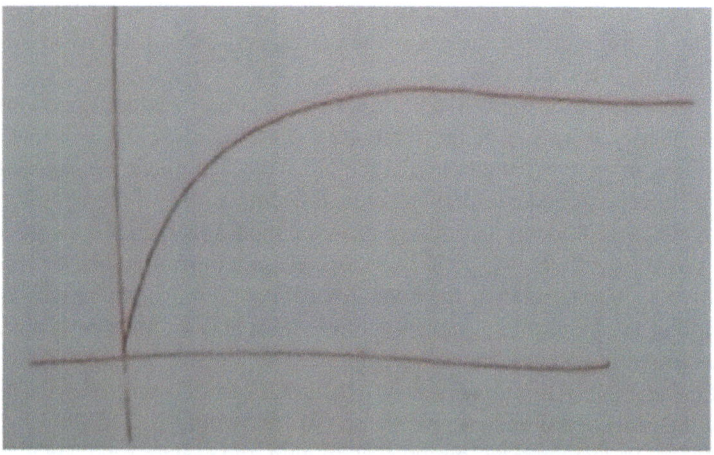

Fig. 12.3 Graph of the output signal behaviour and Transfer function of the plant. (Source: Mendoza and Cordero (2018))

The previous paragraph alludes to that the temperature had not reached its maximum, thus the comment: *it was not stabilized*. Finally, they seek to control the temperature of the bulb by using the ON-OFF controller. Figure 12.4 shows how the curve, which represents the behaviour of the output signal, in certain time intervals exceeds the reference value and in others it does not. This is due to the control mechanism as the teacher expresses it below:

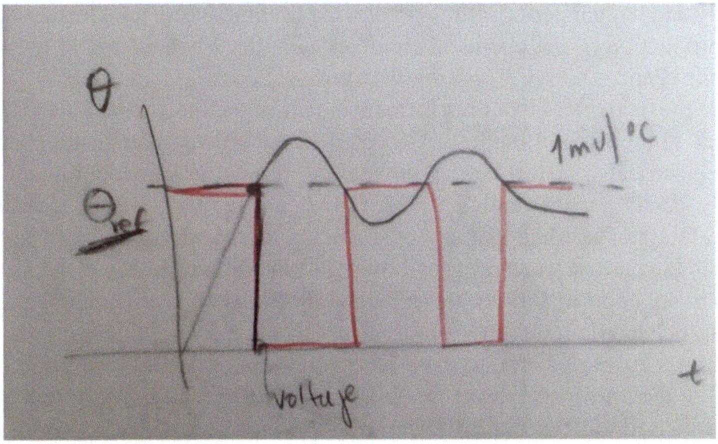

Fig. 12.4 Graph of the output signal using the on-off control. (Source: Mendoza and Cordero (2018))

> (...) they will give a reference value θ ref (...) this would be temperature θ (vertical axis), the temperature will rise and there will surely be a surplus, but at this moment the lightbulb will have been turned off (...) (What is in red means the energy in the lightbulb, or the voltage, (...) or the effect of turning off the lightbulb). The temperature will vary more or less in this way (black curve). The point is that this is going to be oscillating (...) That is, if the temperature is greater than or equal to (...,) then turn on, if not, turn off (...) (Class observation cited in Mendoza & Cordero, 2018).

In the control system design, three moments are observed: *M1*: *System dynamics* ($\theta \rightarrow \theta ref : \theta$ it tends to behave like f); *M2*: *Adjustment of the transfer function* or *behaviour model* ($a\theta' + \theta = \theta ref$); *M3*: *Control of the output signal* and *stability* ($a\theta' = \theta ref - \theta$: *a control*). Each of these moments is subjected to the *Reproduction of a desired behaviour*, that is, the output signal tends to behave as the reference value or input signal (Fig. 12.4).

In this way, stability is *signified* in the behaviour of the input and output signals, causing *procedures* such as the comparison between the output signal and the reference value, modifying the parameters of the differential equation modelling the system and meaning it as an *instrument*, which is responsible for modelling the stability of the output signal and thus achieve that the behaviour initially proposed is reproduced.

12.6.3 Community of Chemical Industrial Engineers

These engineers are industrial chemists and in their day-to-day professional they make diagnostics of electrical transformers; and what they problematize are the concentration behaviours of the chemical elements that make up the transformers, in monthly and annual periods: they predict the lifetime of the transformer according to the reproduction of the chemical elements' graphical behaviours in monthly and yearly units (Pérez-Oxté & Cordero, 2016).

The diagnosis of electrical transformers is part of the daily life of a *CCM* (ChIE). They work in a chemical laboratory located in the Yucatan peninsula. There they diagnose electrical transformers to prevent serious failures in equipment through their early detection, preventing them from being damaged. They constructed a Graphical Diagnostic Method that consist of the concentration history for each of the gases in the electric transformer oil over time. They analyse the concentrations of the gases to see abnormal increases or behaviours that can be indications of possible failures in the electric transformer.

Eight gases are classified into three blocks of analysis, which emerge from the oil contained in the transformer, and the description of the conditions of their graphical behaviour is used to determine whether or not the transformer is in good condition:

1. *Block 1*. Gases that are indicators of possible failures in the transformer: Hydrogen—is an indicator of partial discharges; Ethylene—Hot spots; and Acetylene—Arc. Hydrogen is always present in all faults; however, an increase in this is an indicator of something abnormal in the transformer. It is present because it is easily formed and becomes apparent, for example, by electricity flashing inside the transformer. Ethylene comes in higher concentration and requires twice as much energy as Hydrogen to form, while Acetylene should not be produced since it is a reason to consider a possible failure in the electric transformer and thus, leaving it out of order. There should be no elevated levels of these three gases, they may form but their concentration must be stable.
2. *Block 2*. Gases that must store a ratio 1:10 in their concentrations for the situation to be considered normal: Carbon monoxide and Carbon dioxide. The extraordinary behaviour of these gases indicates a fault called Paper pyrolysis, which means that the paper inside the transformer is burning. The presence of these gases appears naturally due to the wear of the transformer. The ratio 1:10 of these gases is represented graphically as a parallel behaviour.
3. *Block 3*. Gases that indicate a natural wear on the electric transformer: Methane, Ethane and Water. The behaviour of the concentrations of these gases must be stable, that is, with slow and constant increases.

With the Diagnostic Graphical Method, the main question that this community makes is about the assignment of significations to *the graph* in the specific situation of the diagnosis of transformers. These graphs are considered as models of behaviour, which is to say, as tools that allow reading, interpreting, and inferring information on the trending of the concentrations of the gases that the transformer has. Based on this, decisions are made (Pérez-Oxté & Cordero, 2016).

Three uses are identified: *Statistical control*, *Graphic-fault relationship*, and *Graphical diagnostic model*. The diagnosis of transformers is the specific situation, in which the main focus is "behaviour reproduction". These behaviours are re-signified in the moments of each use. This occurs since the situation generates an environment of reciprocal relations between the arguments of Prediction,

Trending Behaviour and Optimization, starting from the re-signification of the graph's uses in which functions and forms are debated.

The core of this environment is the reproduction of behaviours (Pérez-Oxté & Cordero, 2016): for example, to make conjectures about a possible transformer failure, a reading and interpretation of two of the eight graphs of the gas concentrations is enough. In this case, Carbon Monoxide and Carbon Dioxide are indicators that there is Paper pyrolysis and therefore maintenance must be performed. The ratio 1:10 that the gases must present is analysed from a trend behaviour of the concentrations of the gases.

In Fig. 12.5, it can be observed that the graphical behaviour of Carbon Monoxide tends to be about 150 ppm, while the behaviour of the Carbon Dioxide tends to stabilize at 1500 ppm, concluding that the state of the transformer is in good condition.

When the trending behaviour is not stable in the proportion of 10%, this is an indicator that the transformer has a possible failure. By examining this example, it is possible to recognize the variation of the concentrations and stability aspects from analysing the trending behaviour of both gases and predicting whether the transformer requires maintenance or not. In the models, there are increases in the concentrations of gases on certain dates, but these are considered normal, because the increase of one of the gases is similar to the one of the other.

Ideally, in the graphic models, there would be a ratio of one gas to the other of 10%, but that does not happen, and it is enough that the relationship is fulfilled in the trend of the models. In this sense, the moments indicated above play the following role:

1. *Moment 1. Statistical control.* The variation and the stability in the concentrations of gases in a transformer. Discussion: Graphs with similar variations model stable behaviours.
2. *Moment 2. Graphical-fault relationship.* The trending behaviour of the gas concentrations to determine the stability in the graphs. Discussion: graphics show non-similar variations that lead to question the trending behaviour of them.
3. *Moment 3. Diagnostic graphic model.* Prediction in the simultaneity of variations in gas concentrations and the optimization for future trends. Discussion: stability as a quality to discern between normal and extraordinary behaviours (Pérez-Oxté & Cordero, 2016).

In summary, it can be said that the *specific situation* that this community of engineers *CCM* (ChIE) treats in their daily professional routine is to *signify certain concentrations at* certain *periods of time* to analyse and make decisions about the state of the transformer.

Staging. From the specific situation of *CCM* (ChIE) a staging was performed (Pérez-Oxté & Cordero, 2016.) Five students (men) participated. All were in the third semester of the Industrial Chemical Engineering training program in the Chemical Engineering department at the Autonomous University of Yucatan, Mexico.

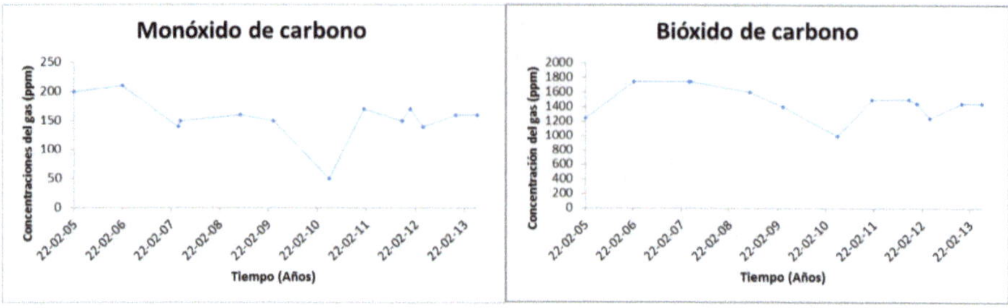

Fig. 12.5 Models of behaviour of Carbon monoxide and Carbon dioxide gases. (Source: Authors' elaboration based on data from electrical transformers)

In the research report they were called community of mathematical knowledge of industrial chemical engineering trainees. For the purposes of this discussion, the learning episodes of re-significations of the uses of mathematics between the three moments indicated previously:

- *The re-signification between Moment 1 Statistical Control and the Moment 2 Relation graphic-failures*. The behaviour of the concentrations of the gases is analysed. The presence of variations in the concentrations and the conditions in which they are present served to characterize those graphical models that predict whether the electric transformer is in good condition or not. For this purpose, arguments such as the prediction of the electric transformer were generated from looking for the reproduction of the behaviours with trends similar to ideal graphic models.

  Debate and new functions and forms: The graphics functioning focused on the behaviour of the concentrations with the form of the graphs considered as *constant behaviours* was discussed. The new way was to assess variation by comparing states, concentration, in-time intervals, and determining growths and decreases in the graphs. The new functioning of the graph was the prediction of the chemical element's concentration state in future time intervals.
- *The re-signification between Moment 2, Graphic-fault relationship, and Moment 3 Graphical diagnostic model.* The qualities of the graphic models are analysed, to make differences or similarities that help to discern normal and extraordinary behaviours. For this purpose, they are subjected to selection situations, where the graphic model would be closest to an ideal behaviour. Arguments were generated to optimize diagnostics for future time intervals.

  Debate and new functioning and forms: The argumentative functioning of the graphic on the trend with the form of extraordinary behaviours that had peaks or not, at a certain time, was discussed. The new functioning of the graphic consisted of characterizing the behaviour, at a certain interval of time and with a certain proportion, as a condition of the appropriate behaviour. The new graphical form consisted in comparing the changes of states, in a determined time.

An important aspect to note is that these training engineers were in the third semester of their studies, and the corresponding school mathematics focuses attention on the concept of function, with its properties, and its graphs, its derivability, and analytical methods for calculating maxima and minima, and for solving differential equations. In the learning of re-significations that are described above, first, this does not have an algebraic formula of the function and, secondly, with numerical data interpreted by the graphs' uses, the variation to predict and the transformation to reproduce behaviours, are valued and selected to optimize behaviours.

This is an example of the modelling category delimited to the data collection that has been presented here. The result shows us that the investigation is the revelation of functional mathematics that emerged from those communities of engineers in the scenarios described. This category of modelling involves reflecting on its meaning in school mathematics. First, that category (*trend behaviour*), a priori, is not in the school treatment of mathematics.

Then, the advancement in characterizing the category of modelling in teachers and students of the different educational levels is desired. This can be a fundamental component of the reference frame that will derive the educational change of mathematics: the decentration of the mathematical object that will allow the entry of the uses and meanings of the objects that emerge in the communities in the different scenarios.

12.7 Efficacy of the Category for Learning Mathematics

The category of trending behaviour of functions ζ(ctf) is related to the derivative and asymptotic behaviour, stability, and optimization. The *trending behaviour of functions* is intrinsic to the graph and generates a development of uses of the graph that re-signifies mathematical knowledge, by making distinctions and forming constructions as an essential part of modelling (Cordero, 2008).

In the following, there are three situations described by Cordero (2008). Each one is formed by three moments that indicate a development of the category.

1. S_1: Variation of coefficients in the transformation of a function. Moments: Changes and slope changes; Comparison between graphs and ratios; And Simulation of graphic behaviours. This situation privileges the conception of function from a relation with its behaviour. The function is understood as the relation between variables where the variability of one of them is represented, that is to say, the ways of behaving concerning the others. In this sense, it is affirmed that *the function is re-signified as an instruction that organizes behaviours*: f is transformed into $F = Af(Bx + C) + D$ (Cordero, 2008; Buendía & Cordero, 2005; Cordero et al., 2010).
2. S_2: Relations between functions through operations. Moments: The Sum of functions in a graphical context; Graphic sequences in the sum of functions

$f + h_k = g_k$; and Transformation of graphs $f + \ ? \ = h$. Here, the operations between the functions through their graphs becomes operating functions by looking for trending behaviour (Cordero, 2008), not from local or point-to-point aspects but through the shapes of the curves.

3. S_3: Stability of differential equations. Moments: (a) Sequences of first order differential equations (EDL) $y' + y = F$, varying F; (b) Sequence of first order linear differential equation, varying coefficients $ay' + y = F$; and (c) Sequence of second-order linear differential equation and generalizations: $y' = F - y$; $ay' + by'' = F - y$; $a_1y' + a_2y'' + \cdots + a_ny^n = F - y$. Here, the differential equation is seen as a relation between functions that determines behaviours. ζ(ctf) favours *identifying* coefficients in the function, *recognizing* patterns of graphical behaviour, *searching* trends in behaviours and establishing *relationships* between functions. The signs "+" and "=" are transformed into notions of tendency about the behaviours (Buendía & Cordero, 2013; Cordero et al., 2016).

The previously mentioned allows recognizing in these uses of mathematical knowledge is a category that produces a continuous material, that is, it has a genesis and a development, for it to reproduce in the educational system. For this, it must be constituted as an argument, in this case, of specific situations where "reproductions of behaviours" are meanings that provide procedures such as variation of parameters, and which involve functional tools to organize behaviours through instructions (Cordero, 2008).

12.7.1 The Transversality T_k

The *transversality of the mathematical knowledge* will be the new epistemological statuses (Fig. 12.6). To achieve the educational impact of these new statuses, it is necessary to establish *programs that will have to be systems* to favour the functionality of mathematical knowledge. These systems should articulate to science, education, and society. The immersion in the Communities of Mathematical Knowledge will be fundamental to achieve the horizontality of knowledge.

12.8 Conclusions: The Educational Change in Mathematics: A Hope

It is necessary to understand mathematics teachers as a community of mathematical knowledge that builds its own mathematical categories, particular to their environment, regulated by the mutual relations between school knowledge and reality. The reference framework (MR) will guide the necessary articulations in autonomous actions in the Mathematics teaching, hence the importance of generating research on the teachers' function (Opazo-Arellano et al., 2018), which will lead to the

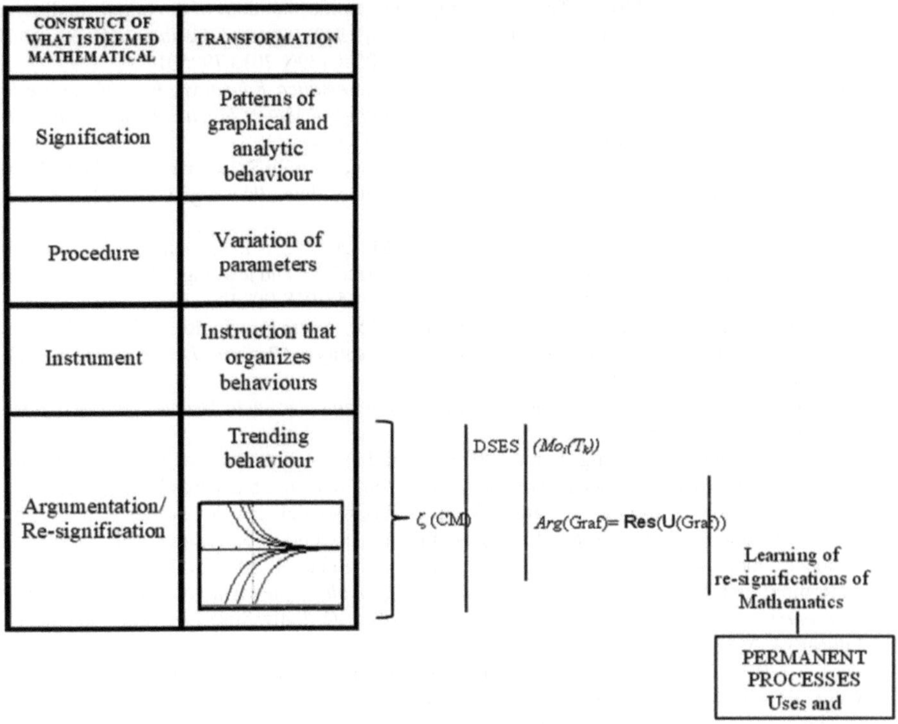

Fig. 12.6 Re-signification of **U**(Graph) and the moments of transversality $Mo_i(T_k)$. Design of School Situations of Socialization (DSES). (Source: Authors' elaboration)

permanence of the environment of the reciprocal relations that happen in the mathematical functionality, and to the educational change of the mathematics decentralizing the object (Cantoral, 2013; Cantoral et al., 2015).

References

Arendt, H. (2005). *La condición humana [The human condition]*. Paidós.

Bissell, C., & Dillon, C. (2012). Ways of thinking, ways of seeing. In *Mathematical and other modelling in engineering and technology*. Springer.

Blum, W. (2011). Can modelling be taught and learnt? Some answers from empirical research. In G. Kaiser, W. Blum, R. Borromeo-Ferri, & G. Stillman (Eds.), *Trends in teaching and learning of mathematical modelling* (pp. 15–30). Springer.

Blum, W., & Borromeo-Ferri, R. (2009). Mathematical modelling: Can it be taught and learnt? *Journal of Mathematical Modelling and Application, 1*(1), 45–58.

Buendía, G., & Cordero, F. (2005). Prediction and the periodical aspect as generators of knowledge in a social practice framework. A socioepistemological study. *Educational Studies in Mathematics, 58*(3), 299–333. https://doi.org/10.1007/s10649-005-2295-5

Buendía, G., & Cordero, F. (2013). The use of graphs in specific situations of the initial conditions of linear differential equations. *International Journal of Mathematical Education in Science and Technology, 44*(6), 927–937. https://doi.org/10.1080/0020739X.2013.790501

Cantoral, R. (2013). *Teoría Socioepistemológica de la matemática Educativa. Estudios sobre la construcción social del conocimiento [Socioepistemological theory of educational mathematics. Studies on the social construction of knowledge]*. Gedisa.

Cantoral, R. (2019). Socioepistemology in mathematics education. In S. Lerman (Ed.), *Encyclopedia of mathematics education* (pp. 1–7). Springer Nature. https://doi.org/10.1007/978-3-319-77487-9_100041-1

Cantoral, R., Moreno-Durazo, A., & Caballero-Pérez, M. (2018). Socio-epistemological research on mathematical modelling: An empirical approach to teaching and learning. *ZDM Mathematics Education, 50*(1–2), 77–89. https://doi.org/10.1007/s11858-018-0922-8

Cantoral, R., Reyes-Gasperini, D., & Montiel, G. (2015). Socioepistemología, Matemáticas y Realidad [Socioepistemology, mathematics and reality]. *Revista Latinoamericana de Etnomatemática, 7*(3), 91–116.

Carraher, T., Carraher, D., & Schliemann, A. (1997). *En la vida diez, en la escuela cero [In life ten, in school zero]*. Siglo XXI.

Common Core Standards Writing Team. (2013, April 11). *Progressions for the common core standards in mathematics high school, modelling*. Common Core Standards Writing Team.

Cordero, F. (2008). El uso de las gráficas en el discurso del cálculo escolar. Una visión socioepistemológica [The use of graphs in the discourse of school calculation. A socioepistemological vision]. In R. Cantoral, O. Covián, R. M. Farfán, J. Lezama, & A. Romo (Eds.), *Investigaciones sobre enseñanza y aprendizaje de las matemáticas: Un reporte Iberoamericano* (pp. 285–309). Díaz de Santos-Comité Latinoamericano de Matemática Educativa. A. C.

Cordero, F. (2016). Modelación, funcionalidad y multidisciplinaridad: El eslabón de la matemática y el cotidiano [Modelling, functionality and multidisciplinarity: The link of mathematics and everyday life]. In J. Arrieta & L. Díaz (Eds.), *Investigaciones latinoamericanas de modelación de la matemática educativa* (pp. 59–88). Gedisa.

Cordero, F. (2022). *La Matemática, sus Usos y Significados. Un Programa Socioepistemológico de la Matemática Educativa [Mathematics, its uses and meanings. A socioepistemological program of educational mathematics]*. Manuscrito en preparación.

Cordero, F., Cen, C., & Suárez, L. (2010). Los funcionamientos y formas de las gráficas en los libros de texto: Una práctica institucional en el bachillerato [The functions and forms of graphics in textbooks: An institutional practice in high school]. *Revista Latinoamericana de Investigación en Matemática Educativa, 13*(2), 187–214.

Cordero, F., & Flores, R. (2007). El uso de las gráficas en el discurso matemático escolar. Un estudio socioepistemológico en el nivel básico a través de los libros de texto [The use of graphs in school mathematical discourse. A socioepistemological study at the basic level through textbooks]. *Revista Latinoamericana de Investigación en Matemática Educativa, 10*(1), 7–38.

Cordero, F., Gómez, K., Silva-Crocci, H., & Soto, D. (2015). *El Discurso Matemático Escolar: La Adherencia, la Exclusión y la Opacidad [The school mathematical discourse: Adherence, exclusion and opacity]*. Gedisa.

Cordero, F., Solís, M., Buendía, G., Mendoza, J., & Zaldívar, D. (2016). *El comportamiento con tendencia, lo estable y las Ecuaciones Diferenciales Lineales. Una Argumentación Gráfica [Trend behavior, stability and Linear Differential Equations. An graphic argumentation]*. Gedisa.

Guber, R. (2001). La etnografía. Método, Campo y Reflexividad. Norma: Bogotá.

Hirsh, C. & McDuffie, A. (2016). Annual perspectives in mathematics education: Mathematical modelling and modelling mathematics. National Council of Teachers of Mathematics.

Huincahue, J. (2017). *Propuesta de modelación matemática en la Formación de Profesores y bases para una variedad de Modelación desde la Teoría Socioepistemológica [Proposal of mathematical Modelling in Teacher Training and bases for a variety of Modelling from*

Socioepistemological Theory; Doctoral dissertation, Pontificia Universidad Católica de Valparaíso]. Repositorio de producción científica asociada a proyectos y becas financiadas por la Agencia Nacional de Investigación y Desarrollo de Chile. Retrieved from http://repositorio.conicyt.cl/handle/10533/232985.

Kaiser, G., & Sriraman, B. (2006). A global survey of international perspectives on modelling in mathematics education. *Zentralblatt fur Didaktik der Mathematik, 38*(3), 302–310. https://doi.org/10.1007/BF02652813

Kant, I. (1998). *Crítica de la razón práctica [Critique of Practical Reason]* (E. Miñana y Villagrasa & M. G. Morente (Trans.), 4th ed.). Ediciones Sígueme. (Original work publisher 1788).

Lim, H., Buendía, G., Kim, K., Cordero, F., & Kasmer, L. (2010). The role of prediction in the teaching and learning of mathematics. *International Journal of Mathematical Education in Science and Technology, 41*(5), 595–608. https://doi.org/10.1080/00207391003605239

Medina-Lara, D., Cordero, F., & Soto, D. (2018). La resignificación de la media aritmética, bajo el análisis gráfico de la representación de los datos [The resignification of the arithmetic mean, under the graphic analysis of the representation of the data]. *Acta Latinoamericana de Matemática Educativa, 31*(1), 655–661.

Mendoza, E. J., & Cordero, F. (2018). La modelación en las comunidades de conocimiento matemático. El uso de las matemáticas en ingenieros biónicos. El caso de la estabilidad [Modelling in the communities of mathematical knowledge. The use of mathematics in bionic engineers. The case of stability]. *Revista Latinoamericana de Etnomatemática, 11*(1), 36–61.

Mendoza, E. J., Cordero, F., Solís, M., & Gómez, K. (2018). El Uso del Conocimiento Matemático en las Comunidades de Ingenieros. Del Objeto a la Funcionalidad Matemática [The use of mathematical knowledge in the communities of engineers. From object to mathematical functionality]. *Boletim de Educação Matemática, 32*(62), 1219–1243.

Morales, A., & Cordero, F. (2014). La Graficación-Modelación y la Serie de Taylor. Una Socioepistemología del Cálculo [Graphing-modelling and the Taylor Series. A socioepistemology of calculus]. *Revista Latinoamericana de Investigación en Matemática Educativa, 17*(3), 319–345.

Opazo-Arellano, C., Cordero, F., & Silva-Crocci, H. (2018). ¿Por qué estudiar la identidad disciplinar en la formación inicial del docente de matemáticas? [Why study disciplinary identity in the initial training of the mathematics teacher?]. *Premisa, 20*(77), 5–20.

Pérez-Oxté, I., & Cordero, F. (2016). Una epistemología basada en la transversalidad de los Usos de la gráfica de una comunidad de ingenieros Químicos industriales [An epistemology based on the transversality of the uses of the graph of a community of industrial chemical engineers]. *Investigación e Innovación en Matemática Educativa, 1*(1), 24–30.

Pollak, H. (1979). The interaction between mathematics and other school subjects. In *New trends in mathematic teaching IV* (pp. 232–248). UNESCO.

Suárez, L., & Cordero, F. (2010). Modelación-Graficación, una categoría para la matemática escolar. Resultados de un estudio socioepistemológico [Modelling-Graphing, a category for school mathematics. Results of a socioepistemological study]. *Revista Latinoamericana de Investigación en Matemática Educativa, 13*(4), 319–334.

Zaldívar, D., Cen, C., Briceño, E., Méndez, M., & Cordero, F. (2014). El Espacio de Trabajo Matemático y la situación específica de la matemática funcional: Un ejercicio de diálogo [The Mathematical Working space and the specific situation of functional mathematics: An exercise in dialogue]. *Revista Latinoamericana de Investigación en Matemática Educativa, 17*(4-II), 417–436. https://doi.org/10.12802/relime.13.17421

Chapter 13
Modelling and Anticipation of Graphical Behaviors in Industrial Chemical Engineering: The Role of Transversality of Knowledge in Learning Mathematics

Irene Pérez-Oxté and Francisco Cordero

13.1 Introduction

This document addresses the relationship between learning mathematics, university engineering and the engineering profession. The purpose comes from the disassociation between school mathematics and reality,[1] which we assume as the main challenge for learning mathematics. We consider that this disassociation is the consequence of the epistemological lack of uses and meanings of mathematical objects in school scenarios.

The aim is to know, from the social construction of knowledge, the function of the mathematical knowledge in professional scenarios, mathematics in the diverse disciplines and, in general terms, the ordinary mathematical knowledge of people. Based on the socioepistemological theory, we build theoretical frameworks to establish the pertinence of the uses of mathematical knowledge at all educational levels (elementary, secondary and higher education).

Therefore, the specific goal of this research is to consider a professional scenario of industrial chemical engineers who diagnose electrical transformers which gives evidence of the emergence of a category of use named *anticipation of behaviors* through the segmentation of times; this is called *periodization.* Based on this

[1] Reality is understood as those functional relationships between school mathematics and daily life.

I. Pérez-Oxté (✉)
Universidad Abierta y a Distancia de México, Ciudad de México, Mexico
e-mail: irene.perezoxt@nube.unadmexico.mx

F. Cordero
Centro de Investigación y de Estudios Avanzados del Instituto Politécnico Nacional, Ciudad de México, Mexico
e-mail: fcordero@cinvestav.mx

M. Rosa et al. (eds.), *Mathematical Modelling Programs in Latin America*,
https://doi.org/10.1007/978-3-031-04271-3_13

category, its transversality of mathematical knowledge in engineering students is formulated through the design of activities triggering re-significations of anticipation. The transversality of uses is confronted with the usual school mathematics to valorize the uses of mathematical knowledge in learning mathematics.

Finally, a foresight of the educational impact is formulated based on the results. For example, the uses of anticipation learning mathematics. It is worth mentioning that the courses in the area of mathematics are a filter for many careers in engineering. The curriculum at the Faculty of Chemical Engineering has many mathematical courses in the first semesters (FIQ, 2014). Nevertheless, little is questioned about learning mathematics based on the reality and the scientific-humanistic approach defined by Freudenthal (1968).[2]

Various educational models currently consider the teaching of mathematics close to the student. Modelling is an option for learning processes. For example, meaningful learning is the basis of the science, technology, engineering, and mathematics (STEM) curriculum. The core of this educational model is focused on the ability to innovate, invent and solve problems. The perception of some upper secondary school students is set on a practical dimension; they claim scientific knowledge as something that provides access to the comprehension of the world.

Similarly, they express interest in studying a STEM career because it allows them to appreciate the employment relationship (Holmegaard et al., 2014). However, the range of conceptions that educators have about this term is also questioned. For example, according to Wong et al. (2016), the lack of clarity surrounding the acronym is an issue among science and mathematics educators in England; it is often interpreted as a form of interdisciplinary work.

One of the topics in the STEM literature is the sequencing of activities of modelling in the mathematics and science curriculum; one of its purposes is to combine it with engineering based on a coherent learning experience (Sokolowski, 2018). The modelling cycle of Blum (2002) has been an important reference for this topic. The common principle of the modelling methods is to determine a solution from a situation problem inductively. The path between the problem and its solution is not trivial in many cases and sometimes the students do not benefit by from the processes of modelling (Sokolowski, 2018).

In spite of these questionings about learning mathematics of thematic areas in the STEM model, the role of mathematics at different educational levels is critical. According to Long, Iatarola and Conger (2009 cited in Loveys & Riggs, 2019), the type of mathematics courses attended during the secondary level is essential to determine the preparation of American students for the mathematics at the university level. For example, Sanders (2009) proposed an integrative STEM education which includes approaches that explore teaching and learning of two or more thematic STEM areas or between a STEM discipline and one or more school subjects.

[2]The mathematics educational approach to other sciences and social practice. In other words, considering mathematics as a human activity, mathematization from contexts, and mathematics for all students (Freudenthal, 1968).

Kang et al. (2019) highlighted the impact of science, in the life of boys and girls, being the clue for helping people and society, and in this sense, it is interesting how scientists balance their work and everyday life without giving up their personal time. The requirements of the socioepistemological vision are to construct epistemologies of uses and meanings of mathematics and establish re-significations of these uses at different educational levels, in discipline where functional knowledge is expressed[3] and in daily life of people.

For example, in Medicine, Cantoral et al. (2018) reported how variation and changes in the context of interpretation of electrocardiograms are used in the practices of cardiologists. The work developed by Mendoza and Cordero (2018) is an example of the use of the stability of movement and signals evidenced in the activity of bionic engineering underlying the Mathematical Work of Lyapunov. They proposed knowledge where stability, under the practice of controlling the temperature of a light bulb, means that the behavior of an output signal tends to reproduce the behavior of an input signal.

13.2 Professional Scenario in Engineering

With an emphasis on the transversality of knowledge, we placed this investigation between professional and academic scenarios: The Dissolved Gas Analysis (DGA) for the activity of maintenance of electrical transformers by industrial chemical engineers and students from the third semester of industrial chemical engineering. The continuity of the courses of maintenance and diagnosis plays an important role in the phases of professional development of these engineers.

In fact, the activities of their professional occupation are the anticipation of failures of electrical transformers. This is why its professional slang is provided with diagnosis techniques, for example, Dissolved Gas Analysis (DGA). However, these techniques may generate false positives, this means, they could indicate that the transformer has a flaw when it doesn't really exist. Figure 13.1 shows the mechanism for the interpretation of gases DGA.

It is important to highlight that Fig. 13.1 also exemplifies the process of diagnosing because:

> There is great uncertainty in the database of gases because of the variety of patterns and the amount of gases generated by the different types of malfunctions that are affected by many factors, among them the type of oil and its temperature, the method of sampling, the characteristics of isolation and the environment stands out. Therefore, different interpretations of failures or conflict between them can be obtained with the different techniques of interpretation of the DGA (Sarria-Arias et al., 2014).

[3]Functional knowledge, in plain terms, means a useful knowledge of people in their worldly and professional life and work (Arendt, 2005). From Socioepistemology, it is the result of the transversality of the use of people's knowledge in different situations that are re-signified.

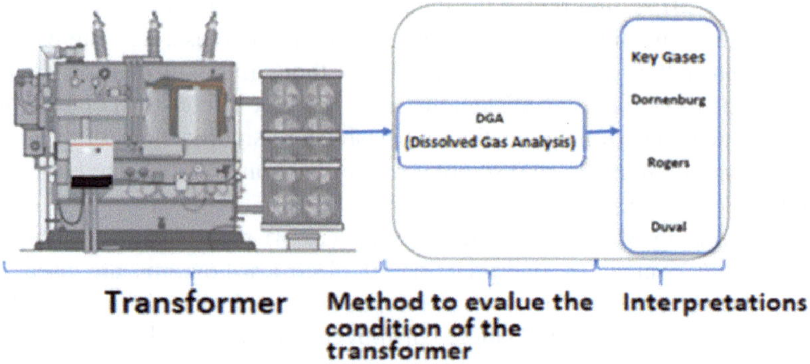

Fig. 13.1 Mechanism for the interpretation of gases DGA. (Source: Taken from Sarria-Arias et al. (2014, p. 106))

Source of the faults	Key Gas	Percentage of the current gas	Graph
Arc	Acetylene	CO: <0.1% H2: 60% CH4:5% C2 H6:1.6% C2H4: 3.3% C2H2: 30%	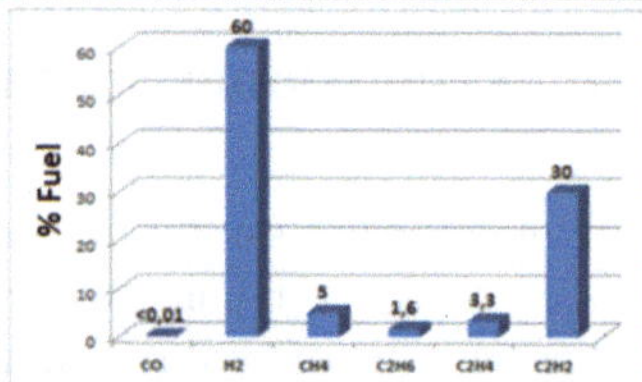

Fig. 13.2 Diagnostic approach of the key gas method. (Source: Taken from Sarria-Arias et al. (2014, p. 108))

The extraction of a sample of the oil in the transformer is required for the analysis of the dissolved gases; this sample is analyzed in the laboratory and gives useful information about the state of the oil and the identification of the type of fault. It is important to notice that different types of failures generate different gases (Pandey & Deshpande, 2012).

The interpretations are consulted in the literature and analyzed during the students' engineering studies. For example, "in the key gas method, the presence of fuel gas depends on the temperature of the oil in the transformer" (Sarria-Arias et al., 2014, p. 108). Figure 13.2 represents graphically the percentages of the current gas when the failure is an arc of the transformer.

In summary, the first step of the procedure is analyzing the different inspections, essays and necessary proofs to diagnose at an early stage the possible problems that could emerge during the life of the transformer (ABB ability, n.d.). This information is analyzed to take decisions about the need for actions on the transformer (conducting advanced essays or corrective actions). Prediction is the main characteristic of their activity. This means they examine if the transformer is fully operational by taking measurements over time. Statistics are taken periodically. The trends of the data are obtained taking into consideration time and international norms of this discipline.

Our community of study was defined by chemical engineers working in the state of Yucatan, Mexico. The issue concerning the diversity of interpretations on the readings of the statistics was attended to by building their own method characterized by the efficiency to diagnose the transformers in this state. Under this professional scenario, we decided to study the functional mathematical knowledge built by this community and its relationship with the educational field.

To this end, some activities were built with the intention of promoting the construction of meanings around the graphical behaviors of the concentrations of specific chemical elements with students of industrial chemical engineering. The transversality of knowledge is analyzed based on the *Socioepistemological Theory* with emphasis on the uses of mathematical knowledge.

13.3 A Category for Modelling and Functional Mathematics: Theoretical Framework in Socioepistemology

The empirical study reported is placed on the social construction of mathematical knowledge, specifically on the reciprocal relationship between mathematics and the industrial chemical engineers. The theoretical framework considers the Socioepistemological constructs of uses, re-significations and transversality of knowledge. They are articulated with the category of modelling whose principle is based on the problematization of the relationship between disciplinary domains of science and daily life.

The epistemology of uses of mathematical knowledge resulting from this principle involves the valorization of uses for the learning of mathematics (Mendoza & Cordero, 2018). The following constructs are defined with this theoretical framework:

- *Uses of mathematical knowledge*. They are the organic functions of the situations (***functioning***) manifested by the "tasks" that conformed the situation and the ***form*** of use will be the class of these "tasks". These tasks can be activities, actions, executions and alternation between own domains of the organism of the situation (Cordero & Flores, 2007).
- *Re-signification of uses*. When the alternation of tasks happens, a new organic function is generated, that will debate with the forms of uses. This *act of use* is the re-signification of the uses of mathematical knowledge (Cordero & Flores, 2007). Re-significations occur in specific situations. These are part of an environment and are made up by the elements: meaning, procedure and instrument which derived the argumentation of the situation (Cordero et al., 2019).
- *Transversality of knowledge*. Transversality is the re-signification of the uses of knowledge between scenarios or knowledge domains (for example, between

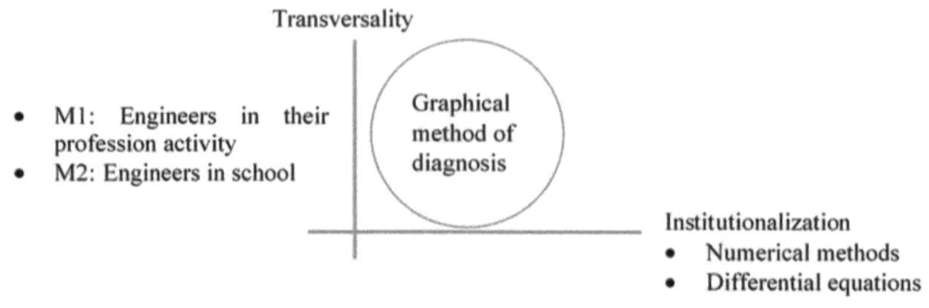

Fig. 13.3 A category of modelling in a specific situation. (Source: Authors own elaboration)

school and work, or between mathematics and engineering). It occurs in moments (M1M2 in Fig. 13.3) which are phases in the situational process (Mendoza & Cordero, 2018).

Functional mathematics is defined as the result of the transversality of use of mathematical knowledge of industrial chemical engineers (community of study). It is made up of an epistemology of uses from the activity of diagnosing electric transformers. The basis for the design of school activities is the contribution to the emergence of uses in engineering students.

The category of modelling, in the context of the diagnosis of transformers, is the re-signification of uses in the transversality from profession to school. This means, between the activity of diagnosing electric transformers and the discussion of the behaviors of graphical models. We reported the re-signification that emerges in this alternation. Figure 13.3 shows a category of modelling in a specific situation.

In this context, Fig. 13.3 also points out the graphical method of diagnosis as knowledge. Its emergence will be attended to at two moments of transversality: Moment 1, Engineers in their professional activity and Moment 2, Engineers in school (M1 and M2 shown in Fig. 13.3). The emergence of knowledge is confronted with school mathematics: on the one hand, the professional activity, a category of *periodization-anticipation* and on the other hand, school, algorithmic optimization methods.

The graphical method of diagnosis is a method built by industrial chemical engineers given the complex activity of diagnosis. The method is recognized as a functional knowledge because it is an efficient process for maintenance; it is grounded on a database registry maintained by engineers for many years. This method leads to a record named historic, it refers to the concentration of eight gases (chemical elements) represented in parts per million (*ppm*[4]) in the oil of the transformer; for example, Fig. 13.6 shows 16 records over the course of 7 years.

The use of the emergent mathematical knowledge inferred in the investigation involves dividing a portion of time into periods to discuss graphical behaviors, which is called action of periodization. These actions were carried out on the

[4]Parts per million (ppm) is a unit of measure for measuring concentration. It refers to the number of units of a specific substance for every million units of the whole set.

interpolation instrument that reproduces graphical behaviors through procedures that compare different periods based on the established behaviors.

In this regard, the periods are compared to conclude if there are stable or extraordinary behaviors. Ideally, stable behaviors are reproduced during the different periods. This could mean that the transformer is in good shape. Then, *anticipate the behaviors* is an argumentation that is built with the idea of periodization. This fact is named *re-signification of graphical behaviors in a periodization situation* in our theoretical framework.

We will now give evidence that this re-signification occurs in chemical engineers and its transversality in engineering students.

13.4 Method of Research: From Profession to School

This is qualitative research. The decisions taken in the investigation allowed the emergence of a functional knowledge to be seen, that will be described next. It was decided to examine research focused on revealing the uses of mathematical knowledge in an engineering community in their professional activities in order to identify a functional mathematical knowledge. The work selected was Torres (2013) that focused on the use of graphs to diagnose electrical transformers and its results showed that graphs are models of the behavior of chemical elements that define the life of a transformer.

Participants were industrial chemical engineering students (five students). The students' productions were analyzed based on the theoretical framework. An epistemology of uses that allowed the design of activities was built with these results. Finally, a triangulation of data between these two scenarios was conducted: interviews with chemical engineers and the productions of the engineering students. The following scheme of implementation (Fig. 13.4) shows the path followed in the work and the variables observed empirically.

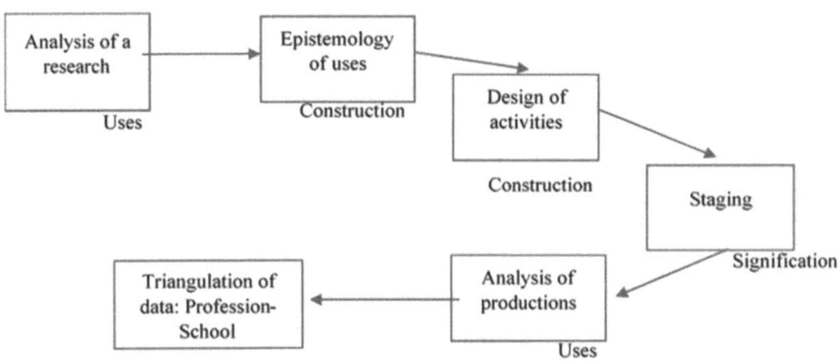

Fig. 13.4 Scheme of implementation of the research. (Source: Authors own elaboration)

The first variable observed was the uses of mathematical knowledge. The information from the interviews (of professional engineers) focused on the implementation of its method of diagnosis. Attention was paid to the mathematical instruments that allowed the construction, the meanings associated to the graphical behaviors and the processes induced by the meanings were characterized. Characterizing the graphical behaviors into regular or extraordinary was the guideline to make decisions on the transformer. We will now define the elements used in the research based on the Socioepistemological theory.

The epistemological structure underlying the characterization of the use of mathematics in a scenario, like engineering, is called specific situation (Cordero, 2001). The elements that composed this structure are (Buendía & Cordero, 2005):

- *Meanings*. Students engage in interactions with their teachers and classmates, which are reflected through situational arguments. We believe these meanings can be based on two connections that are consistent with definitions and properties as long as personal interpretation is full of images and metaphors (Bishop 1999 cited in Buendía & Cordero, 2005).
- *Procedures*. Operations induced by meanings.
- *Instruments*. System of resources to build meanings in the context of interactions (Cordero, 2001).
- *Argumentation*. It refers to the re-significations of uses that express modelling of a situation.

The conformation of an epistemology of uses (periodization situation) and the design of activities that was put into play with engineering students were conducted based on the information collected. The second variable observed was the productions of the students. The intention was to characterize the uses that emerged with the activities built.

13.4.1 Periodization Situation

The disciplinary activity of the industrial chemical engineers led them to the construction of a graphical method of diagnosis that allowed them to minimize the false positives in the electrical transformers. According to Torres (2013), the professional activity of the engineers is summarized in the following paragraph as the industrial chemical engineers anticipate fails in electrical transformers. They built a graphical method of diagnosis to fulfill this purpose. This method is composed of the history of the concentrations of eight chemical elements registered over time. The registry was built for approximately 2 years.

Graph readings were taken to achieve this activity. In this sense, the use of a graph as a statistical control is at the core of their activity. The chemical elements divided in three blocks are associated with fails that can occur in the electrical transformer. Figure 13.5 shows the relationships between chemical elements and possible fails in the electrical transformers.

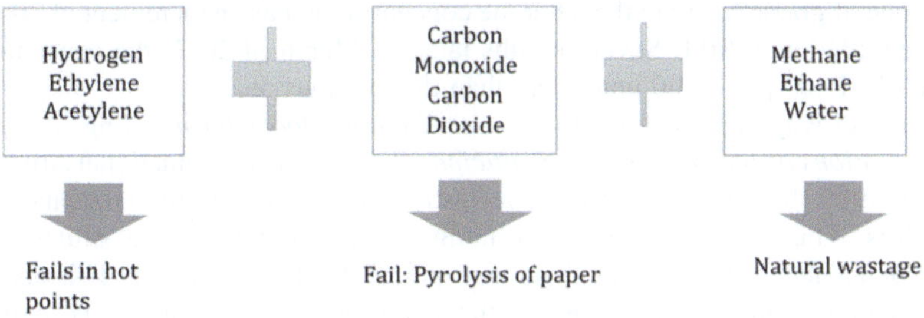

Fig. 13.5 Relationships between chemical elements and possible fails in the electrical transformers. (Source: Adapted from Torres (2013))

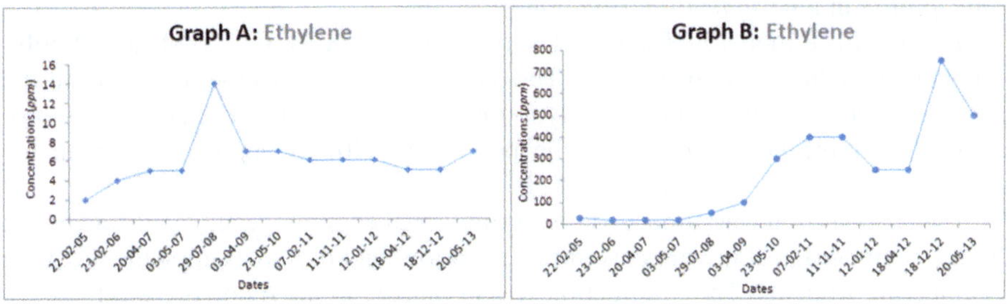

Fig. 13.6 Concentrations of ethylene and its presence is indicative of hot points. (Source: Authors own elaboration based on data from electrical transformers)

The graphical models are characterized by considering the dates when the concentrations of the chemical elements are registered. On the y-axis are the concentrations measured in parts per million (*ppm*) and the Time-axis does not have a constant scale. This is because the concentrations are registered every 6 months approximately, but if there are extraordinary behaviors it is more likely to take consecutive readings to monitor the state of the transformer. Figure 13.6 shows an example of the model of the behavior for the element Ethylene with the mentioned characteristics.

A model of deterioration in ideal conditions would imply the null existence of concentrations of the chemical element at any time. But this is not always possible due to the natural wastage and the oil reactions in the electrical transformer.

Figure 13.6 shows two graphical behaviors of Ethylene. The *presence* of this element indicates hot points in the oil of the electrical transformer. A stable behavior can be appreciated in model A because "it shows 7 ppm of Ethylene since 2009; a year later, in 2010, it still shows 7 ppm; then a year later, 2011, it shows 6 ppm, a year later, 2012, it is between 6 and 5 ppm and in the last test, May 2013, it shows 7 ppm" (free translation from Torres, 2013, p. 106).

Graph B exemplifies an extraordinary behavior. In this model, "the increase of the presence of Ethylene gas in the final analysis in the month of March is showing a hot point" (free translation from Torres, 2013, p. 111). The analysis of graph B is similar to the one of graph A, but in the latter the concentration has an increment of 200 ppm between 2010 and 2011. Seven months later, in March of 2012, the concentration has reached 400 ppm (100 ppm more than the previous record).

In a brief way, the graphical method of diagnosis *looks for meanings of specific concentrations at certain periods to anticipate behaviors.* The situational structure is composed by the interpolation of data (instrument); this is, to organize and define behaviors. This instrument gives meanings in the context of the situation, like reproducing an ideal behavior from one period to a later period. This meaning leads to a procedure which consists of the action of periodization; this is segmentation by periods and comparison of behaviors. Altogether, the instrument, its signification and the procedure generate an argumentation named *anticipation.*

In this regard, Table 13.1 shows that the activities are justified under a *category of anticipation* in a *periodization situation.*

The graphical model in Table 13.1 shows 13 data (A, B, C, ..., M) and 2 polynomials modelling the behaviors. The action of periodization is taken if the non-desirable predictive value of polynomial 1 (VP_{P1}) occurs. The aim is to reproduce a desirable behavior (P_2) in the period [13, 20] as opposed to obtain the polynomial that best fit to the data (A, B, C, ..., M).

The analysis of the arguments of the professional engineer regulated the activities built (methodological instruments of investigation) and that were implemented with engineering students. The information was organized in the following way:

Table 13.1 Periodization situation that generates an argumentation of anticipation. (Source: Pérez-Oxté (2021))

Elements of construction	Periodization situation
Meanings	Reproduction of graphical behaviors
Procedure	Comparison of periods
Instrument	Interpolation $P(x_i) = y_i$ is the polynomial that fits to the points $(x_1, y_1), \ldots, (x_n, y_n)$ n pairs of points for different x_i
Argumentation/re-signification	Anticipation Reproducing a desirable behavior (P_2) in a period 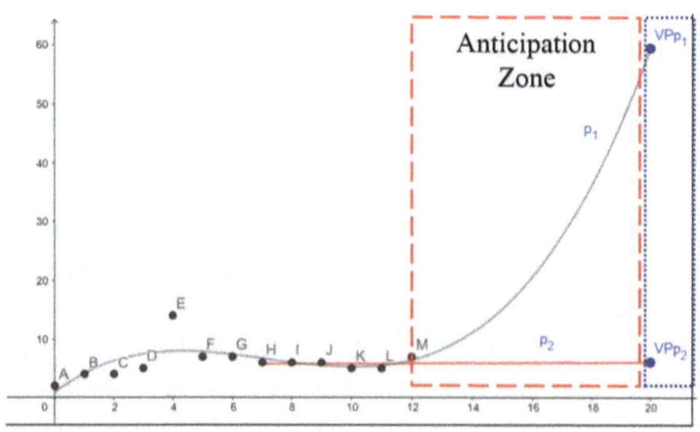

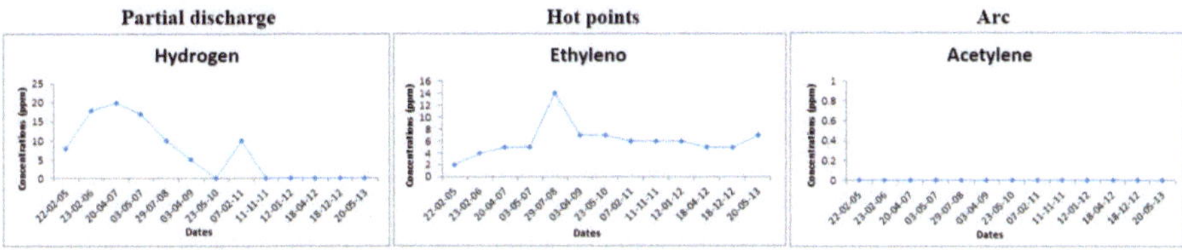

Fig. 13.7 Block 1: Gases associated to an index of failure in the electrical transformer. (Source: Authors own elaboration based on data from electrical transformers)

- *Activity 1*: Use of statistical control. Variation and stability of the concentration of gases in a transformer.
- *Activity 2*: Use of the relationship graph-failure. The trend behavior of the concentration of gases in order to determine the stability of graphs.
- *Activity 3*: Use of the graphical model of diagnosis. Prediction of the simultaneity of the variations of the concentrations of gases and the optimization of future trends.

The proposed situation for future engineers is the following:

1. Determine the state of the electrical transformer through the graphical method.
2. Indicators in the activities.

Five indicators that structured the activities are listed below. For practical effects, their nomenclature is I_i. Two compound blocks for different gases were shown to engineering students. Three gases were analyzed in block 1: hydrogen, ethylene and acetylene. Carbon monoxide and dioxide of two different transformers were shown in block 2. Figure 13.7 shows block 1 that deals with gases associated to an index of failure in the electrical transformer.

- I_1. Describe the variations of the concentration of gases (Block 1).
- I_2. Anticipate the behaviors under the following: *If Ethylene decreased from 14 to 7 ppm from year 2008 to 2009, how does the concentration of gas decrease after 2009?*
- I_3. Determine the graphical characteristics needed to conclude that the transformer is in good shape for a period of 5 years.
- I_4. Build a graph *Subtraction* that expresses the subtraction of the concentrations of hydrogen and ethylene and indicate the behavior of the resulting graph with respect to the graphs of hydrogen and ethylene.
- I_5. Analyze the behaviors simultaneously (Block 2), identify the similitudes and differences and identify the trends (Condition $\frac{y_{BCi}(t_i)}{y_{MCi}(t_i)} = 10$).

Figure 13.8 shows Block 2 that deals with graphical models of gases that should keep a ratio of 10%.

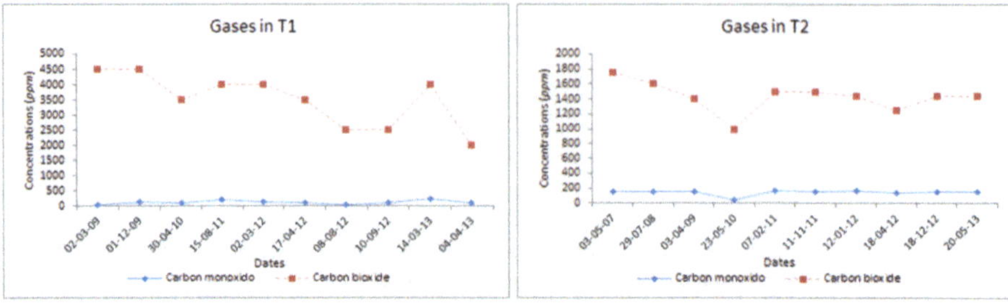

Fig. 13.8 Block 2: Graphical models of gases that should keep a ratio of 10%. (Source: Authors own elaboration based on data from electrical transformers)

13.5 Data Analysis

The resulting scientific knowledge of this research was built from the specific situation of the diagnosis of the transformers. A functional knowledge that gives meanings to the graphical behaviors emerged. The analysis of data considered the descriptions and narratives given by the engineering students.

The information was first organized by the procedures used by the engineering students: analysis of the variation by comparing states, varying or distinguishing between graphical behaviors. Then, the meanings associated were questioned: movement, patterns of behavior or adaptation. After the codification of the empirical data, the analysis was refined in terms of the *functioning* and *forms* of use of the knowledge of the situation.

13.5.1 Context of the Study

The study involved five male students was formed in the third semester of industrial chemical engineering at the Faculty of Chemical Engineering at the Autonomous University of Yucatan, Mexico. The engineering students were taking courses like Differential Equations, Phases of Equilibrium, Organic Chemistry, Analytical Chemistry and Instrumental Analysis, Numerical Methods 1 and Research Workshop 1.

Experimentation was carried out for 2 h outside of their regular courses; the activity was divided in three moments. Notes were taken during the staging and it was video recorded. The nomenclature $\boldsymbol{EI_i - E_j}$ was used to make reference to the engineering student *i* in team *j*. The aim of the staging was the characterization of the meanings that participants gave to the graphical behaviors.

Finally, the productions of the engineering students were triangulated with the arguments of the professional engineers (information collected in Torres, 2013). The hypothesis was that the mathematical arguments of professional engineers are the same in students.

13.5.2 Constructions of Knowledge in Industrial Chemical Engineering

Three moments were characterized during the three activities, each with the respective elements of construction of the engineering students. They are summarized in Table 13.2.

13.5.2.1 M1: Prediction of Graphical Behaviors for Diagnosis

The arguments of variation expressed by the engineering students focused on two aspects, the local and global variation, as a first approach to give meaning to the behaviors of the concentrations of the elements.

For example, one engineering student (EI_i) described the variation of the concentrations of Ethylene gas (transcription 1). He gives meanings regarding a permanent state by highlighting a period with no significant changes in the concentrations during a considerable period of time. He identifies a period with great changes in the concentration in parts per million to give an explanation in terms of the associated failure.

> $EI_3 - E_2$: It can be observed that from 22-02-05 to 03-05-07, there was a slight increase in the concentration of ethylene gas (from 2 to 5 ppm). After these dates, an abrupt change in the concentrations was observed; it was reflected with a considerable increase and a later decrease in a period of two years. It could be caused by a failure (a hot point). (Transcription 1: Student answer).

The global analyses were expressed in terms of stable behaviors to answer the posed questions. The transcription of the student $\boldsymbol{EI_2 - E_1}$ gives evidence of $\mathbf{I_1}$ (transcription 2), the graphical models define a predictive behavior linked with the state of the electrical transformer. In this way, a connection between behaviors and trends is evidenced with a situation of variation.

> $EI_2 - E_1$: With regard to hydrogen, we could talk in general terms about a decrease until the concentration reaches zero, so there would be NO problems of partial discharges by the end of the process. As for the inconvenient hot points associated to ethylene, we could talk in a

Table 13.2 Elements of construction of industrial chemical engineering students. (Source: Pérez-Oxté (2021))

Elements of Construction	*M1*: Prediction of graphical behaviors for diagnosis	*M2*: Trend behaviors for stable behaviors	*M3*: Selection of ideal behaviors
Meanings	Movement Permanent states	Patterns of graphical behavior	Pattern of adaptation
Procedures	Comparison of two states	Variation of parameters	Discrimination of attributes
Instruments	Amount of continuous variation	Instruction that organizes behaviors	The stable[a]

[a] What we term *the stable* is formed by an ideal object in a selection situation

general way about a constant behavior with a slight tendency to increase (Transcription 2: Student answer).

Indicators $\mathbf{I_2}$–$\mathbf{I_3}$ highlighted the confrontation between the situation of diagnosis and the prototypical graphical models in this first moment. This was expressed by the comparison between the graphical model of the ethylene gas and a lineal model in a certain period (transcription 3).

> ***EI$_2$–E$_1$***: After the fall in 2009, the concentration of ethylene was expected to continue decreasing in an almost constant way (under a lineal model); however, according to the graph this does not happen. From this year on, the concentration remained quite stable at an interval of between 6 and 8 ppm. In terms of the transformer, this is interpreted in that the presence of hot points from 2009 is the same despite the passing of time over years, which is good because the transformer remained in good shape (Transcription 3: Student answer).

Prediction of behaviors underlies in the identification of regular behaviors. The action of association of data composed by the prototypical models indicates a link between known mathematics and the situation itself.

13.5.2.2 M2: Trend Behaviors for Stable Behaviors

The students established relationships between the three graphs when they were asked to sketch the graph obtained from the subtraction (comparison) of the concentrations of the graphical models of hydrogen and ethylene ($\mathbf{I_4}$—Block 1). These relationships are translated in the following expression:

$$\textbf{\textit{Reduction of Concentration}} + \textbf{\textit{Absence of Peaks}} = \textbf{\textit{Regular Behavior}}$$

Figure 13.9 shows evidence of the production of the community and its respective justification.

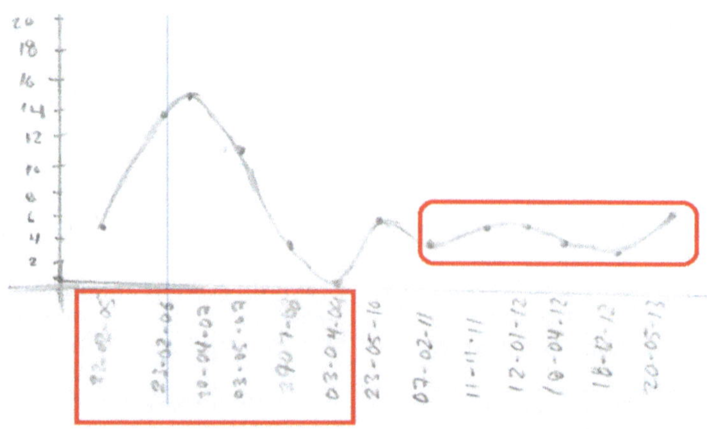

Fig. 13.9 Graphic model of subtraction of hydrogen and ethylene. (Source: Student's work)

> ***EI_I–E_I***: This is how I was taught. For example, you can say that the complete reduction on the part of the hydrogen with regard to the ethylene, cannot be seen at the beginning. If you see the graph of ethylene, it is still present in a practical sense but... once the hydrogen is reduced, the concentration of Ethylene is maintained at "zero", but in these graphs, the aim is that these peaks do not appear because if they do appear then it means that they are failing even with the reduction; the concentrations are failing because it is a comparison. In this case, the behavior of the hydrogen and ethylene are shown but no peak is shown so this means that the reduction they are making is correct (Transcription 4: Student answer).

By this time, it was concluded that the engineering students identified behaviors with trends when comparing graphical models. This also means distinguishing extraordinary behaviors.

13.5.2.3 M3: Selection of Optimal Behaviors (Ideal)

Indicator 5 (**I_5**), under the analysis of two gases, carbon monoxide or dioxide (Fig. 13.10), and the task of determining whether the gases meet the 10% relationship, give evidence of the contrast between the meaning of a *proportion of 10%* and the real data analyzed. Therefore, this specific situation demands a re-signification of their knowledge in such a way that the existence of behaviors with a tendency to this *proportion* is enough.

This trend led them to look for *similar variations* in the graphical models. The ideal is that both graphs modeled parallel straight lines, but this does not happen. A debate about the **ideal model** emerged in moment 3; they justified it by associating a model of a constant function at zero to claim that those were the best concentrations to declare that the transformer is in good condition. However, they recognized how difficult it was due to the natural wastage of the electrical transformers.

In this way, these models are also immediately discarded despite being constant models. Lineal models were considered the most suitable and closer to the reality of the life of electrical transformers because it is desirable to consider wear represented by increasing constant and non-exponential behaviors. The following excerpt (transcription 5) gives evidence of the existence of ideal graphical models like *the best that could happen*.

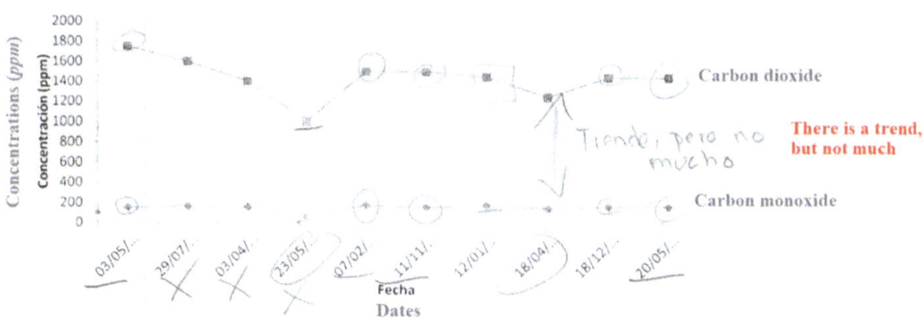

Fig. 13.10 Gases that should keep a ratio of 10%. (Source: Student's work)

> ***EI_2-E_1***: There is a trend of increase or being maintained in range; so, it could not be possible to bring the concentrations of ethylene gas down to zero; maybe it could reach a minimum and be kept there (Transcription 5: Student answer).

Engineering students idealized behaviors and compared them with the models of the solved activities, the comparison of behaviors over periods of time is among their procedures.

13.5.3 Re-significations Between Moments

In the following sections there is a discussion about the re-significations between moments.

13.5.3.1 Re-signification Moment 1–Moment 2

Discussion. Description of the variation of the concentrations through the comparison of states in a local or global way. The graph was considered as a *model of behavior* instead of the representation of a function.

The presence of variations in the concentrations and the current conditions were used to characterize the graphical models that predicted if the electrical transformer was in good shape or not. Argumentations, like the prediction of the state of the electrical transformer from the reproduction of behaviors with similar trends of ideal models, were created to this end.

Debate and New Functioning and Forms. Table 13.3 shows the use of graph as statistical control.

13.5.3.2 Re-signification Moment 2–Moment 3

Discussion. The qualities of the graphical models are analyzed; this means, identifying normal behaviors with ideal trends or extraordinary behaviors that could indicate a failure in the transformer, from where it is possible to predict behaviors. The analysis of the qualities of the graphical models is carried out by identifying differences and similarities to discriminate between normal and extraordinary behaviors. To this end, they are submitted to situations of selection, where the graphical model is the closest to the ideal model. Argumentations were generated to optimize diagnosis for future intervals of time.

Table 13.3 Use of graph as statistical control. (Source: Students' work)

Fu	Fo
Predict stable behaviors.	Comparison between the concentration of gases in the oil in the electrical transformer.

Table 13.4 Use of the graph in the relationship graph-failures. (Source: Students' work)

Fu	Fo
Identify trend behaviors with a specific condition.	Analysis of trend behaviors from its attributes.

Table 13.5 Use of graph as a model of diagnosis. (Source: Students' work)

Fu	Fo
Build graphical models with trends.	Distinction of attributes in graphical models.

Debate and New Functioning and Forms. The argumentative *functioning* of the graph on the trend of the *form* of extraordinary behaviors that showed peaks or not in a certain period of time was discussed. The new *functioning* of the graph consisted in the characterization of the behavior in a certain period of time and with specific proportion as a condition for the behavior. And the new graphic *form* consisted in the comparison of changes in the states in a specified time. Table 13.4 shows the use of the graph in the relationship graph-failures.

Table 13.5 show the use of graph as a model of diagnosis.

The argumentation of anticipation was generated with the *functioning* and *forms* identified in the productions of the students. A process of periodization where ideal graphical behaviors are proposed for reproduction in immediate periods emerged in the situation. Part of the structure of the situation was the analysis of behaviors and its comparison of different periods.

13.6 Confrontation of Uses of Anticipation and Usual School Mathematics[5]

By way of conclusion, the category of mathematical knowledge built from the emergence of uses of mathematical knowledge is not focused on the mathematical object; in fact, its focus is on the uses and meanings of the mathematical objects. The category of uses is a plural epistemology (professional and school scenarios) and of transversality of knowledge. The focus is on the re-signification of uses. Then, the uses of anticipation are confronted with the usual school mathematics.

Engineering students did not use the mathematical knowledge from their mathematics courses when facing the periodization situation in spite of their knowledge of the concepts of function and derivative and optimization methods to determine

[5]Understand it by the usual mathematical content that appears in textbooks. For example, in the mathematical texts used in engineering, there is a prevalence of the basic mathematical concepts of calculus and statistics as mathematical objects from which properties, definitions or theorems, in general terms, are derived without considering the meanings of the uses of the mathematics of engineers (Cordero et al., 2019).

derivatives, critical points, maximums and minimums. Instead, in an autonomous way, they re-signified the variations to determine regular or irregular behaviors, and then, deduce an anticipated behavior. The instruments were the numerical data and the graphical behavior and not the functions (formulas) nor the derivatives (algorithms). Table 13.6 exemplifies the previous discussion.

The impressions of the engineering students of the activities carried out were captured as a result of the methodology of *focus group*. Trends and regularities of their points of view were classified in *two different aspects of seeing mathematics* and *mathematics in engineering*.

13.7 Reflections and Conclusions

The category of modelling was characterized to valorize the functional relationship between school mathematics and a specific engineering situation. This relationship provided the meanings associated to the graphic behavior in use. We highlight the following aspects:

1. Absence of peaks in the graphical models with regular behaviors.
2. Constant behavior in periods as the ideal behaviors in spite of variations.
3. Trends of graphical behaviors in the reduction of the concentration of gases.

The focus of the modelling approach in this article is the function of mathematics as an organic incorporation (Mendoza & Cordero, 2018). There were also two aspects that professional engineers discussed explicitly in their graphical method that the engineering students did not consider:

1. Discussion of the *Constant*

    EI_1-E_1: During the period 2005–2007, a considerable increase in the concentration of hydrogen is shown; then a decrease in the levels of concentration is shown for the period 2008–2010 and it spiked again in 2011 reaching a concentration of 10 ppm and regulated remaining at zero, namely constant, by the end of the same year.

2. Discussion of Behaviors

    On one side, in the first moment of the activity and explained by the inheritance of school mathematics, the industrial chemical engineering students made reference to the *constant* although there was no explicit constant function in the model (Transcription 3). This reference is re-signified when the students made reference to *behaviors that vary at an interval*, *behaviors with trends*, *normal behaviors*. On the other side, there is a functional justification to idealize behaviors in the activities, for example, by sketching the graph that comes from the subtraction between two other graphs.

    This is, what the ideal *subtraction graph* would be, students re-signified the idea of *constant* to establish a functional relation for this specific situation: ***Reduction of the Concentration*** + ***Absence of Peaks*** = ***Regular Behavior***. With these antecedents, we claim that meanings were given to graphical behaviors and that the constant function to monitor behaviors was decentralized, given the posed activities.

Table 13.6 Examples of confrontation between the uses of anticipation and school mathematics. (Source: Authors own elaboration)

School mathematics	Uses of anticipation
Linear models of approximation	*Irregular/regular behaviors*
The straight line that has the best fit to the points (x_1, y_1), (x_2, y_2), (x_3, y_3) is of the form $y = c_1 + c_2x$ so the problem focuses on finding the best values of c_1 and c_2 that substituting in the model gives: $c_1 + c_2(x_1) = y_1$ $c_1 + c_2(x_2) = y_2$ $c_1 + c_2(x_3) = y_3$	A regular behavior is to avoid exponential increases, contrary case of maintaining it constant or with few variations. It can be said that there is a modelling $P_1(t_i) = 0$ for $0 \leq t_i \leq 7$ in graph B of Fig. 13.6. However, there is an increasing behavior of the data $P_2(t_i) < P_2(t_{i+1})$ in the period $8 \leq t_i \leq 11$. 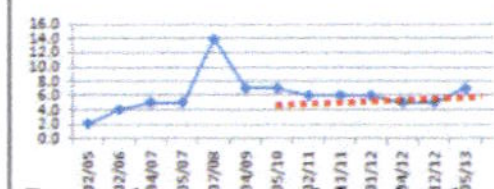 It tends to $y = 6$ in the last 5 years.
Definition of limit	*Behaviors with a tendency in a period*
Let f be a function defined in an open interval that contains c (with the possible exception of c) and let L be a real number. The statement $\lim_{x \to c} f(x) = L$ means that for every $\varepsilon > 0$ there is $\delta > 0$ such that if $0 < \lvert x - c \rvert < \delta$, then $\lvert f(x) - L \rvert < \varepsilon$.	Tendencies in the numerical and graphical analysis are sought. The values in the graph oscillate around *5 ppm* in the last six dates. The trend of the behavior to stabilization and the possibility of repeating on the following dates are elements of construction of the use of anticipation. 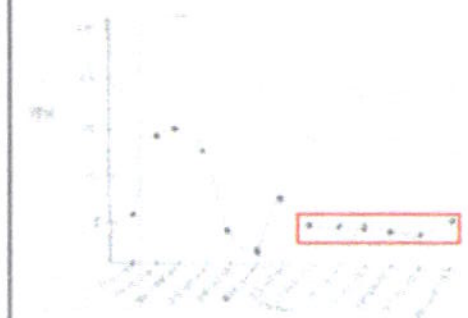 $8 \leq t_i \leq 13$ $y_i \in (5 - \varepsilon, 5 + \varepsilon)$

(continued)

Table 13.6 (continued)

School mathematics	Uses of anticipation
Second derivative criterion	*Ideal behaviors and their reproduction*
Determine the relative extreme points of a function f such that $f'(c) = 0$ and whose second derivative exists in an open interval that contains c. If $f''(c) < 0$, then there is a maximum at the point $M(c,f(c))$ If $f''(c) > 0$ then there is a minimum $m(c,f(c))$	The first action is periodization. Afterwards, the behaviors can be modeled and select the ones that explain more. For example, a polynomial of 12th degree fits with the 13 data of the graph in Fig. 13.7 but it does not give a good explanation of its behavior. The polynomial $P_1(x) = 0.03x^3 - 0.63x^2 + 3.82x + 0.96$ gives a better explanation in general terms. But the linear model $P_2(x) = 0.03x + 5.53$ in the period [7,12] is the one that describes the desired behavior in the following periods. Predicting the value at the increments is confirmed after predicting $x = 20$ in both models $P_1(20) = 59.56$ and $P_2(20) = 6.16$ confirms the increments. The first one confirms a drastic increment with regard to the data that oscillated between 6 and 7 ppm in the last period. On the other hand, the model in P_2 shows a slight increase so it can be accepted as the model for a desirable behavior.

Acknowledgements This research has been supported by CONACYT with the Project A category of mathematics modelling. The epistemological plurality and the transversality of knowledge: learning the meanings of mathematics in engineering and at different educational levels. Code 0284259.

References

ABB Ability (n.d.). Retrieved from https://new.abb.com/service/es/training/abb-university/spain/formacion-sobre-sistemas-electricos-de-potencia

Arendt, H. (2005). *La condición humana*. Ediciones Paidós Ibérica, S.A. ISBN: 9788449318238.

Blum, W. (2002). ICMI Study 14: Applications and modelling in mathematics education. Discussion document. *Educational Studies in Mathematics, 51*(1/2), 149–172.

Buendía, G., & Cordero, F. (2005). Prediction and the periodic aspect as generators of knowledge in a social practice framework. A Socioepistemological study. *Educational Studies in Mathematics, 58*(3), 299–333.

Cantoral, R., Moreno-Durazo, A., & Caballero-Pérez, M. (2018). Socioepistemological research on mathematical modelling: An empirical approach to teaching and learning. *ZDM, 50*(1), 77–89. https://doi.org/10.1007/s11858-018-0922-8

Cordero, F. (2001). La distinción entre construcciones del Cálculo. Una epistemología a través de la actividad humana. *Revista Latinoamericana de Matemática Educativa, 4*(2), 103–128.

Cordero, F., Del Valle, T., & Morales, A. (2019). Usos de la optimización de ingenieros en formación: El rol de la ingeniería mecatrónica y de la obra de Lagrange. *Revista Latinoamericana de Investigación en Matemática Educativa, 22*(2), 185–212. https://doi.org/10.12802/relime.19.2223

Cordero, F., & Flores, R. (2007). El uso de las gráficas en el discurso matemático escolar. Un estudio socioepistemológico en el nivel básico a través de los libros de texto. *Revista Latinoamericana de Investigación en Matemática Educativa, 10*(1), 7–38.

Facultad de Ingeniería Química [FIQ]. (2014). *Propuesta de modificación del plan de estudios de la licenciatura en ingeniería química industrial*. Recuperado el 30 de Mayo de 2015 de http://www.ingquimica.uady.mx/documentos/planes/iqi_mefi_plan_2014.pdf

Freudenthal, H. (1968). Why to teach mathematics so as to be useful. *Educational Studies in Mathematics, 1*, 3–8.

Holmegaard, H. T., Madsen, L. M., & Ulriksen, L. (2014). To choose or not to choose science: Constructions of desirable identities among young people considering a STEM higher education programme. *International Journal of Science Education, 36*(2), 186–215. https://doi.org/10.1080/09500693.2012.749362

Kang, J., Hense, J., Scheersoi, A., & Keinonen, T. (2019). Gender study on the relationships between science interest and future career perspectives. *International Journal of Science Education, 41*(1), 80–101. https://doi.org/10.1080/09500693.2018.1534021

Loveys, B., & Riggs, K. (2019). Flipping the laboratory: Improving student engagement and learning outcomes in second year science courses. *International Journal of Science Education, 41*(1), 64–79. https://doi.org/10.1080/09500693.2018.1533663

Mendoza, J., & Cordero, F. (2018). La modelación en las comunidades de conocimiento matemático. El uso de las matemáticas en ingenieros biónicos. El caso de la estabilidad. *Revista Latinoamericana de Etnomatemática, 11*(1), 36–61.

Pandey, R., & Deshpande, M. T. (2012). Dissolved gas analysis (DGA) of mineral oil used in transformer. *International Journal of Application or Innovation in Engineering & Management, 1*(2), 208–212. Retrieved from http://www.ijaiem.org/volume1Issue2/IJAIEM-2012-10-30-069.pdf, ISSN 2319-4847

Pérez-Oxté. (2021). *Anticipar-periodizar: Una socialización de saberes matemáticos entre la ingeniería y la docencia*. Tesis de doctorado no publicada, CINVESTAV—IPN, México.
Sanders, M. E. (2009). STEM, STEM Education, STEMmania. *The Technology Teacher, 68*(4), 20–26.
Sarria-Arias, J., Guerrero-Bello, N., & Rivas-Trujillo, E. (2014). Estado del arte del análisis de gases disueltos en transformadores de potencia. *Revista Facultad de Ingeniería, 23*(36), 105–122. https://doi.org/10.19053/01211129.2716
Sokolowski, A. (2018). *Scientific inquiry in mathematics—Theory and practice, a STEM perspective* (1st ed.). Springer. https://doi.org/10.1007/978-3-319-89524-6_4
Torres, L. (2013). *Usos del Conocimiento Matemático. La Simultaneidad y Estabilidad en una Comunidad de Conocimiento de la Ingeniería Química en un Escenario de Trabajo*. Tesis de Maestría no publicada. CINVESTAVIPN, México, D.F.
Wong, V., Dillon, J., & King, H. (2016). STEM in England: Meanings and motivations in the policy arena. *International Journal of Science Education, 38*(15), 2346–2366. https://doi.org/10.1080/09500693.2016.1242818

Chapter 14
Categories of Modelling and Reproduction of Behaviors in Other Disciplines: Teaching Mathematics in Engineering

E. Johanna Mendoza-Higuera, José Luis Morales-Reyes, Falconery Giacoleti-Castillo, and Francisco Cordero

14.1 Introduction

Mathematics has been a compulsory subject in the basic cycles of engineering education. It has a defined status and function that is even visible from the annals of engineering at the École Polytechnique in France and the Escuela Real de Minas in Mexico. The curricular programs, in general, are organized into three groups: basic sciences, engineering sciences, and professional courses. In that way, the fundamental knowledge is initially offered, which will be applied later in the subjects specific to each program and hence allow the development of knowledge within engineering (Cajas, 2009 cited by Mendoza-Higuera, 2020).

In this regard, Mathematics Education, in the last four decades, has been concerned with reflecting on the teaching of mathematics in as engineering, by seeking answers to questions such as: *What mathematics should future engineers learn?*, *What mathematical knowledge do engineers use in their professional practice?*, and *What is the role of mathematics in engineering education?*

In our case, we recognize a reciprocal and horizontal relation between mathematical and engineering knowledge (Mendoza-Higuera, 2020; Giacoleti-Castillo,

E. J. Mendoza-Higuera (✉)
Unidades Tecnológicas de Santander, Bucaramanga, Santander, Colombia
e-mail: edithmendoza@correo.uts.edu.co

J. L. Morales-Reyes
Universidad de Costa Rica, San José, San José, Costa Rica
e-mail: jose.moralesreyes@ucr.ac.cr

F. Giacoleti-Castillo · F. Cordero
Centro de Investigación y de Estudios Avanzados del Instituto Politécnico Nacional, Ciudad de México, Mexico
e-mail: falconery.giacoleti@cinvestav.mx; fcordero@cinvestav.mx

M. Rosa et al. (eds.), *Mathematical Modelling Programs in Latin America*,
https://doi.org/10.1007/978-3-031-04271-3_14

2020; Morales, 2020): one favors the development of the other and vice versa. This view leads to the recognition that there are uses of mathematical knowledge in engineering that make it meaningful from different knowledge communities (academic, school, and professional) and that they are excluded in school mathematics. In this sense, we are interested first in showing the reciprocal relation between mathematics and engineering as scientific disciplines.

Subsequently, we will delve into the problems of teaching mathematics in the training of future engineers. In addition, we will show that in the study of the uses of mathematical knowledge in engineering communities—for example, in electronics—the categories of modelling and reproduction of behaviors typical of Socioepistemological Theory in Mathematics Education emerge (Cordero, 2022) to account for the transversality of knowledge. Finally, we report on some results of implementing a design of a socialization school situation where those categories are the guiding thread of the proposed tasks.

14.2 The Teaching of Mathematics and Engineering

If we consider the engineer as a builder of artifacts that improve the quality of life of humans and make possible what humans cannot do with their bodies and their strength, in the terms of Aracil (2017), engineering has been developing for thousands of years. The invention of artifacts and systems such as the wheel, the steam engine, the computer, the internet, among others, are responses to this need to improve their lives. The Neolithic Revolution began with the sedentary lifestyle and the origin of agriculture (purely technical activities), but also the concern to protect and organize their settlements to live better. Later, with people already established in cities and kingdoms, the need to build bridges, roads, fortifications, artillery, and other instruments needed for warfare arose.

Likewise, it was necessary to establish an economy that would allow facing the needs of society, such as the construction of public works (activities where the progress of technology is perceived). In this sense, different situations can be described that account for the emergence of mathematical knowledge and engineering from practical activity. For example, G. Monge, an innate engineer who, while studying to become a master builder, solved a problem of the location of a fortification that consisted of preparing works so that no part of them would be exposed to direct fire from the enemy. Monge solved it through a procedure of his invention that included geometric representation, breaking with the arithmetic operations traditionally used (Mendoza-Higuera, 2020).

Another example is the case of Diego de Guadalaxara and Tello, who, without having a scientific background, developed instruments for the exploitation of mines. According to Cházaro (2011 cited by Mendoza-Higuera 2020):

> (...) his questions were not aimed at explaining nature or finding the causes of the natural; his interests in mathematics were the same as those of his creators: to measure the land to calculate mine shots, to avoid landslides, and to dewater them (p. 745).

In the two cases presented, the mathematics they developed was based on engineering, either in the construction of works or in mining. On the other hand, from the scientific activity, works such as those of Newton, Cauchy, Fourier, Riemann, Lebesgue, Euler, Lagrange, among others, show the relation that has existed since the beginning between engineering and mathematical scientific activity. For example, Fourier tackled the phenomenon of heat propagation, since this was a relevant problem for the metallurgical industry of the time.

It was necessary to know the interior temperature of the Earth and the way it changes with time and depth. Thus, he studies the behavior of heat propagation in solids through the stable and permanent flow of time. The mathematical solution to the problem is an infinite trigonometric series that represents a system of temperatures, which by its physical character establishes the convergence of the series as they cannot be infinite (Farfán, 2012). Similarly, Cordero (2003) in the works of Cauchy (1882), Riemann (1898) and Lebesgue (1926) identified a pattern that expresses the accumulation of what flows in a region through the comparison of two states which leads to the formulation of the Fundamental Theorem of Calculus (Cordero, 2003).

And the last among a variety of examples that we could present here, Newton, who sought to model, anticipate, and predict natural phenomena, which would finally help to solve technological problems, ends up formulating what is now known as the Taylor Series (Cantoral, 2019). In this way, the relation between mathematics and reality is explicit, exemplified in the study of natural phenomena, which not only provided scientific knowledge but also practical knowledge, hence the link with engineering and technology.

In summary, we emphasize two points. First, mathematics and engineering have lived and still live a reciprocal relationship in terms of their use and construction of knowledge; both have been favored by the development of the other, which ends up providing meanings for mathematics from both disciplines; secondly, the nature of engineering is based on practical or technological work, where knowledge emerges that responds to functional justifications and that account for the use of mathematical knowledge as a tool.

However, in engineering education, there has been, from the beginning, a discussion that focuses on establishing what knowledge engineers need to know: is it scientific or theoretical knowledge that should prevail, or is it practical knowledge? For example, the École Polytechnique was founded in France to train engineers closer to the sciences than to craftsmanship. Romo-Vázquez (2014) based on the work of Belhoste (1994), analyzes the educational models proposed at the École Polytechnique and indicates that it moves from an encyclopedist model to an analytical one and then to one focused on applications.

The first seeks to show the alliance between the sciences and the arts. The second focuses on training in mathematics, especially in mathematical analysis as a basis for physics, mechanics and geodesy. The third focuses on content that was useful for its application. Whatever the model, it generally reflects the search for a balance between the theoretical and the practical. Currently, when analyzing the curricula of various universities, it is observed that the contents are organized into three groups: basic sciences, engineering sciences, and professional courses.

Thus, initially, basic or fundamental knowledge is offered, which will later be applied in the subjects of each program and which in turn allow the development of knowledge within engineering (Cajas, 2009 cited by Mendoza-Higuera, 2020). In most cases, in the second cycle subjects, the lecture is strengthened with laboratories or practical activities that allow the student to apply the knowledge learned.

However, in the first cycle, specifically in mathematics, its teaching is not distinguished according to the specificities of each of the engineering careers or programs (Cajas, 2001), better yet, in some cases, the teaching only focuses its attention on meanings, procedures, and arguments from mathematics but not from its use in other disciplines, from the meanings that emerge in the situations that an engineer or engineering student faces according to his or her specialty.

All of the above seems to respond to a philosophy of mathematics teaching where the academic knowledge proper to mathematical activity prevails and ignores the potential of the knowledge that emerges from doing, proper to human activity. It is assumed that the knowledge produced from doing does not have the same character as scientific knowledge, perhaps that is why basic sciences are required to provide a reasoned logic to the knowledge produced from engineering practices. Mathematics has even been called a service discipline for engineering in that it provides basic tools to be applied later in the solution of engineering problems (Howson et al., 1988).

From *Socioepistemological Theory in Mathematics Education* (TSME for its acronym in Spanish) (Cantoral, 2019; Cordero, 2022) we recognize that school mathematics is goes through a duality rather than a confrontation. In the development and innovation of techniques (either from science, with its support or without it) is assumed that there is a construction of mathematical knowledge, certainly not in the form that mathematical activity demands, but yes, a functional knowledge that includes meanings, arguments, and procedures typical of engineering.

The dual nature of school mathematics consists of understanding that there are scenarios where mathematics is the object of study and others where it is not. That is to say, school mathematics treats mathematics as an object of study, but in other areas, it is an *instrument* (Cordero, 2016). There are professional users of mathematical knowledge, who are not mathematicians and use mathematics but not as an object of study. Figure 14.1 shows the duality of school mathematics.

In these scenarios where school mathematics is taken as an instrument, functional justifications prevail, i.e., all knowledge that accounts for the meanings, procedures, and arguments, comes from its disciplinary practice and not from mathematics per se; in any case, these justifications are different to the reasoned justifications proper to mathematical activity. The TSME, as a model of practices, articulates, and formulates epistemologies from functional and reasoned justifications; and thus clarifies the dual nature of mathematics (Cordero & Flores, 2007; Cordero et al., 2010).

The role of school mathematics in the development of engineering and the training of its human resources, responds to a character of the instrument, where

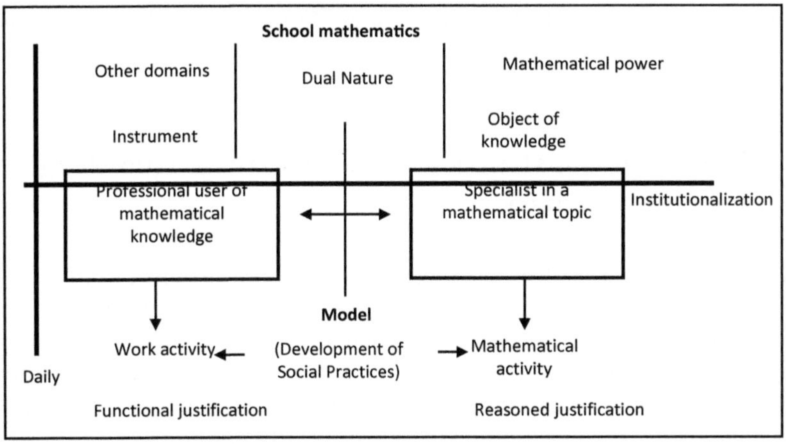

Fig. 14.1 The duality of school mathematics. (Source: Cordero, 2016, p. 70)

the mathematical knowledge that is built there is the expression of its intimacy and corresponds to a functional nature.

This knowledge is built by the engineer in his practice, in his condition of a situated subject, who belongs to a culture, to a community, to a way of understanding and developing the work of his profession, to a way of producing knowledge. Thus, our emphasis is on functional mathematics and on accounting for those categories that emerge in these practices (Mendoza-Higuera, 2020).

In short, school mathematics has the status of fundamental basic science in the training of engineers, and its function is that of a tool at the service of other disciplines. This status and function reveal a problem: the teaching of mathematics in engineering is devoid of the meanings of mathematical knowledge that emerge from engineering practice and is far from the reality of the learner.

Finally, this status and function have caused phenomena in school mathematics, which are expressed in the construct *mathematical school discourse* (dME for acronym in Spanish) (Soto & Cantoral, 2014). This specifies that school mathematics does not have a frame of reference that integrates uses and meanings to mathematical objects inside and outside mathematics.

In this regard, Cordero (2001) points out that the reconstruction of meanings provides categories of mathematical knowledge to form the frame of reference, and, for this reconstruction, the source is the human activity. In 2008, Cordero says that the use of graphs is a category that confronts the dME and in 2016 manages to articulate the frame of reference with a type of modelling that emerges in certain communities.

The dME provokes ignoring the dual nature of school mathematics and the construction of situational, and therefore functional, knowledge in human groups. Mendoza-Higuera et al. (2018) give empirical evidence of that fact with diverse communities of engineers; Giacoleti-Castillo (2020), reveals the *category tendential behavior* as a resignification of the Laplace Transform in a community of electrical engineers, and Morales (2020) offers school situations in higher education to *re-signify* the concept of derivative.

14.3 The Social Construction of Mathematical Knowledge and Its Categories

The TSME seeks to establish a horizontal and reciprocal relationship of school mathematics within mathematics and outside it, specifically in communities of other disciplines such as engineering. Therefore, epistemological plurality and transversality of knowledge, which are promoted by the Forgotten Subject and Transversality of Knowledge Program (SOLTSA for acronym in Spanish) developed within TSME, are unavoidable constructs for this task.

When *epistemological plurality* is assumed, diverse mathematics is admitted, establishing meanings by disciplinary communities, scenarios, and specific situations experienced by people, according to their work and needs. For this reason, we agree to call this knowledge mathematical, which in turn encompasses what we call *functional mathematics*. Its dimension, at the educational level, will have to relate knowledge with institutional knowledge and with reality or realities.

Functional mathematics comes in when people bring their mathematical knowledge into play to face the situations they live in, which entails the use of mathematical knowledge and, in turn, the resignification of these uses. When they move through situations, domains, and scenarios, they show the transversality of knowledge (Cordero, 2022).

In this sense, the SOLTSA works on two lines of research: in the first one, the categories of mathematical knowledge are problematized in different domains where attention focuses on revealing the uses of mathematical knowledge and its resignifications that are condensed in the categories of mathematical knowledge; and, in a second line, multiple factors are configured that contribute to the teaching of mathematics such as identity, inclusion, socialization, and others, through the design of socialization school situation (Cordero, 2022).

So, to address the problem of teaching and learning mathematics for engineering students, the conformation of a frame of reference (FR) is proposed that recognizes the functionality through transversality and the plurality of mathematical knowledge (multidisciplinary domains of knowledge), all in specific situations.

Thus, in a dialectic between mathematics and reality (habitual in people's daily lives), both pieces of knowledge are mixed, and become one; rather, they are transformed into a unity of knowledge, of knowledge in use by people.

The transformation decentralizes the object and the uses are re-signified between situations and between scenarios: the school-academic; the work-profession; and the city-everyday life. With the school situations of socialization, the learning of mathematical resignifications will take place in permanent processes (uses and meanings) as opposed to terminal objects (concepts and definitions) (Cordero, 2022).

The action of functional mathematics consists of re-signifying the uses of mathematical knowledge in the transversality of knowledge proper to the communities of knowledge. These actions take place in specific situations. Specificity rules the uses of mathematical knowledge. It is agreed to call these uses categories of mathematical knowledge.

In this context, Cordero (2022) and Mendoza-Higuera (2020) explain that a category of mathematical knowledge is sui generis due to the epistemological, ontological, and educational stance of the research approach, and state that:

> (...) the category is a process that accompanies the epistemological plurality and the transversality of knowledge that define the mathematical functionality of the communities of mathematical knowledge that take place at school, at work, and in the city (Mendoza-Higuera, 2020, p. 63).

As a whole, the epistemology of uses and its resignifications shape the categories of mathematical knowledge as a social material product to be favored in teaching and learning, which become visible in the design of socialization school situation. In order to shape elements that contribute to the alliance of quality with the teaching of mathematics and that in turn specify the relevance of the new program under development and its educational impact, it has been considered to evidence the socialization to counteract the exclusion of the social construction of knowledge that is visualized from the problematic.To begin to materialize this impact, we seek to constitute processes that transform de school mathematics discourse; we call them design of socialization school situation (DSES). This is based on epistemologies that favor the uses of mathematical knowledge, and also on a perspective which, on the one hand, allows counteracting the phenomena provoked by the *mathematical school discourse* (dME) and, on the other hand, allows the analysis of the participants' resignification process (Morales, 2020; Morales & Cordero, 2020).

These are aimed at the valorization of the epistemology of uses. To this end, the valorization processes are explained with the dialectical perspective of exclusion-inclusion (Soto & Cantoral, 2014). That is to say, people necessarily confront the mathematical school knowledge with their own mathematical knowledge.

This dialectic then composes a model in which the transition between the dME and the *social construction of mathematical knowledge* (CSCM for acronym in Spanish), plays an indispensable role, the latter usually excluded in school mathematics.

At this point, it is worth distinguishing that CSCM is not restricted only to the interaction between people, but as mentioned by Cordero et al. (2015):

> We consider the interactions between individuals (we care about which individuals and their historical processes), the processes of debates and negotiations that the community undergoes to institutionalize knowledge (institutional process), and the functionality of this knowledge in a specific context and situation (functional process), characteristic of social practice (p. 69).

By understanding the processes of exclusion and inclusion as a dialectical relationship, it is pointed out that one does not live without the other, and that exclusion will

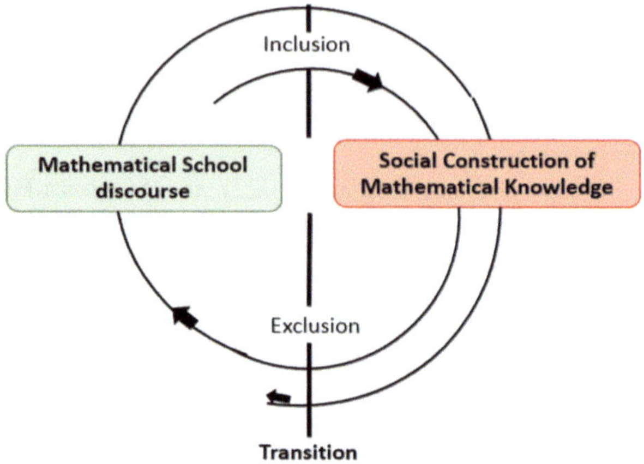

Fig. 14.2 Model of the exclusion-inclusion dialectics. (Source: Soto, 2014)

be characterized by the elements of the dME and inclusion by the CSCM. Figure 14.2 shows the model of the exclusion-inclusion dialectics.

Three laws are considered in the dialectical process, namely: the confrontation of opposites, unity, and change (Soto, 2014). The dialectic is a duality that confronts two contrary categories (dME and CSCM), which as pointed out could not live one without the other but coexist in permanent struggle, which produces unity.

14.3.1 Category of Modelling and Reproduction of Behaviors

The SOLTSA has managed to constitute some specific situations in which categories such as prediction, analyticity of functions, and tendency behavior of functions emerge. These categories have been revealed in historical epistemological studies and, in turn, in the emergence in specific situations of knowledge communities that make use of mathematical knowledge. With the research of Mendoza-Higuera (2013, 2020) and Giacoleti-Castillo (2020), the extension of the category tendency behavior of functions, proper of the transformation situation, to the reproduction of behaviors has been promoted.

Each category is composed of meanings, procedures, arguments, and instruments that make up an epistemology of the uses of mathematical knowledge. Thus, the transformation situation is composed of meanings as patterns of graphical and analytical behaviors that entail procedures of parameter variation to the extent that the function is taken as an instruction that organizes behaviors, maintaining as a guiding thread the search for tendencies when reproducing a known or given behavior[1] (Cordero, 2001; Mendoza-Higuera, 2020; Giacoleti-Castillo, 2020).

[1] In Chap. 12, of this book, the epistemological moments of this category are extensively mentioned.

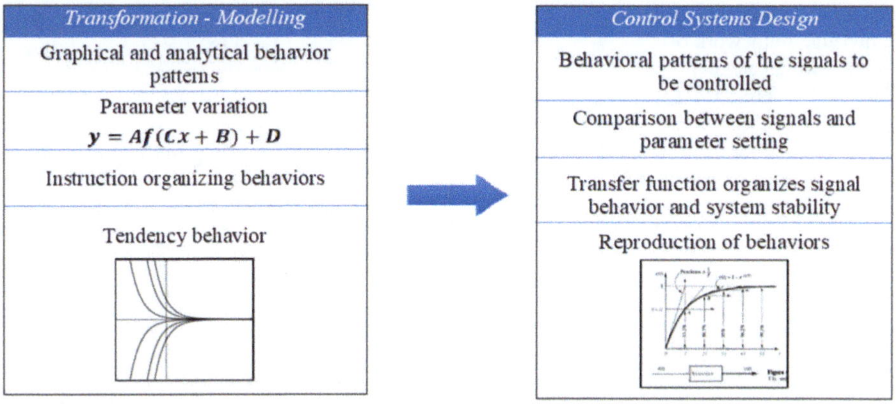

Fig. 14.3 Transformation situation and design of control systems. (Source: Mendoza-Higuera, 2020, p. 190)

In addition, although it will not be extensively discussed here, in Mendoza-Higuera (2020) there is evidence of the reproduction of behaviors in the resignification of uses of linear differential equations with constant coefficients as stability models in the situation of control system designs (Fig. 14.3).

In addition, together with these categories, the category of modelling is established as the process that accompanies the construction of mathematical knowledge as humans transform their reality, in contrast to other modelling proposals that seek to understand reality by relating it to mathematics through a model that represents it. Thus, this category of modelling starts from a principle P′ that corresponds to the functional of the reciprocal relationship between mathematics and everyday life; in this sense, P′ generates the uses of the mathematical knowledge (U(CM)) of people.

When these uses go through different situations (S_{ij}) and domains (D_j), they constitute epistemologies (E_r) and transversalities that re-signify the uses of that mathematical knowledge. And, the category corresponds to this resignification of uses (Cordero, 2022) as shown in Fig. 14.4.

14.4 Category Reproduction of Behavior: Emergence and Staging in Engineering Knowledge Communities

As already mentioned, to address the problem of teaching and learning mathematics in engineering, it is proposed to build a frame of reference that problematizes institutional mathematical knowledge and complements it with functional mathematics (put into use by communities of knowledge). In this way, we will show the emergence of the category reproduction of behaviors in a situation of design of

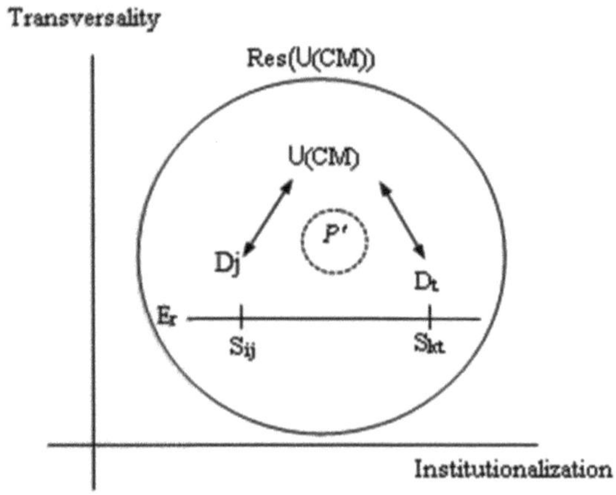

Fig. 14.4 Mathematical knowledge framework of the modelling category. (Source: Cordero, 2022)

control systems, typical of school scenarios of electronic engineers in training, where the Laplace Transform is resignified.

In turn, we present the staging of a design of socialization school situation where the category is the guiding thread for the resignification of the derivative with students of chemical engineering. At the same time, we account for the transversality of the category in situations of different domains such as electronic engineering and chemical engineering.

14.4.1 Control Systems Designs: A Resignification of the Laplace Transform

Giacoleti-Castillo (2020) and Giacoleti-Castillo and Cordero (2020) exhibited the functional factors that relate to continuous and discontinuous behaviors in a specific situation of control system design, and how these factors re-signify the Laplace Transform. In the approach of this research, two elements are of vital importance: control systems and the Laplace Transform. The first element defines the specific situation of the study community; the second is the mathematical object re-signified in the specific situation of the community of engineers in training.

The design of control systems was constituted as a situation, given that it is a central activity in the daily work of electronic engineers, both in their professional scenario and in their training (Mendoza-Higuera & Cordero, 2018). Control systems are devices designed by engineers where, given a certain desired behavior, several control actions (procedures) are executed to reproduce it. This alludes to the category reproduction of behavior since it is sought to reproduce desired or pre-established characteristics (behaviors).

In the signals and actions executed in a control system, behaviors occur that are typically modeled by functions that are defined discontinuously (e.g., the unit step function); however, the reproduction of the behaviors in the system output is continuous. The purposes of a control system can be summarized as follows: detect an error and correct the error. Given the disturbances that occur in a system, there are times when the desired behavior is not being achieved. In other words, an error is occurring that is detected by the system and to correct it, the same system is fed back by carrying out certain control actions.

Control theory uses transfer functions to perform these control actions, which characterize the input-output relationships of systems described by linear differential equations. The transfer function of a control system is defined as the quotient between the Laplace Transform of the output signal (response function) and the Laplace Transform of the input signal (excitation function), where the initial conditions are zero (Ogata, 2010).

$$Transfer\ Function\ G(s) = \frac{L\{output\ signal\}}{L\{input\ signal\}}$$

That is, it is in the transfer function where the control actions are applied to reduce the difference between the output signal and the input signal so that the desired behavior of the system is obtained. The control system designed by the mathematical knowledge community of engineers in training (CCM(IEF) for its acronym in Spanish), in this study consists of controlling the behavior of the temperature of the water contained in a bottle, in which edible algae (Spirulina) are grown.

The optimum temperature for algae growth is between 31 and 39 °C. But the temperature of the water in the container has a behavior that is frequently disturbed by the ambient temperature, among other factors. This causes the water temperature to be out of the optimal range at times. To avoid this problem, the community designs a system that controls the behavior of the water temperature in the desired range of 33–37 °C. The main objective of the control system is, then, to always maintain the water temperature in the desired range (Castro et al., 2019). Figure 14.5 shows the physical equipment of the control system.

The control system (Fig. 14.5) works in such a way that the sensor takes the temperature of the water contained in the bottle. The Arduino software receives the temperature data provided by the sensor (it processes the data every 5 s). If the water temperature in the container is out of the desired range, then the system generates an error signal and activates a feedback procedure: a control action is executed to correct the error, which consists of activating the pumps for recirculation of water from another container to the bottle containing the algae, in order to return the water temperature to the desired range in the shortest possible time (Castro et al., 2019). When the water temperature returns to the desired range, recirculation stops.

In the behaviors of the water temperature being controlled, the community identifies the following components of the control system: desired temperature range (input signal) and temperature obtained in the container (output signal). The reproduction of behaviors in this community is defined through the following two epistemological factors: timing, and the tendency in a range.

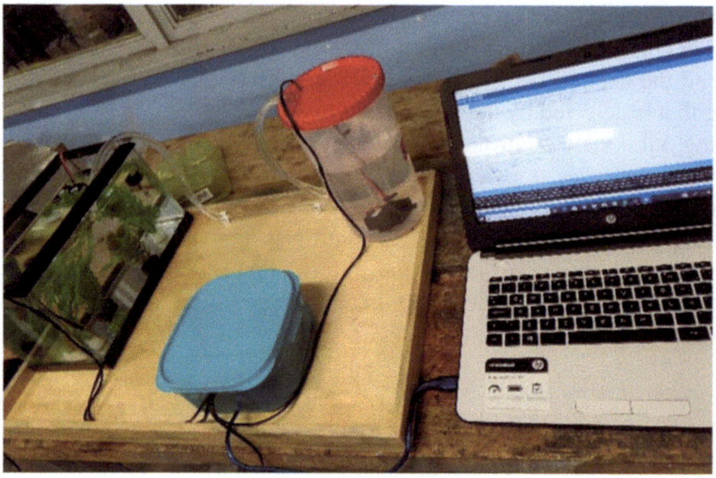

Fig. 14.5 Physical equipment of the control system. (Source: Giacoleti-Castillo, 2020, p. 71)

Timing. Under control theory, timing is defined as the role of the time domain in control systems. This consists of problematizing the phenomena or behaviors of a system in a temporal domain so that control actions can be executed (Hernández, 2010). This role of time has several connotations: one of them is its non-linear consideration; time is not considered as a domain where phenomena occur in which no intervention is possible. On the contrary, in the design of the control system, the engineer schedules, in advance, different actions to be executed at certain moments (of time) so that the system controls behaviors over time.

This treatment of time in the specific situation of the community allowed the system components to be built in advance so that the control actions would be activated at certain future moments (m_1, m_2, m_3...), where control errors will occur (temperature outside the desired range). In designing the control system, they were anticipated in time to control the effects of disturbances that would occur at those future times, so that the behavior of the water temperature is ideal at all times (Fig. 14.6).

The tendency in a range. This tendency refers to the behavior in a region, in which the reproduction of the behavior is desired to occur in the control system. As mentioned above, given the desired behavior, the system was built to control/maintain that behavior, in such a way that at all times the tendency of the signal is in the desired range (33–37 °C). Figure 14.6 shows a graph displayed by the system showing the output signal, which has a trending behavior in the desired range (we have drawn the blue lines to indicate the desired range of temperature).

In this specific situation, the behaviors of the control system became problematic over time (horizontal axis of the graph). This is per what was pointed out by Ogata (2010), who indicates that in the operation of a control system it is necessary to interpret the signals in the time domain. In the case of the CCM(IEF), this

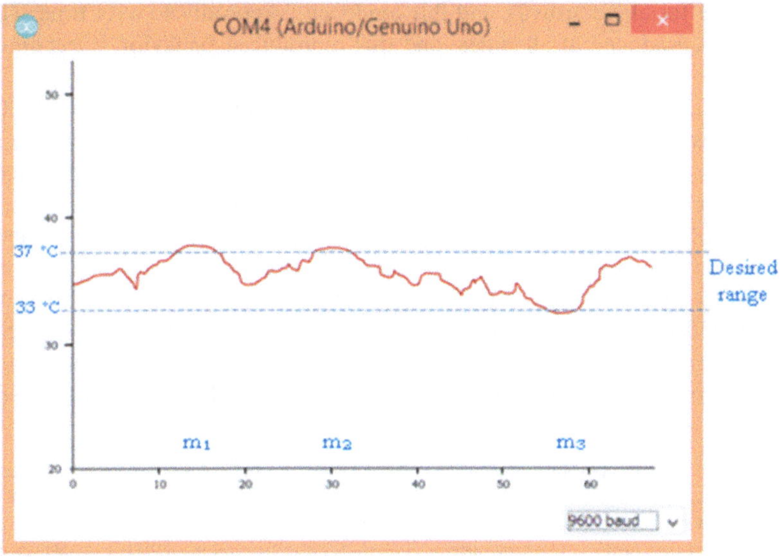

Fig. 14.6 Temperature behavior in the output signal of the system. (Source: Giacoleti-Castillo, 2020, p. 77)

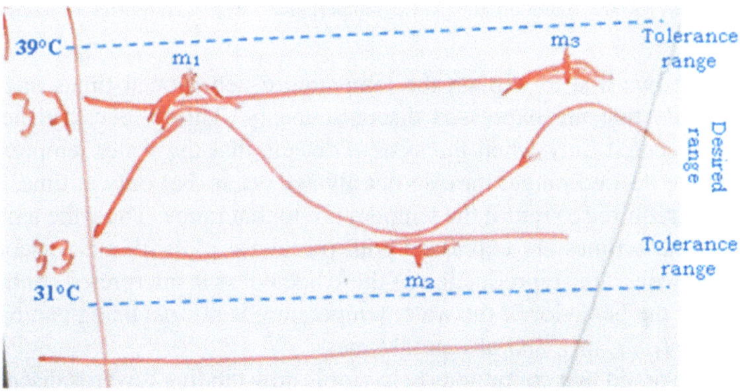

Fig. 14.7 Continuous behavior from discontinuous behaviors. (Source: Giacoleti-Castillo, 2020, p. 78)

interpretation allowed them to problematize the behavior of the water temperature, which occurs with a tendency in a range-, to reproduce the desired characteristics.

The two factors described above give functional meaning to the continuous reproduction from discontinuous behaviors in the specific situation of the control system design. In this way, then, the reproduction of behaviors is interpreted with a continuous graph. Figure 14.7 shows a graph constructed by the community to explain the operation of the control system. This figure shows that the behavior of the output signal graph describes a tendency in the 33–37 °C range, which is the desired range for algae growth, and this tendency is reproduced over time. That is, behavior is reproduced at all times, and this reproduced behavior is trending in a range.

In controlling the behavior of water temperature for algae harvesting, CCMIF problematizes the tendency of that behavior: it is desired to control that the water temperature is in the desired range at all times. To this end, timing reproduces the desired behavior, which is interpreted as continuous behavior.

The role of time as timing occurs when the community schedules in advance the control action to be executed at future times (m_1, m_2, m_3, ...) where control errors will occur (temperature out of the desired range). The control action executed at those moments is the recirculation of the water by pumping, with the purpose that the water temperature returns to the desired range in the shortest possible time and thus prevents it from crossing the tolerance range. Regarding this, community members state the following:

> If at any time this temperature leaves the desired range, the system starts working, either to heat the water or to cool it, depending on the temperature at which it is. Because what we want is for it to always be in that range. When [the temperature] at some point reaches higher or lower than that range, the pumps are activated, and the first pump delivers the water; the second pump takes out a quantity of water, within a few seconds. So, the water is recirculated. Say, you are going to supply 2 or 3 seconds of water; then it waits 5 seconds, it takes the data back from the temperature sensor and, if it is or is not already in the range, it activates or does not activate the pump. And it is from that that we have it, from a certain number of cycles we maintain the water temperature range (Transcript of community interview, 2019).

Figure 14.7 shows that to explain the temperature behavior at times m_1, m_2, m_3, traces are made that are expressed discontinuously. This is because the control actions are executed only when the system detects that the water temperature has gone out of the desired range; they do not always occur, but only at times (m_1, m_2, m_3) when it is required to return the temperature to that range. Thus, the temperature behaviors at these times are expressed with piecewise plots, that is, discontinuous behaviors. However, the reproduction of these behaviors is interpreted continuously; this is because the behavior of the water temperature is always being controlled, not only at times m_1, m_2, m_3, and also at all times.

It should be noted that continuous behavior is now tending towards discontinuous behavior. In other words, the behavior of the output signal, which is expressed continuously, is a reproduction resulting from discontinuous behaviors. This is because, at times when the system signals an error signal, the water temperature reproduced has a behavior within the tolerance range before returning to the desired range; thus, this behavior reproduced in the output tends to the discontinuous behaviors of the times m_1, m_2, m_3. That is, continuous behaviors are reproduced from discontinuous behaviors.

Finally, the functionality of the Laplace Transform (LT) in this specific situation refers to the two purposes of a control system: to detect an error and to correct it. Precisely, the uses of the LT are put into operation in the control system when, at various times, the water temperature behavior is determined, and when the system's

Table 14.1 Construction of the mathematical in the specific situation of the knowledge community

Construction of the mathematical	**Core situation**	**Specific situation**
	Transformation	Design of water temperature control system for algae harvesting
Meanings	Graphical and analytical behavior patterns	Behavior of the system signals (continuous and discontinuous) Timing Trend in a range
Procedures	Parameter variation	Comparison of signals (input and output) and recirculation of water in the containers (Feedback in the transfer function to achieve a desired behavior)
Instrument	Instruction that organizes behaviors	Instruction that organizes continuous behavior $Transfer\ Function = \frac{LT(output\ signal)}{LT(input\ signal)}$
Argumentation/ resignification	Trend behavior/behavioral reproduction	Continuous reproduction from discontinuous behavior 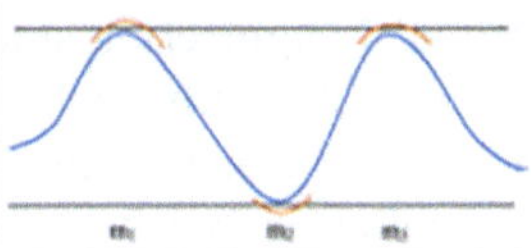

Source: Giacoleti-Castillo (2020, p. 84)

feedback procedure is executed, to reproduce the behavior of the temperature in the desired range. In addition, this functionality is endowed with meanings referring to the two epistemological factors presented above: the timing and the tendency in a range. Table 14.1 shows the construction of the mathematical in the specific situation of the knowledge community.

In the problematization of this specific situation, elements alluding to a functional justification of the Laplace Transform are put into operation; in other words, knowledge is constructed that corresponds to the mathematical, that is, to that which is useful to the community in its daily work and activities. This is in contrast to school mathematics, which focuses its attention on the object and ignores the uses of mathematical knowledge in everyday life.

From these functional aspects, the Laplace Transform is re-signified as the instruction that organizes continuous behavior in the control system. All of the above composes an epistemology of uses of the Laplace Transform (LT) in the specific situation. This epistemology confronts the school mathematics of the LT, which privileges its algorithmic and utilitarian character is privileged as a method to solve a differential equation, leaving aside its functional value that responds to the daily work of the communities in specific situations.

14.4.2 Transformation of Mathematical School Discourse: Staging the Reproduction of Behaviors

The uses of mathematical knowledge in other disciplines and the categories they form are the basis for elaborating of school designs that strengthen the meanings usually provided by school mathematics. When starting from recognizing certain uses of the derivative, specifically in the daily professional life of engineering, which is usually excluded in school mathematics, a specific scenario is visualized to promote a transformation of the dME.

For this purpose, what is reported by Pérez-Oxté and Cordero (2016, 2020), who analyze how a community of chemical engineers study the state of an electrical transformer from the graphical analysis of its chemical compounds. In this context, uses of the derivative emerge that are not normally part of school calculus: prediction, tendency behavior (reproduction of behaviors), and analyticity.

Thus, from these uses, Morales (2020) elaborated a school design, which was socialized with chemical engineering students, to investigate the processes of valorization of the uses of the derivative, which arise in the student, in the transition from signifying it as the slope of a tangent line to re-signifying it either as a prediction, or a tendency behavior, and or analyticity.

Thus, a contribution is made, on the one hand, by providing the mathematics teacher with an environment of uses and meanings of the derivative, and on the other hand, a design of school situation to improve and strengthen the learning of the derivative with chemical engineering students. These designs allow moving from the focus on the mathematical object to the valuation of uses of mathematical knowledge (Medina, 2019; Morales, 2020).

Object decentering is considered as a *sine qua non* condition to propitiate the interaction between two contrary categories: the dME and the CSCM (as already deepened in previous sections). The perspective that guided this design was the exclusion-inclusion dialectic (Soto & Cantoral, 2014), which establishes that to make transit between the dME and the CSCM there must be a confrontation between these elements, subsequently, a unity and finally a change would occur. Table 14.2 shows how the laws of dialectics and the elements that make up the specific situation are articulated in the construction of the design.

In particular, the DSES is based on three situations: approximation, variation, and transformation, and on the exclusion-inclusion dialectic perspective, as can be seen in the general scheme of Fig. 14.8.

The composition of Morales' design (Morales, 2020) allows observing a transit and dependence between these situations; that is, the approximation works as the prelude to reaching the situation of variation, and these, in turn, promote the emergence of the generation of behaviors through the analysis of tangent lines. In this way, the aim is to find out whether the role of the tangent line in the generation of behaviors is superimposed on the algorithmic search for its equation, which is what is commonly done in the dME.

Table 14.2 The perspective of the design of socialization school situation

Laws of the exclusion-inclusion dialectics	Articulation of the laws of dialectics in the construction of the school situation design for socialization
Confrontation of opposites	Activities are developed in which the arguments arising from the specific situation take precedence over those stipulated in the school mathematical discourse. The aim is to break with the usual application given to the mathematical object.
Unity	The situation is composed of four elements: meanings, procedures, instruments, and arguments. These elements do not necessarily act in a linear order; any one of these may occur first and the others may follow. A particular feature of this interaction is that when the participant focuses on the definition of the derivative, they will not move to the CSCM, whereas when they value the environments of use and meanings, they will have made the proposed transition.
Change	It will appear when those who participate in the resolution of the design, consider, in a horizontal relationship, the arguments/resignifications that emerge from the specific situations.

Source: Morales (2020, p. 47)

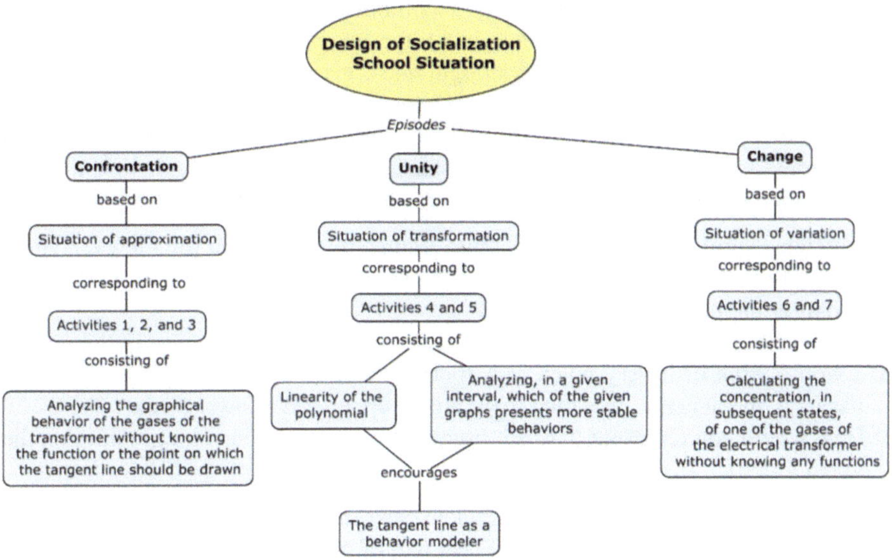

Fig. 14.8 The general outline of the design of socialization school situation. (Source: Morales, 2020, p. 48)

This design was staged with five students from the Bachelor of Chemical Engineering from a Costa Rican university, who had already passed calculus courses in one and several variables, in addition to linear algebra and differential equations. This allowed analyzing how the use of a design based on the uses of mathematical knowledge led to the resignification of the uses of the derivative of this group of students.

In the following, three episodes are presented that give an account of the implementation of the DSES: confrontation, unity, and change. For this, the concrete case of one of the participating students will be evidenced. In the transcripts, the student's comments/reasonings/justifications are denoted with A, B, ..., the interviewer's interventions with E, and the interviewer's clarifications of the students' behavior in square brackets.

The first of the situations presented in the DSES is the approximation. This situation aimed to confront school mathematics, in which a function is usually given, a point is indicated, and then a tangent line is asked to be determined and graphed. In the first part of the activity, the tangent line shows whether the electrical transformer shows stable or extraordinary behavior. In other words, the drawing of the tangent line does not only respond to the fact of the following instruction but has an intrinsic functionality in the situation presented. In this sense, the student will signify the behavior of the tangent with that of the curve that models the concentration of a certain gas inside the transformer from the variation between these two, alluding to ideas of tendency.

The second part of the activity is a combination of the approximation and transformation situations. On the one hand, the inverse problem of school mathematical discourse is presented, i.e., a straight line is given, and a curve must be drawn on it for which the given line is its tangent, and, in order to achieve this curve, the variation of parameters and their disciplinary knowledge about stable behavior is used. Draw a line that exhibits stable behavior on the given graph in the initial context, such that it is locally tangent at some point on the given graph.

A: And what would be the equation of that graph? [remains silent for a few seconds]. The only thing I can think of is to do the tendencies piecewise, before the peak and after the peak.
E: And then how would you draw a line that shows a stable behavior in the graph?
A: I would think of it very much as a control chart, in which there is a line that is basically the average, in which there are an upper and a lower limit.
E: How would you represent that? [Subsequently starts drawing straight lines on the given graph].

In this first part of the activity, the student can recognize that at $x = 0$ the equation of the tangent line corresponds to the linear part of the polynomial, although he does not see it in the first graph. Figure 14.9 shows the activity one solving.

E: What would be the equation of the red line?
A: It would be equal to $y = 6$
E: Is that red line tangent to the graph? [At this point the student is silent and unsure].
A: A tangent line represents the derivative of a function. Well, the slope.
E: So, the red line would be?
B: No because it touches it at more than one point.
A: No, it can only be at one

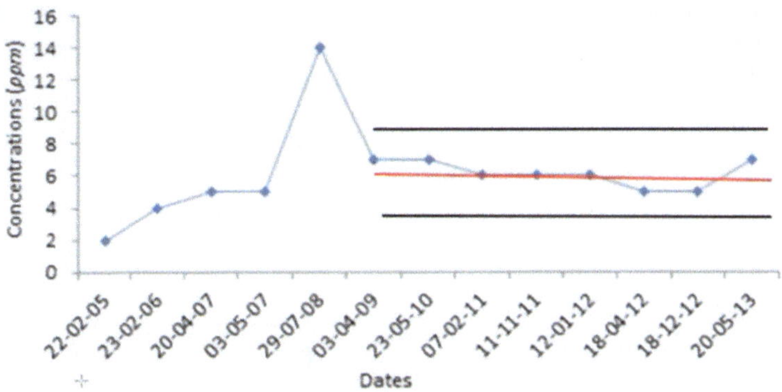

Fig. 14.9 Activity one solving. (Source: Morales, 2020, p. 68)

As the student starts to be asked questions about the relationship he observes between the given graph and that of the tangent line, he begins to notice that both have the same behavior in a neighborhood centered on the tangent point.

Thus, the student begins to affirm that the graph of the tangent line allows modelling the graphical behavior of the gas. In this section we can see how the student looks for the graph to take a certain tendency, which he considers stable when it takes more and more the shape of the straight line, that is, he looks for the differences between them to be minimal and in this way the tangent line is the instruction that organizes its behavior.

The above discussion corresponds to the preamble of the second part of the activity, which aims to see the unit, in this sense, it is intended to know if they use the tangent line as a generator of the behavior of a curve in a neighborhood around zero. For this, the graph of a polynomial is provided, and the signs of the linear part are modified, after which the students are asked to determine what the graph would look like as a result of this change.

The episode of change is the last of the exclusion-inclusion dialectic, in which only functional arguments of the derivative are expected to be used. The activity consists of making a prediction of the subsequent state of the transformer, knowing only the behavior at a given time and the variation that occurs in it.

In Fig. 14.10, A and B represent the behavior of ethylene at different times. Assume that you only know A and the change from A to B (but not B). Build a model that allows you to predict B from these data.

A: I can think of that [draws the red line], just look at the tendency of A and B, and draw a linear regression, which in very engineering words is a rule of three. [Figure 14.11 shows student's work in this stage].

E: How would you do it?

A: What I would do is try to associate the height of A to a value [plots a vertical and a horizontal axis]. And since it is at different times it is time-dependent [label the horizontal axis with t] and the behavior of ethylene is concentration [label the

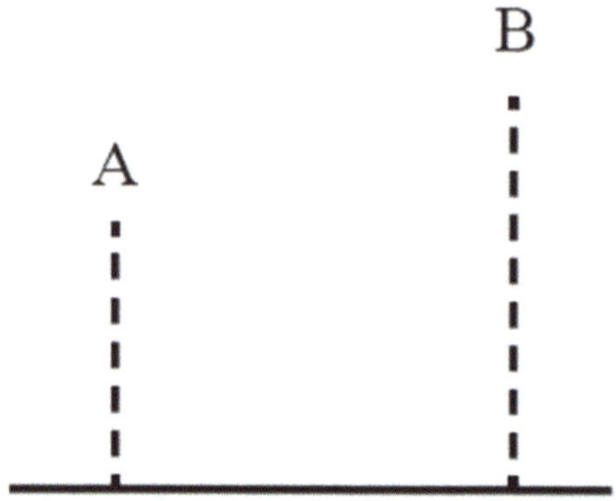

Fig. 14.10 Diagram of third stage behavior. (Source: Morales, 2020, p. 79)

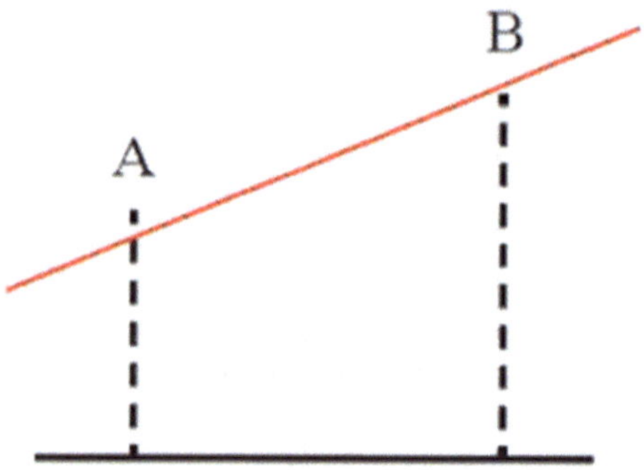

Fig. 14.11 Student solving in the third stage. (Source: Morales, 2020, p. 79)

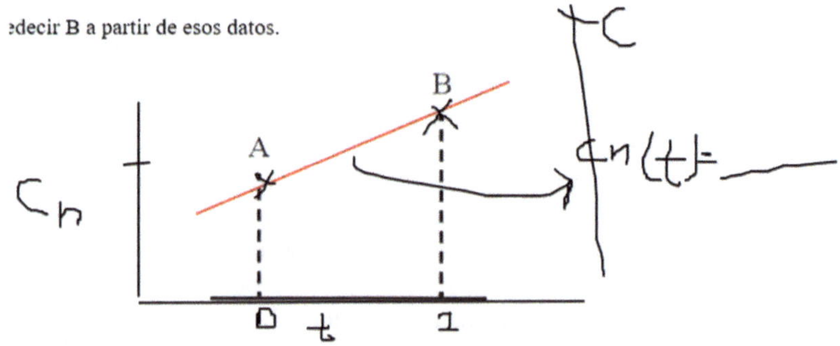

Fig. 14.12 Student solving in the third stage. (Source: Morales, 2020, p. 79)

vertical axis with C] so I would have ordered pairs, and if I name A time zero and B time one, with that I could get the concentration as a function of time. [Figure 14.12 shows student's work in this stage].

Assume that the graphical behavior of ethylene is now unknown. Knowing that the data were taken every 6 months, that one of the concentrations was 6.02 ppm and

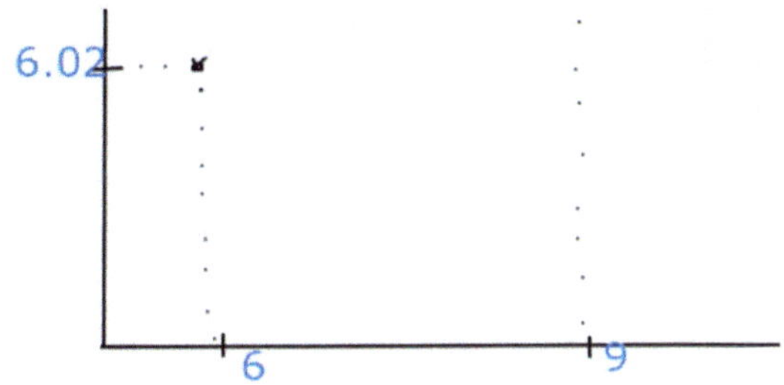

Fig. 14.13 Student solving in the third stage. (Source: Morales, 2020, p. 80)

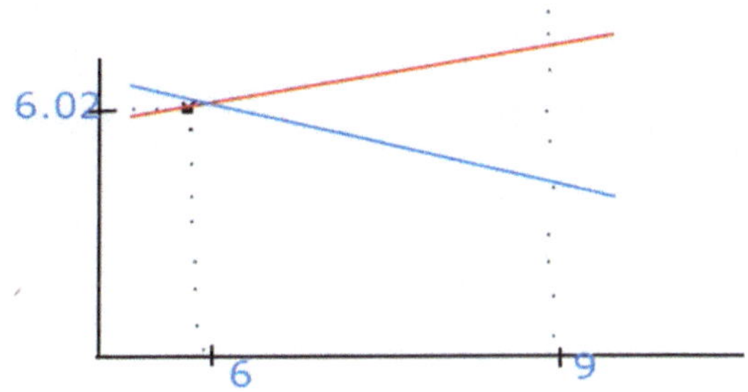

Fig. 14.14 Student solving in the third stage. (Source: Morales, 2020, p. 81)

that the variation from that point to another is 2.7, then predict what concentration would be 3 months later.

A: Well, the first thing is that since you only have one piece of data, you would have to assume a tendency.
E: And how do you mean?
A: For example, by now, in the previous example, we assume a linear tendency. Because if I had several data, I would know how that concentration behaves over those 6 months, but otherwise I would have to assume one.
E: And then what could be done?
A: At the moment what comes to mind is to try to see it graphically [makes a graph]. This point is for the 6 months [and makes the asterisk in Fig. 14.13] and the one for the 9 months is here, whether it goes up or down.

A: What I still don't understand is how to interpret the value of 2.7. I imagine it must be like what we were doing now: knowing the orientation of the tangent line, then perhaps I might intuit something.
E: Ok, so it is clear that you don't know what will happen at 9 months if there will be a concentration of more than 6.02 ppm or less than 6.02 ppm. What would happen if you knew the slope of the tangent, what would happen?
A: I would know the sign of the slope, and then I would draw a line. Either something like this (...) or like this [plots two lines seen in Fig. 14.14].

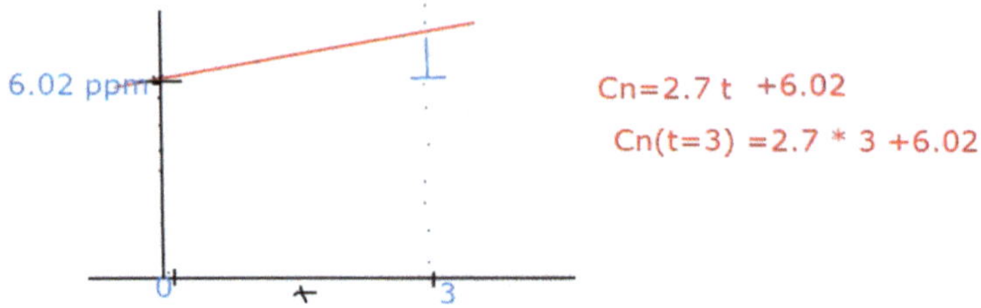

Fig. 14.15 Student solving in the third stage. (Source: Morales, 2020, p. 81)

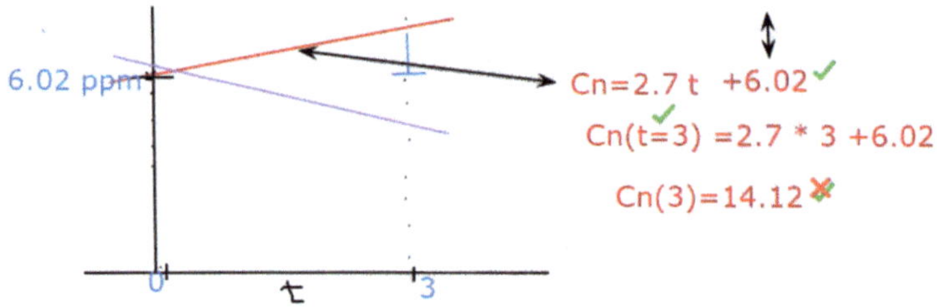

Fig. 14.16 Student solving in the third stage. (Source: Morales, 2020, p. 82)

E: And how do you get the slope of a tangent line?
A: I'm going to construct the nicer graph again but starting at zero. The slope of the tangent line is the derivative, so I know that it would be increasing [makes another graph which is seen in Fig. 14.15].

E: Could you explain a little bit about what you did [as written in Fig. 14.16]?
A: What I did was to express the equation of that tangent line. Before I had put measures 6 and 9, but it was more cumbersome, so it was easier to put the measure at zero. As we know that at time zero I have a concentration of 6.02, then that 6.02 is going to be the intercept of the equation of the tangent line and the 2.07 is the change, that is, it is going to be the slope, so that's how I did that equation [marks with a green "check" the equation $C_n = 2.7t + 6.02$]. And then, as they say, that I evaluated the concentration 3 months later, I simply evaluated at t = 3 [makes another green check, this time on $C_n(t = 3)$]
E: What would happen if the slope were −2.7?
A: I would assemble the same equation only with a negative slope. And in that case, there would be a lower concentration after 3 months.

This section corresponds to the last part of the design, that is, to show the change. From the beginning, the student puts into play re-significations of the tangent line and does so by resorting, for the solving of the activity, to the linear approximation. As it was pointed out in the previous activities, the tangent line is the guide for the generation of stable behavior.

Moreover, although the students had already studied the Taylor polynomial $f(x+h) = f(x) + f'(x)h + \frac{f''(x)}{2!}h^2 + \frac{f'''(x)}{3!}h^3 + \ldots$ and that they could resort to this using the fact that h corresponds to the 3 months that have elapsed since the given shot, making a linear approximation and taking the first datum as zero time, with which one would have $f(0 + 3) = f(0) + 2.7 \cdot 3 = 6.02 + 8.1 = 14.12$, this is not the resource they used.

From the beginning, the student resorts to the plot of a tangent line to predict the behavior of the gas in the requested later state, as well as to the slope formula to construct an equation that allows him to make the prediction. In other words, knowing that the data collection is every 6 months and that the prediction is 3 months later, the student is using the fact that the tangent line behaves similarly to the graph that would model the real behavior of the gas.

The above show how the transversality between situations enables the emergence of the re-significations of the derivative. In the latter case, as a predictor of subsequent gas behaviors based on the search for tangent-tending behaviors.

14.5 Final Reflections

Entering the problem of teaching and learning mathematics in engineering education from TSME and the SOLTSA, in particular, entails several challenges. One of them is to implement a reciprocal and horizontal dialogue with the other, in this case with engineering community.

Immersion in knowledge communities to identify situations where there are uses of mathematical knowledge and to infer the emergence of categories of mathematical knowledge imply precise methodological resources that justify the theoretical constructs that underlie these inferences. In this regard, the knowledge community and its situations are fundamental to describing the functional mathematics that is put into use there and thus forms the frame of reference that will contribute to the redesign of the dME for the training of engineering students.

In this paper, we account for the category of behavior reproduction and its emergence in electronic engineering, in an academic environment, where when facing a specific situation of control system design, functional justifications arise (specific to the situation and its discipline) that account for the uses and re-significations of the Laplace Transform.

The latter goes from being a strategy for solving linear differential equations in the algebraic form to becoming an instruction that organizes continuous behaviors through procedures where the comparison of signals controls the feedback mechanism so that the output signal behaves as required. On the other hand, in chemical engineering where a chemical engineer faces a specific situation of analysis of a transformer, different categories emerge, being the reproduction of behavior (or tendency behavior) one of them.

To determine that there are no indications in the transformer gases, the behavior of each of the graphical models of the dissolved (or real) gases is analyzed, comparing it with a graphical model of the ideal gases (the one required to affirm that the transformer has no faults). In this way, the tendency of gas behavior is

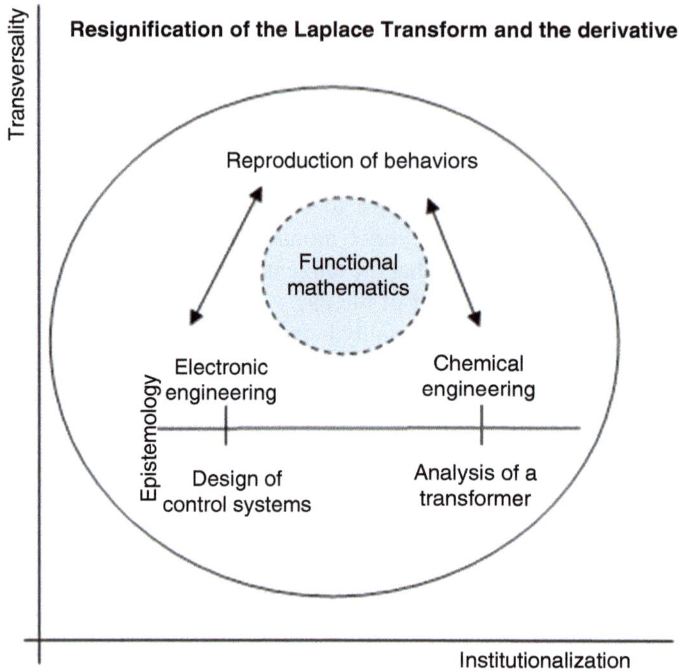

Fig. 14.17 Category of modelling in engineering scenarios. (Source: Authors own elaboration)

analyzed according to the variation between the models, which is characteristic reproduction of behavior (Torres-Burgos, 2013).

The category reproduction of behaviors emerges in situations of electronic engineering (Giacoleti-Castillo, 2020) and chemical engineering (Pérez-Oxté & Cordero, 2016, 2020), pointing out transversality of uses that re-signify the knowledge of the Laplace Transform and the derivative, respectively. In this way, we account for the category of modelling as that action that makes visible the transversality of uses of mathematical knowledge and the re-signification of mathematical knowledge.

In human activity, modelling emerges in a natural way to transform the reality of the engineer who faces situations that need to be solved and, while providing a solution, makes use of reasoned and functional justifications that constitute this functional mathematical knowledge. Figure 14.17 shows category of modelling in engineering scenarios.

As indicated by Mendoza-Higuera and Cordero (2018), the category of modelling does not appear in the usual school mathematics for engineering education, however, it does appear in everyday situations of mathematical knowledge communities, for example, of bionic, electronic, or chemical engineering students.

Finally, to intervene punctually and forcefully in the mathematics classroom, it is necessary to build instruments that contribute to the re-signification of knowledge by the students, in this sense, the designs of the school situation of socialization, where the categories of mathematical knowledge are implemented, will be those instruments.

Then, to disrupt and transform the dME, which will have an educational impact, we offer a design where the student permanently confronts the traditional meanings of the derivative with the use of prediction and tendential behavior. Here the category of behavioral reproduction appears when the tendency is the instrument that allows him to respond to the prediction task.

Consequently, it is convenient to incorporate the reciprocal and horizontal dialogue between mathematics and engineering, where the categories of mathematical knowledge are identified and their implementation can be established in classrooms to promote the educational impact and establish permanent programs that contribute to this path.

References

Aracil, J. (2017). Las matemáticas y la Ingeniería [Mathematics and engineering]. In A. Durán-Guardeño & J. Ferreirós- Domínguez (Eds.), *El valor de las matemáticas* (pp. 11–36). Editorial Universidad de Sevilla.

Belhoste, B. (1994), Un modèle à l'épreuve. L'Ecole Polytechnique de 1794 au Second Empire. In B. Belhoste, A. Dahan-Dalmedico & A. Picon (Eds.), *The Formation Polytechnicienne*, 1774–1994, Dunod, pp. 9–30.

Cauchy, A. (1882). *Oeuvres Complétes D'Augustin Cauchy*. Académie des Sciences (Ed.). Gauthier-Villars, (II série. Tome IV).

Cajas, F. (2001). The science/technology interaction: Implications for science literacy. *Journal of Research in Science Teaching, 38*, 1–15.

Cantoral, R. (2019). Socioepistemology in mathematics education. In S. Lerman (Ed.), *Encyclopedia of mathematics education* (pp. 1–7). Springer Nature. https://doi.org/10.1007/978-3-319-77487-9_100041-1

Cajas, F. (2009). *El conocimiento de ingeniería como conocimiento escolar [Engineering knowledge as school knowledge]*. En P. Lestón (Ed.), Acta Latinoamericana de Matemática Educativa, 22, 77–84. Comité Latinoamericano de Matemática Educativa.

Castro, J., Hernández, A., Natarén, A., & Solís G. (2019). *Automatización de un sistema de control de temperatura para la cosecha de algas por el método de hidroponía* [*Automation of a temperature control system for algae harvesting by the hydroponics method*]. (Internal document). Instituto Tecnológico de Tuxtla Gutiérrez.

Cházaro, L. (2011). Los instrumentos matemáticos en la Nueva España: circulación, usos y transformaciones de la medición [Mathematical instruments in New Spain: Circulation, uses and transformations of measurement]. *La Gaceta de la Real Sociedad Matemática Española, 14*(4), 739–752. Sección: Historia.

Cordero, F. (2001). La distinción entre construcciones del cálculo. Una epistemología a través de la actividad humana [Distinction between constructions of the calculus. An epistemology through human activity]. *Revista Latinoamericana de Investigación en Matemática Educativa, 4*(2), 103–128.

Cordero, F. (2003). *Reconstrucción de significados del Cálculo Integral: la noción de acumulación como una argumentación [Reconstructing meanings of Integral Calculus: The notion of accumulation as an argumentation]*. Grupo Editorial Iberoamérica.

Cordero, F. (2008). El uso de las gráficas en el discurso matemático escolar. Una visión socioepistemológica [Use of graphs in mathematical school discourse. A socioepistemological view]. In R. Cantoral, O. Covián, R. Farfán, J. Lezama, & A. Romo (Eds.), *Investigaciones sobre enseñanza y aprendizaje de las matemáticas: Un reporte iberoamericano* (pp. 265–286). Díaz de Santos-CLAME A.C.

Cordero, F. (2016). Modelación, funcionalidad y multidisciplinariedad: el eslabón de la matemática y el cotidiano [Modelling, functionality and multidisciplinarity: The link between mathematics and everyday life]. In J. Arrieta & L. Díaz (Eds.), *Investigaciones latinoamericanas de modelación de la matemática educativa* (pp. 59–88). Gedisa.

Cordero, F. (2022). *La Matemática, sus Usos y Significados. Un Programa Socioepistemológico de la Matemática Educativa* [*Mathematics, its uses and meanings. A socioepistemological program of educational mathematics*]. Manuscrito en preparación.

Cordero, F., & Flores, R. (2007). El uso de las gráficas en el discurso matemático escolar. Un estudio socioepistemológico en el nivel básico a través de los libros de texto [The use of graphs in mathematical school discourse. A socioepistemological study at the elementary level through textbooks]. *Revista Latinoamericana de Investigación en Matemática Educativa, 10*(1), 7–38.

Cordero, F., Gómez, K., Silva-Crocci, H., & Soto, D. (2015). *El discurso matemático escolar: la adherencia, la exclusión y la opacidad [Mathematical school discourse: Adherence, exclusion and opacity]*. Gedisa.

Cordero, F., Mena, J., & Montalto, M. E. (2010). Il ruolo della giustificazione funzionale in una situazione di risignificazione dellàsintoto [The role of functional justification in a situation of re-signification of the symptom]. *I'insegnamento della Matematica e delle Scienze Integrate, 33B*(4), 457–488.

Farfán, R. (2012). *Sociepistemología y ciencia. El caso del estado estacionario y su matematización [Sociepistemology and science. The case of the steady state and its mathematization]*. Gedisa.

Giacoleti-Castillo, F. (2020). *La temporalización y la tendencia como factores funcionales de la reproducción de un comportamiento continuo a partir de discontinuos. Una resignificación de la Transformada de Laplace en un sistema de control* [*Timing and tendency as functional factors in the reproduction of continuous behavior from discontinuous. A resignification of the Laplace Transform in a control system*]. Unpublished master's thesis, Department of Mathematics Education, Cinvestav-IPN.

Giacoleti-Castillo, F., & Cordero, F. (2020). Reproducción continua de Comportamientos Discontinuos. De la Transformada de Laplace a la Continuidad de la Reproducción de Comportamientos [Continuous reproduction of discontinuous behaviors. From the Laplace transform to the continuity of the reproduction of behaviors]. *Acta Latinoamericana de Matemática Educativa, 33*(2), 615–626.

Hernández, R. (2010). *Introducción a los sistemas de control: Conceptos, aplicaciones y simulación con MATLAB*. Pearson Educación. ISBN: 978-607-442-842-1.

Howson, A. G., Kahane, J. P., Lauginie, P. Y., & dev Turckheim, E. (Eds.). (1988). *Mathematics as a service subject. ICMI study series*. Cambridge University Press.

Lebesgue, H. (1926). *Lecons sur l'intégration et la recherche des fonctions primitives*. Gauthier-Villars. Deuxiéme édition.

Medina, D. (2019). *Transformación educativa del docente de matemáticas. Un episodio: El uso de la compensación como una resignificación de la media aritmética* [Educational transformation of the mathematics teacher. An episode: The use of compensation as a re-signification of the arithmetic mean]. Unpublished doctoral thesis, Department of Mathematics Education, Cinvestav-IPN.

Mendoza-Higuera, E. J. (2013). *Matemática funcional en una comunidad de conocimiento: el caso de las ecuaciones diferenciales en la ingeniería* [Functional mathematics in a knowledge community: The case of engineering differential equations]. Unpublished master's thesis, Departamento de Matemática Educativa, CINVESTAV-IPN.

Mendoza-Higuera, E. J. (2020). *El uso de la matemática en la Ingeniería Biónica. De la estabilidad a la reproducción de comportamientos en un sistema de control. Una Categoría de Modelación* [*Use of mathematics in Bionic Engineering. From stability to reproduction of behaviors in a control system. A modelling category*]. Unpublished doctoral thesis, Department of Mathematics Education, Cinvestav-IPN.

Mendoza-Higuera, E. J., & Cordero, F. (2018). La modelación en las comunidades de conocimiento matemático. El uso de las matemáticas en ingenieros biónicos. El caso de la estabilidad [Modelling in mathematical knowledge communities. The use of mathematics in bionic engineers. The case for stability]. *Revista Latinoamericana de Etnomatemática, 11*(1), 36–61.

Mendoza-Higuera, E. J., Cordero, F., Gómez, K., & Solís, M. (2018). El uso del conocimiento matemático de los ingenieros. Del objeto a la funcionalidad matemática [The use of mathematical knowledge by engineers. From object to mathematical functionality]. *Boletim de Educação Matematica, 32*(62), 1219–1243.

Morales, J. L. (2020). *Resignificación de los usos de la derivada en un diseño escolar con perspectiva de dialéctica exclusión-inclusión: predicción, comportamiento tendencial y analiticidad* [*Resignification of the uses of the derivative in a school design with a perspective of exclusion-inclusion dialectic: Prediction, tendency behavior and analysis*]. Unpublished master's thesis, Department of Mathematics Education, Cinvestav-IPN.

Morales, J. L., & Cordero, F. (2020). Resignificación de la derivada en una situación escolar con perspectiva de dialéctica exclusión-inclusión: un estudio socioepistemológico [Resignification of the derivative in a school situation with a perspective of dialectic exclusion-inclusion: A socio-epistemological study.]. *Acta Latinoamericana de Matemática Educativa, 33*(1), 453–461. Retrieved from https://www.clame.org.mx/documentos/alme33_1.pdf

Ogata, K. (2010). *Ingeniería de Control Moderna* (5th ed.). Pearson Educación, S.A.

Pérez-Oxté, I. & Cordero, F. (2016). Una epistemología basada en la transversalidad de los usos de la gráfica de una comunidad de ingenieros químicos industriales [An epistemology based on the transversality of the uses of graphics in a community of industrial chemical engineers]. *Investigación e Innovación en Matemática Educativa*, 1, 23–30.

Pérez-Oxté, I. & Cordero, F. (2020). *Modelling and anticipation of graphical behaviors in industrial chemical engineering. The role of the transversality of knowledge in learning mathematics*. Manuscript submitted for publication.

Riemann, B. (1898). Oeuvres *Mathámatiques Riemann*. Gauthier-Villars. (Trans. L. Laugel).

Romo-Vázquez, A. (2014). La modelización matemática en la formación de ingenieros [Mathematical modelling in engineering education]. *Educación Matemática* (Edición especial), 314–338.

Soto, D. (2014). *La dialéctica Exclusión-Inclusión entre el discurso Matemático Escolar y la Construcción Social del Conocimiento Matemático* [*The exclusion-inclusion dialectic between mathematical school discourse and the social construction of mathematical knowledge*]. Unpublished doctoral thesis, Department of Mathematics Education, Cinvestav-IPN.

Soto, D., & Cantoral, R. (2014). El discurso Matemático Escolar y la Exclusión. Una visión Socioepistemológica [Mathematical school discourse and exclusion. A socioepistemological view]. *Bolema—Boletim de Educação Matemática, 28*(50), 1525–1544.

Torres-Burgos, L. (2013). *Usos del conocimiento matemático. La simultaneidad y estabilidad en una comunidad de conocimiento de la ingeniería química en un escenario del trabajo.* [*Uses of mathematical knowledge. Simultaneity and stability in a chemical engineering knowledge community in a work setting*]. Tesis de Maestría no publicada. DME, Cinvestav-IPN, Mexico.

Chapter 15
Prospective Mathematics Teacher Discipline Identity and the Modelling Category: The Value of the Learner's Knowledge

Claudio Opazo-Arellano, Sindi Marcía-Rodríguez, Henry Chávez-Martínez, Eleany Barrios-Borges, and Francisco Cordero

15.1 Introduction

To transform education demands to value people's knowledge, however, to achieve this aspiration implies to formulate permanent programs which impact the initial teacher training. The impact will define a vision change about the meaning of learning to teach where a consequence is the *discipline identity* construction (Opazo-Arellano et al., 2018; Opazo-Arellano, 2020). An initial teacher training principle is *to learn to teach* since it defines pedagogical and discipline knowledge of the prospective teacher (Blanco & Mercedes, 2005; Contreras et al., 2010; Cornejo, 2014).

However, which is the knowledge nature that is learned by the prospective teacher? And moreover, what nature is the knowledge that the teacher teaches during his initial teacher training? In order to answer these questions, two work lines are opened. On one hand, some problematize the construction of the educational frameworks in the country where the teacher works. On the other hand, the discipline knowledge that joins and distinguishes the teaching within the different educational levels (Mercado, 2002).

To build the meaning of teaching is something more solid that just acquiring specific knowledge about how to teach or how to organize what it is taught. Both aspects are relevant nevertheless; the most important aspect is within the knowledge

C. Opazo-Arellano (✉)
Universidad Austral de Chile, Puerto Montt, Los Lagos, Chile
e-mail: claudio.opazo@uach.cl

S. Marcía-Rodríguez · H. Chávez-Martínez · E. Barrios-Borges · F. Cordero
Centro de Investigación y de Estudios Avanzados del Instituto Politécnico Nacional, Ciudad de México, Mexico
e-mail: sindi.marcia@cinvestav.mx; henry.chavez@cinvestav.mx; eleany.barrios@cinvestav.mx; fcordero@cinvestav.mx

M. Rosa et al. (eds.), *Mathematical Modelling Programs in Latin America*,
https://doi.org/10.1007/978-3-031-04271-3_15

nature that is learned to teach in the initial teacher training. This research describes the prospective mathematics teacher' knowledge finding that what is learned is the hegemonic character of the school mathematics.

In Cordero's words (2022) a zoom to the mathematical objects (definitions and concepts) over the permanent processes (uses and meanings). These categories express a different relationship with the mathematical knowledge since in the first one is standard and utilitarian and the second one favors a horizontal and reciprocal relationship within the mathematical knowledge. Therefore, to value the knowledge that builds and spreads who learns (Cordero, 2016) in this initial teacher training privileges the first relationship with the mathematics knowledge.

A consequence of this event is the phenomenon of adhesion to the school mathematics discourse (Cordero & Silva-Crocci, 2012) that is why the prospective teacher adopts the school mathematics knowledge as the unique referent for his teaching. However, the social construction of the learner of the mathematics knowledge is excluded. The adhesion construct has been interpreted and developed with basis on theoretical-methodological elements of the socio-epistemology forgotten subject and knowledge mainstreaming (Cordero, 2022).

To emphasize that to adopt a culture, a problem or some knowledge without questioning or disrupting its nature contributes to the adhesion phenomenon (Cordero & Silva-Crocci, 2012). In this sense, Opazo-Arellano et al. (2018) and Opazo-Arellano et al. (2021) explain this phenomenon extent in the initial teacher training of the prospective mathematics teacher process. They affirm that when this knowledge community adopts the school knowledge through a hegemonic way, the teaching of mathematics causes the exclusion of autonomous argumentations of the learner. This leaves in a second place the epistemological pluralism[1] and the knowledge mainstreaming[2] within the mathematics classrooms.

As it was described above, the authors indicate that in order to avoid the adhesion phenomenon within the initial teacher training it is a condition to construct the discipline identity. In other words, a resistance to the school mathematics discourse (Opazo-Arellano, 2020). In this sense some factors that contribute to the prospective teacher to face the adhesion to the school mathematics discourse were identified finding *the re-signification of the mathematics knowledge*[3] as a concrete demonstration of the teaching of mathematics transformation. Besides this construct defines epistemologically and ontologically the discipline identity (Opazo-Arellano & Cordero, 2021).

---

[1]Epistemological pluralism is opposed to hegemony in the sense of the consideration of different arguments, meanings and procedures that exist and that are associated with mathematical knowledge in a specific situation and context (Cordero et al., 2015).

[2]Mainstreaming is the resignification of the uses of knowledge between scenarios or knowledge domains, for example: between school and work; or between mathematics and engineering (Cordero et al., 2019).

[3]Re-signification expresses the mobility of the uses and meanings of mathematical knowledge in different specific situations, typical of other domains of knowledge and everyday life (Mendoza-Higuera et al., 2018).

The factors that were found are results from making the mathematics knowledge problematic in terms of the definite integral and asymptote of a function. The collection of this data was developed in communities of prospective mathematics teachers in Honduras highlighting two autonomous argumentations through analysis*: the accumulation and tendency behavior* (Marcia-Rodríguez & Cordero, 2021; Cordero & Domínguez, 2001; Chávez-Martínez, 2022). These argumentations form the socio-epistemological corpus of the teaching of mathematics where the common thread is the *modelling category* (Cordero, 2022).

The modelling category transforms into the epistemological and ontological justification that develops a vision about the teaching of mathematics. That means it defines the construction of the discipline identity. Since the prospective teacher plans, delivers and evaluates the teaching of the mathematics knowledge since *the epistemological pluralism (Two expressions are used in the text: epistemological pluralism and Epistemological Diversity. Please unify all by epistemological diversity.) and transversality of mathematical knowledge.* Factors that are demonstrated when socialization school situations designs are applied since in this process the school knowledge is faced and people' mathematics knowledge emerge (mathematics knowledge uses).[4]

What derives into that the prospective mathematics teacher avoids the adherence to the school mathematics knowledge because it legalizes the mathematics knowledge diversity, it resists the mathematics knowledge put in practice is teaching and learning scenarios and projects the re-signification of the mathematics knowledge within the learner's autonomous argumentations.

A summary about what it was stated before is the discipline identity because it expresses the teaching vision and the mathematics knowledge that is constructed by the prospective teacher when he prepares the teaching within the initial teacher training. Figure 15.1 shows the discipline identity within the initial teacher training.

15.2 Adhesion Phenomenon and the School Mathematics Discourse: A Latino-American Perspective

Within the school-academic production and spreading of scientific knowledge scenarios there are cultural behaviors that overshadow the knowledge that does not come from of the countries known as "central". That means supremacy of the dominant cultures thinking is given. At the same time, a pejorative load to the Latino-American cultural expressions facing the world and within the same region (Fig. 15.2). These behaviors that weak the different fields of the discipline task of the

[4]The uses of mathematics knowledge are understood as the organic functions of the situations (working) that are demonstrated in the *tasks* that form the situation. These tasks types are the form of the math knowledge use. The tasks can be: activities, actions, own rotation of domains of the organism of the situation (Cordero & Flores, 2007).

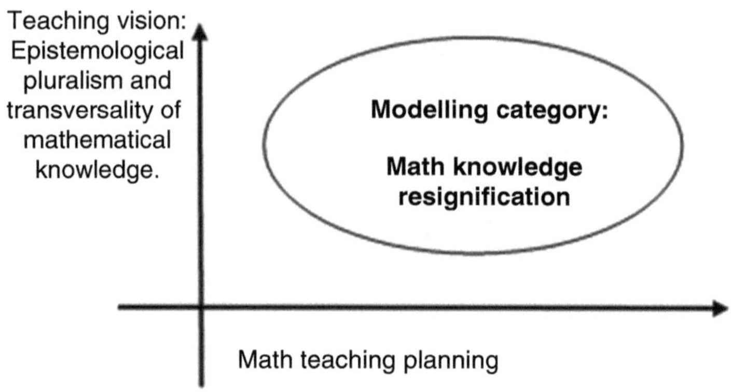

Fig. 15.1 Discipline identity within the initial teacher training. (Source: Opazo-Arellano, 2020)

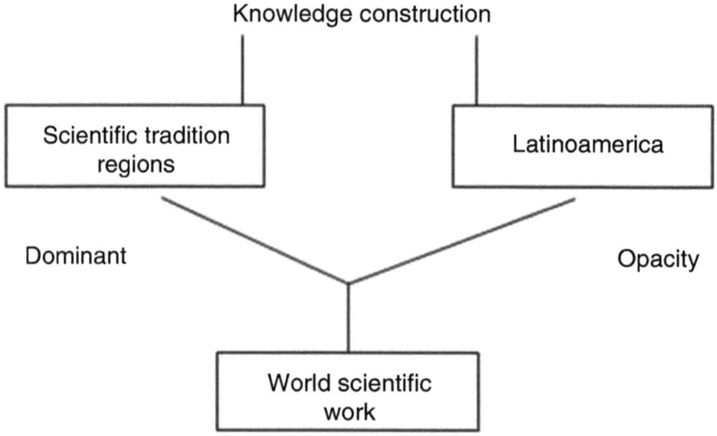

Fig. 15.2 Socio-historical aspects that suggest the knowledge. (Source: Cordero et al., 2015)

mathematics education in Latino-America were known as the adhesion phenomenon (Cordero & Silva-Crocci, 2012; Silva-Crocci, 2021).

The adhesion phenomenon:

> Creates superiority slogans to do a good job in case some practices from central countries are adopted. Some examples are the thesis titles in English in peripheral countries no matter its natural language is the Spanish, to adopt concepts of dominant-occidental origin, to name academic programs of peripheral countries, to work around problems initiated by mentors or programs from central countries, among other behaviors (Cordero et al., 2015, p. 33).

What stated before describes the scenario that suggests and gives meaning to the adhesion phenomenon. One characteristic is that this phenomenon reduces the process of setting up of endogenous theoretical varieties in Latino-American research programs. In these terms, Cordero and Silva-Crocci (2012) refer to the disadvantage that Latino-American live about the knowledge construction. They

Table 15.1 A prospective mathematics teacher's answer

Section a: What can be the values of a and b?
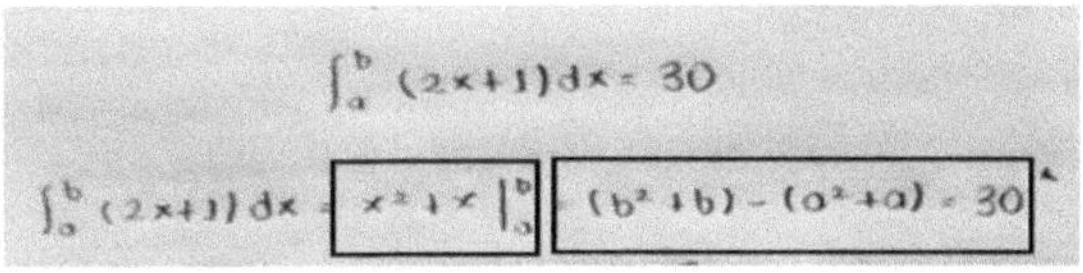
Participant's notes:
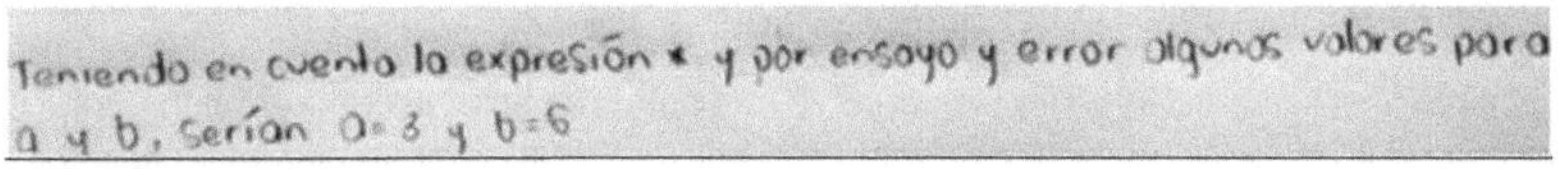
Participant's notes: Considering the expression * and by trial and error some values for a and b would be a = 3 and b = 6.

Source: Opazo-Arellano et al. (2020)

declare that if there are no resistances facing these facts that cause disadvantage they will not vary. In this context what the authors defend is to reject to be adherent (Cordero et al., 2015).

It is important to declare that this phenomenon has an effect on the mathematics teaching and learning problem where the school mathematics discourse plays an essential role: its supremacy over the teachers and students' cultural thinking provokes adhesion what implies attitudes that do not criticize the mathematics contents that are taught such as the exclusion of its construction and its working opacity (Cordero & Silva-Crocci, 2012; Silva-Crocci, 2021).

It is relevant to mention that the adhesion to the school mathematics discourse reaches the initial teacher training because the prospective teacher adopts the school knowledge as the unique referent since the mathematics knowledge construction's situational characteristic. This derives in to focus the attention on the concepts and definitions over the uses and meanings that people create daily.

About this Opazo-Arellano et al. (2020) show how the prospective mathematics teacher adopts an absolute loyalty to the definite integral concept. This avoids the *accumulation* as an autonomous argumentation that emerges from who learns to teach. To identify the adhesion to the school mathematics discourse some arguments, procedures and meanings that result from the mathematics knowledge of the integral being problematic (Table 15.1).

Table 15.1 was developed from the following an excerpt of the audio transcription related to the interview conducted with a prospective mathematics teacher, which is reproduced below (Opazo-Arellano et al., 2020).

> I already solved the integral without reading the questions... I simply saw this and the first I thought was to solve the definite integral. Logically leaving the variables "a" and "b" that area values I do not know. Then in the first question I was asked which would be the value for "a" and for "b" then I simply put the expression in the calculator which showed me 3 and 6 so that was what I put and I did not try with anything else.

Table 15.2 An prospective mathematics teacher's answer in Honduras

Which of the functions do have asymptote behaviors and which do not? Explain

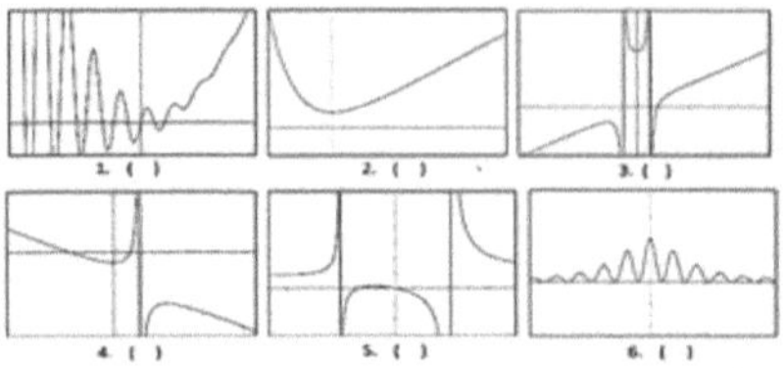

Answer	Answer transcription
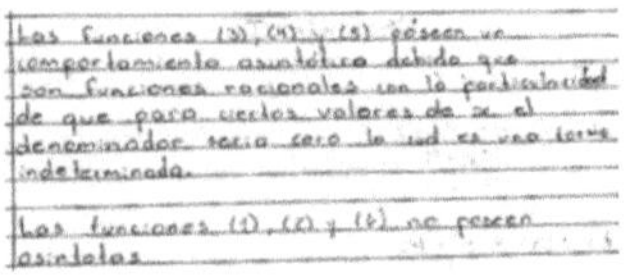	The functions (3), (4) and (5) have a symptote behavior since they are rational functions. For some values of *x* the denominator would be 0 which is an indefinite form. The functions (1), (2) and (6) do not have asymptotes.

Source: Students' work

To solve this definite integral is in terms to apply the procedure that norms what is usual at school mathematics where many of these procedures are acquired and reproduced in routine problems proposed by calculation texts (Valencia Álvarez & Valenzuela González, 2017). This makes to adopt the school knowledge in a hegemonic way causing that the prospective mathematics teacher does not participate of the social mathematics knowledge construction (Opazo-Arellano & Cordero, 2021).

Another example of adhesion to the school mathematics discourse is the asymptote as it is usual within the teaching of mathematics where it is privileged the graphic representation of a function $f(x)$ accompanied by a straight line that is close to $f(x)$ without cutting it. This instruction is not questioned within the initial teacher training (Chávez-Martínez, 2022) what derives into the school knowledge as the unique referent to plan, deliver and evaluate teaching (Opazo-Arellano & Cordero, 2021).

As an example, some demonstrations of this school knowledge during the initial teacher training are present. These are part of a preliminary analysis where the adhesion phenomenon was identified when a prospective mathematics teacher made problematic asymptote behaviors (Table 15.2). In this case the adhesion is in the reproduction of the mathematics object asymptote as straight line.

In the example of the Table 15.2 the prospective mathematics teacher refers to the vertical asymptotes which relate with the values of the rational functions where the denominator makes indefinite forms. This logic favors prototypes of functions that have the graphic behavior of the Fig. 15.3 (Domínguez, 2003).

It is important to state that this prototype of functions are what the school math discourse privileges about the asymptote teaching making the graphic representations hegemonic as the straight line for example.

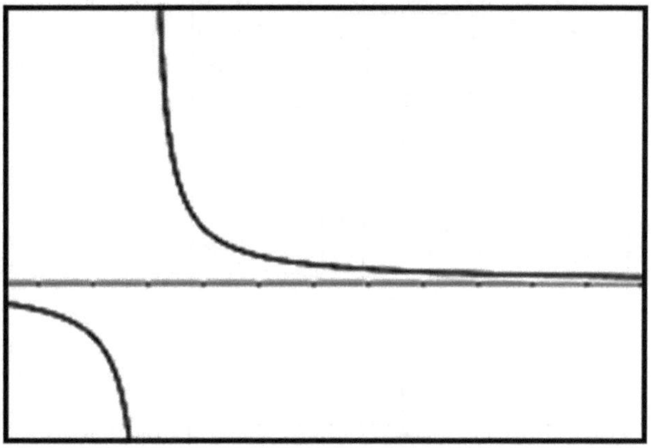

Fig. 15.3 Common asymptote teaching during the initial teacher training at UPNFM (Universidad Pedagógica Nacional Francisco Morazán). (Source: Authors own elaboration)

15.3 Discipline Identity and Socialization School Situations Design

The two examples presented in the previous section show the reach of the school mathematics discourse where what the prospective mathematics teacher learns is defined by this dominant epistemology leaving aside the math knowledge that constructs and spread the teacher who learns to teach through the initial teacher training. Then, the demand itself makes to avoid the adhesion to the school mathematics discourse. Making the school knowledge problematic is relevant since this mechanism makes tense the nature of the knowledge that is learned by the prospective mathematics teacher.

At the same time, it opens a gap to legitimize the diversity of knowledge, to resist the mathematics knowledge put into practice and project the re-signification of the knowledge from the people's autonomous argumentations. It is convenient to declare that to make the mathematics knowledge problematic supposes an epistemological change what is supported from a frame of reference[5] that favors people's mathematical knowledge re-significations. In this sense, a consequence of this process is to value the accumulation (Fig. 15.4).

In other words, the autonomous argumentations that show the uses of math knowledge of the learner, as a result of the debate between functions and forms, which recognize people's experience. It is remarkable that this frame of reference has been systematized in an ontological and epistemological structure where the synthesis is in the situations derived from the systematic study of people' knowledge in their diverse communities of mathematics knowledge as shown in Table 15.3.

[5]To build the frame of reference it is necessary to know, reveal and value the use of mathematical knowledge of the work, the school, the work and the people. All this placed in a horizontal and reciprocal relationship. The frame of reference will articulate the functional and the everyday, promoted by contemporary socialization studies, since they express people's own knowledge and environment, the dialectic between academic and native knowledge.

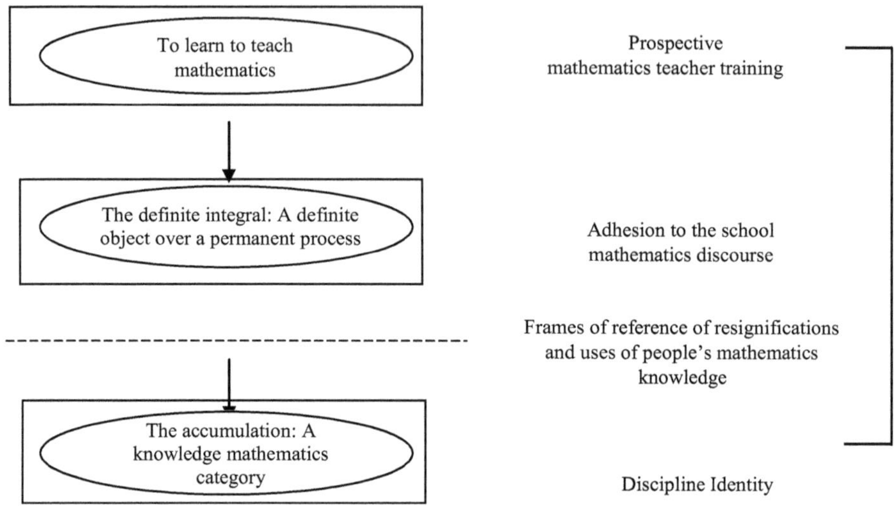

Fig. 15.4 Synthesis of the problems and role of disciplinary identity in initial mathematics teacher training. (Source: Opazo-Arellano et al., 2020)

The question that suits in this context is: *How is the vision of mathematics used in the initial teacher training?* The answer is in the *Socialization school situation design* since it is important for the educational impact. Therefore, its essential function is to value the mathematics knowledge and its re-significations in the knowledge teachers and students' communities. It is also necessary to keep the reciprocity and horizontality between the school and people's habits. That means, through the socialization school situation design contributes to disrupt and transform the school mathematics since the diversity of mathematics knowledge (Cordero, 2022).

The construction of socialization school situations design is compound by two elements: an epistemological basis and a perspective. The epistemological basis roots in each of the situations that compose the mathematical epistemology (Table 15.3). The perspective is one of the multifactors that have been built as the discipline identity (Fig. 15.1). Then the discipline identity and the socialization school situation design articulate in the initial teacher training with basis on making the mathematics knowledge problematic.

The educational impact is necessary however, it is not enough to add or eliminate a mathematics subject from the school curriculum. On the other hand, to promote an epistemological change that increases the learner's knowledge to teach. In this way, when the prospective teacher plans, delivers and evaluates his teaching he will legitimize, resist and project a mathematics classroom that avoids the exclusion of the mathematics knowledge social construction (Opazo-Arellano, 2020). What favors to value knowledge which participate within the social construction of the mathematics knowledge (Cordero et al., 2015). Next, two cases where mathematics knowledge being problematic and the socialization school situation design participate will be shown. The summary is in the epistemological pluralism and the knowledge mainstreaming.

Table 15.3 Calculation and analysis socio-epistemology

Situations							
Mathematics construction	Variation	Change	Transformation	Approximation	Selection	Weight	Periodization
Meanings	Flow Movement Accumulation State Permanent	Area under the curve Position of a mobile Movement of a fluid Thermal constant	Behavior and analytical graphs patterns	Border Derivation Integration Convergence	Adaptation pattern	Behavior distribution	Behaviors reproduction
Procedures	Two states comparison	Comparison of two states	Parameters variation	Formal logic operations (quotient)	Qualities distinction	Equate	Periods comparison
Instruments	Continuous variation quantity $f(x+h)-f(x)=\alpha h$ $\alpha = f'(x)$	Quantity of continuous variation $F(b)-F(a)=\int_b^a f dx$	Instruction that organizes behaviors $y = Af(Bx + C) + D$	Analytical forms $\lim_{h\to 0}\frac{f(x+h)-f(x)}{h} = f'(x)$	What is stable	Balance point $\sum_1^n (x_i-\bar{x})=0$	Interpolation 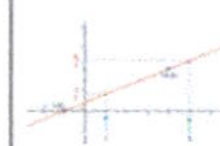
Argumentation Re-signification	Prediction $E_0 + Variación = E_f$	Accumulation	Tendency behavior 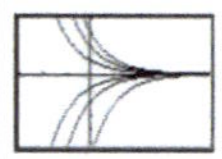	Functions analyticity $f(x+h)=f(x)+f'(x)h$ $+f''(x)\frac{h^2}{2!}+\cdots$	Optimization 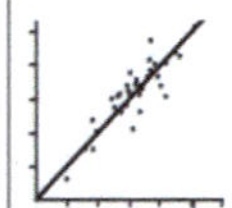	Compensation 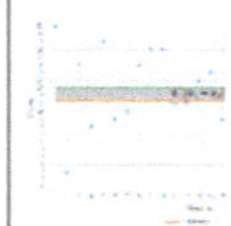	Anticipation 

Source: Cordero (2022)

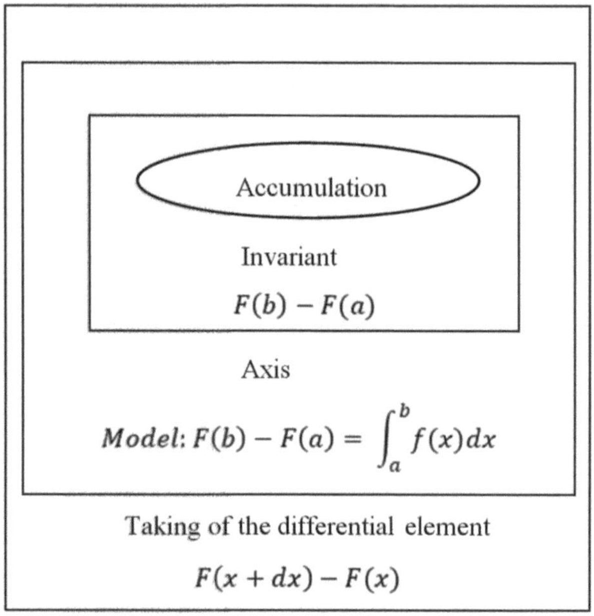

Fig. 15.5 Definite integral new structure. (Source: Cordero, 2022)

15.4 Epistemological Diversity: The Case of Accumulation in the Prospective Mathematics Teachers

The category of accumulation is the expression of the Integral mathematics knowledge function that people have. In Cordero (2003), we find that in mathematics, physics and engineering (Students and Teachers) the structure of the definite integral $\int_a^b f(x)dx = F(b) - F(a)$ is in action. This use made a change of the orientation of the structure: Analyze the model from the root that means: $F(b) - F(a) = \int_a^b f(x)dx$. The change of orientation in the definite integral is the pattern of the theory of integration construction which results into a new structure of the definite integral together to the differential element (Fig. 15.5).

The notion of accumulation is the nucleus of the definite integral structure. The invariant (independent to the context) is the comparison of two states $F(b) - F(a)$. The model is $F(b) - F(a) = \int_a^b f(x)dx$ the main point of the structure and the differential element $F(x + dx) - F(x)$ is an action that relates in a systemic form the progressive phases of the structure and remains invariant to the contexts of the quantities that stream (Cordero, 2022).In order to understand deeper what was explained before Cordero (2003) perceives potentiality in this new structure in the definite integral and suggests a treatment of the argumentation of the integral. Because of its peculiarity he decides to call it category of argument which defines an argumentative plan: the variation. In the variation plan, the integral is associated to the notion of quantity through which form conceptions about the integral ("$\sum dP = P$"), where the objects result to be interrelated variables ($y = f(x)$) and the procedures base on ideas to compare.

Then the differential element been taken plays the role of process in this argumentation field. The conceptions are influenced by the understanding of the local situation to know the worldwide situation (Cordero et al., 2002; Cordero, 2003).

In this regard, we find in Mota (2019) how a community of bio-mathematicians analyzes the specific situation of the thermal constant calculation (how to make the accumulation of the degrees-days). The objective of the calculation proposed by the community is to measure the accumulation of degrees-days each hour. In order to do that the accumulation of degrees-days from the $i - 1$ hour to the i hour are described such as:

$$\frac{max\{(g_{ij} - T_b), 0\}}{24}$$

According to the author the previous expression determines how the degrees-days vary in the local form that means $max\{(g_{ij} - Tb), 0\}$ represents how the temperature varies and the $\frac{1}{24}$ represents to the differential of time, 1 h. The representation of this accumulation as the local state of the situation is the following:

$$F(t + dt) - F(t) = F'(t)dt$$

That results in:

$$F(t + dt) - F(t) = \frac{max\{(g_{ij} - T_b), 0\}}{24}$$

In this context, Mota (2019) mentions that in the previous example it is observed how the quantification of the phenomenon allows to recognize some aspects for example, how the phenomenon varies (temperature) and the time interval in that type of variation remains (1 h). In this study was identified how the differential element be taken contributes with recognizing the accumulation of the local element. This author declares that the category of accumulation emerges such a *functional knowledge*[6] that helps to solve the problem exposed by the bio-mathematicians mathematics knowledge community.

It is relevant to note that the accumulation shows the knowledge of a specific community therefore in the case of the bio-mathematicians the accumulation give the meaning of the thermal constant. What was stated before creates an epistemology different to what it is taught through the school mathematics. Figure 15.6 shows the accumulation category frame.

This epistemology is excluded or overshadowed in calculation lessons as a habit since the classic treatment of the integral in the texts of calculation is generated from the *approximation* category. The argumentative plan is the approximation. In this sense, the integral is associated to the number notion through which establishes

[6]That means where the uses of mathematics knowledge are expressed.

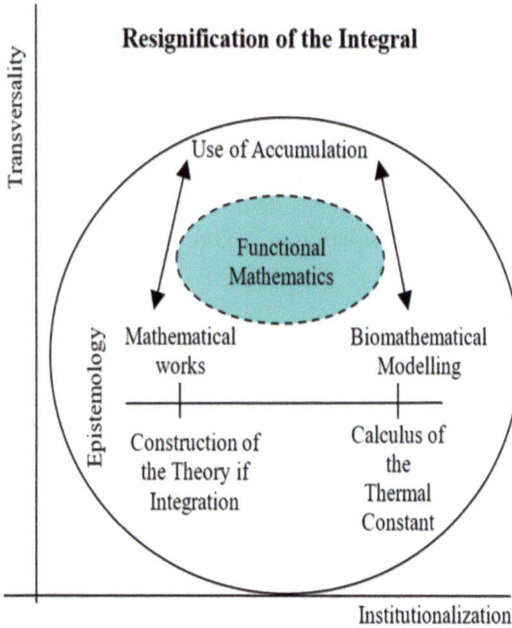

Fig. 15.6 Accumulation category frame. (Source: Marcía-Rodríguez, 2020)

integral definitions "$\lim_{n\to\infty} \sum_{i=1}^{n} f(x_i)\Delta x_i$") where the objects are functions (*f: [a, b]* → ***R***) and the procedures based on the ideas to achieve (interval division, rectangle addition that cover a given region). In this regard, the border is the process used in this argumentation plan. The properties are external to the mathematical object and component elements (Cordero, 2003).

In order to recognize the epistemology effect within Fig. 15.6 which motivated the design of the construction of a socialization school situation where prospective mathematics teachers are. As a result, a phenomenon that is varying continuously is recognized: a constant quantity through time interval. What was exposed before was decided in a research reference frame; eight prospective Mathematics teachers (6 men and 2 women), in their third year of career at Universidad Pedagógica Nacional Francisco Morazán, at Nacaome Campus, at Valle, Honduras participated in this research. This career curriculum has 12 periods (each year has 3 periods) with 49 subjects in total.

It is important to mention that two of the situations that were constructed within the design of the socialization school situation are under the context of the consumption of degrees-days of a pandemic (unit of measurement for the heat accumulation in an organism) and the filling and the emptying of a container. One of the characteristics is the use of the graph when an algebraic expression is absent when it is discussed how something is changing.

In terms of the results of the socialization school situation design, they show how it is considered constant in a small space the phenomenon that is varying continuously within the indicated contexts which are represented by the temperature or water quantity. This development allows the participant to recognize accumulated quantities within short periods of time. In the same sense, this allows to know the total accumulation they add each quantity accumulated in the local place that means the variation within the local place represents small quantities accumulations (differential element) since the adding of those quantities defines the total accumulation (Table 15.4).

It is relevant to mention that the community of prospective teachers does not think about generating areas to approximate to a curve but they face a phenomenon that is varying continuously where each variation accumulates some quantity. In order to value the accumulation as an autonomous argumentation of the participants favors to develop procedures which lead to re-significate the definite integral as something usual within the teaching of mathematics at school since the construction elements as a result of an empirical research.

In this sense the elements of the change situation are relevant as shown in Table 15.3. In other words, the prospective mathematics teachers do not emulate a procedure of something that was previously taught (Cordero, 2022; Marcia-Rodríguez & Cordero, 2021). Therefore the epistemological diversity implies to value other epistemologies that are developed in people's habits (accumulation) and if they are incorporated in the school mathematics that allows to give meaning to the mathematics object.

15.5 Transversality of Mathematical Knowledge: The Case of Behavior with Tendency to the Prospective Mathematics Teachers

The forgotten subject program and transversality of knowledge recognizes in what is usual in the teaching of the school mathematics the absence of people's mathematics knowledge. It is true that this knowledge lives in scenarios such as at work, the city and at school. On the other hand, transversality is promoted of mathematics knowledge such as an element that will transform the school mathematics including what is functional from the knowledge and promoting a connection between the school and routine knowledge (Cordero, 2022).

Transversality of mathematical knowledge is the re-signification of the uses of knowledge in at least two domains or different scenarios. For example: between the school and work or between mathematics and engineering (Mendoza & Cordero, 2018). This means that mathematics knowledge adopts meanings in specific situations (Se_i), according to its use. In particular, the asymptote of a function $f(x)$ is described in the school as straight line that comes closer to $f(x)$ without cutting it.

However, in domains (D_k) such as the bionic engineering (D_1), biology (D_2) and the epidemiology (D_3) the asymptote builds an environment of meanings referred to tendency behaviors among functions: the temperature of the reference of a focus

Table 15.4 Students' answers where they recognize the accumulations in a local form

Second situation activity

1. The following graphs show the forecast of the atmosphere temperature for Friday (blue) and the minimum temperature for the degrees-days[a] consumption of the virus (red).

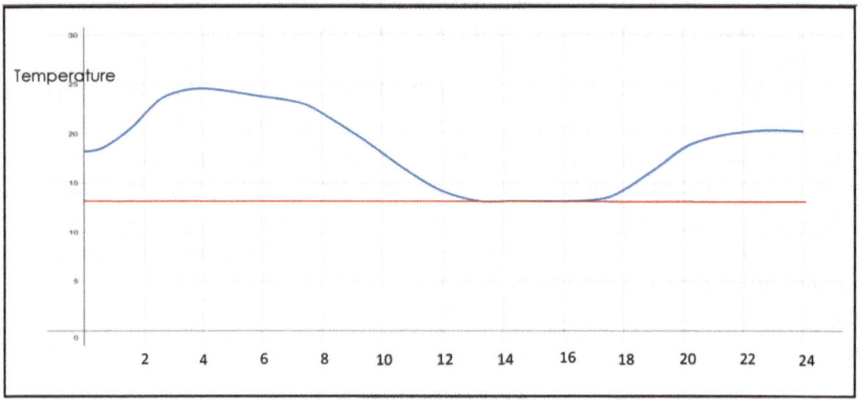

(a) How many degree-days will the bug consume between 0 and 8 h that day

Student	Evidence
AHN	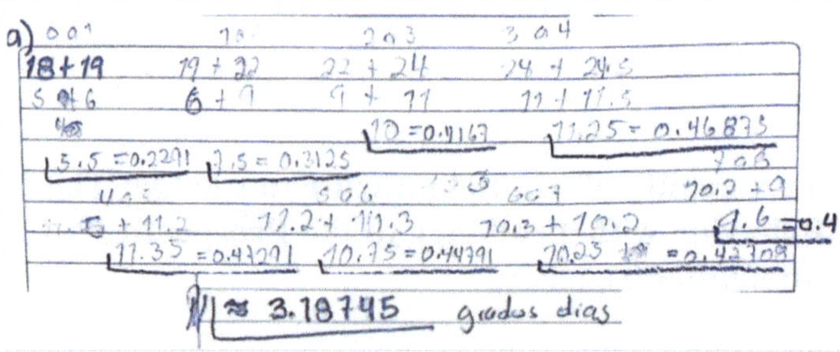
	I took what was in 0, we had said 18, then 19 [. . .] then the result was 11. Then 11/2, resulted in 5.5. **To this 5.5 such it was for 1 h, I divided it between 24 and resulted in 0.2291. Then I did the same with the others and I added all of those results and it was 3.18.**
KHN	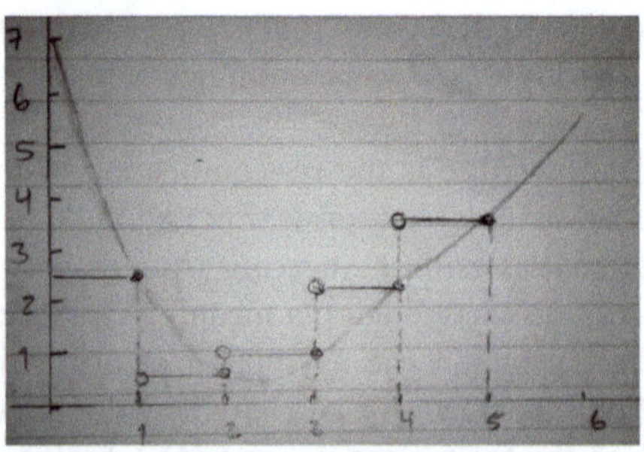
	Graphic representation of the accumulated quantities of the local form developed in one of the activities of the DSES.

Source: Marcia-Rodríguez and Cordero (2021)

[a] The activity is within a context of an insect growing, degrees-days is the unit of measurement to measure the quantity of heat *consumed* or *accumulated.*

(Se_1) (Mendoza-Higuera et al., 2018; Mendoza-Higuera, 2020), the ability of population support (Se_2) (Soto & Vilches, 2018) and the speed of tendency among recovery rates and infection in a pandemic (Se_3) (Chávez-Martínez, 2022).

With this own mathematics knowledge epistemology at work scenario, the socialization school situation design was built and put into practice with prospective mathematics teachers. The students gave new meanings to the asymptote through the value of the mathematics knowledge uses on other domains showing arguments related to the tendency of functions behavior. When this was put into practice 11 prospective mathematics teachers participated (8 men and 3 women)[7] who were in their fourth year of career at Universidad Pedagógica Nacional Francisco Morazán, at Nacaome campus, at Valle, Honduras.

15.5.1 Tendency Behavior of Functions: A Learner's Autonomous Argumentation

Euler, in his translation of *Introducción al Análisis del Infinito* (1748) offered a context of asymptote as an intrinsic property of a straight line which characterizes the behavior through the form of the infinite branch of the straight line (Domínguez, 2003, p. 50). In Domínguez's words (Domínguez, 2003) this refers to the behavior has form and in consequence this fact could help in our contemporary to re-significate the asymptote straight line such as curved asymptote.

This aspect is relevant because the epistemological status from the asymptote in the school mathematics discourse privileges the asymptote of a function such as a straight line (Cordero & Domínguez, 2001). Our concern is to incorporate this epistemology in the socialization school situations design but also that epistemology of uses that people believe.

It is convenient to mention that the Socio-epistemology worries about to recover people's autonomous argumentations in specific situations.

This mathematics knowledge is functional since who learns to teach focus his attention to routine problems. Some examples are the accumulation and the tendency behavior. In terms of the tendency behavior Cordero (1998) affirms that "it is an argument that establishes relationships between functions and it is composed by a coordinated collection of concepts and situations of calculation where some variation worldwide are discussed" (p. 56).

The construction of this argument gives meanings to the asymptote from patterns of graphic and analytical behavior where the subtraction and the quotient are the criteria to allow building asymptote functions $\boldsymbol{f(x)}$ to a given function $\boldsymbol{g(x)}$. The asymptote function is recognized as an instruction that organizes tendency

[7] In the examples that are exposed, to protect the identity of the participants we have decided to name (H) to each of the prospective mathematics teachers. Moreover, a number is present to distinguish them.

Table 15.5 The situation of the transformation and the asymptote

Mathematics construction	Transformation		Asymptote: uses of the asymptote
Meanings	Graphic and analytical behavior patterns		Construction patterns: subtraction and quotient $\lim_{x \to \infty} [f(x) - g(x)] = 0$ $\lim_{x \to \infty} \left(\frac{f(x)}{g(x)}\right) = 1$
Procedure	Variation parameters		Vary the asymptote function $f(x)$ from a given asymptote function $g(x)$
Instrument	Instruction that organizes behavior		The subtraction and quotient are instructions that organize asymptote behavior
Argumentation/ re-signification	Tendency behavior 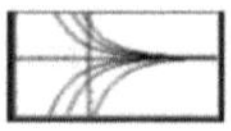		Asymptote behavior

Source: Chávez-Martínez (2022)

behaviors. These elements are summarized in Table 15.5 as a transformation nucleus situation[8] that generates a wider knowledge: the asymptote.

15.5.2 Socialization School Situation Design

The design has three specific situations Se_1, Se_2 and Se_3. These allow an alternation of domains that generate new uses of the asymptote which are given by establish a mutual relationship to achieve and keep a desired temperature, keep the fluctuation of a population around the supportive capacity and determine a positive recovery fee in a pandemic.

15.5.3 Mainstreaming Examples and Evidence in the Prospective Mathematics Teachers

In the Se_1 it is expected that the participants describe the temperature function behavior of a focus $T(t)$ when the time passed by knowing that it must to achieve

[8]The nucleus situation creates a calculation Socio-epistemology that promote the knowledge emergency made by math construction elements (knowledge put into practice), signification, procedure, instrument and argument (Cordero, 2001; Cordero et al., 2019).

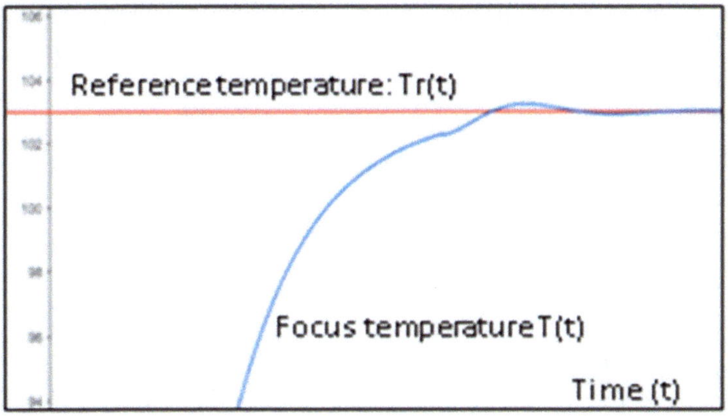

Fig. 15.7 Temperature $T(t)$ of a focus and reference temperature $Tr(t)$. (Source: Chávez-Martínez, 2022)

and keep closer to the reference temperature $Tr(t)$. Figure 15.7 shows the temperature $T(t)$ of a focus and reference temperature $Tr(t)$.

It is known that an on-off control reestablishes the temperature of a focus $T(t)$ comparing it with the reference temperature $Tr(t)$, determining an error $\varepsilon(t)$ that controls the on and off of the focus and with this the temperature. Some of the tasks are:

(a) Write an expression that relates $T(t)$, $Tr(t)$ and $\varepsilon(t)$. Besides, it answers the following: which should be the behavior of $\varepsilon(t)$ to keep $T(t)$ closer to $Tr(t)$?

(b) Express $T(t)$ in terms of $Tr(t)$ and $\varepsilon(t)$.

In this sense in terms of the alternative (a) prospective mathematics teachers *compare* $T(t)$ and $Tr(t)$ through the subtraction to determine the error $\varepsilon(t)$ modelling a construction pattern that caused patterns variation procedures and to organize asymptote behavior. In terms of the alternative (b) of the tasks what were given above prospective teachers present a model that will significate the asymptote as the reproduction of the desired temperature considering that it should be achieved and kept (see H1's answer within Table 15.6).

To vary $\varepsilon(t)$ it was achieved to construct asymptote functions to $Tr(t)$,that means to organize behavior. In order to do this, some specific conditions are necessary. For example, in terms of the behavior $\varepsilon(t)$, the prospective teacher H3 promoted a decreasing exponential function while the prospective teacher described a condition to reproduce a desired temperature that means that $\varepsilon(t)$ should have a tendency towards 0 when the time tend to the infinite. Table 15.7 shows H3 and H4 students' answers about the behavior of $\varepsilon(t)$.

(c) How is the $T(t)$ behavior when the time passes by?

Through the description of $T(t)$ in terms of $Tr(t)$, the future teacher H1 turns into the *tendency form* to highlight the global behavior of $T(t)$ from comparing its

Table 15.6 H1 student's answer to the task (b)

Result	Answer's transcription
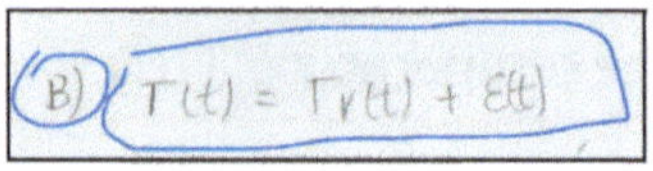	$T(t) = Tr(t) + \varepsilon(t)$

Source: Students' work

Table 15.7 H3 and H4 students' answers about the behavior of $\varepsilon(t)$

H3 student

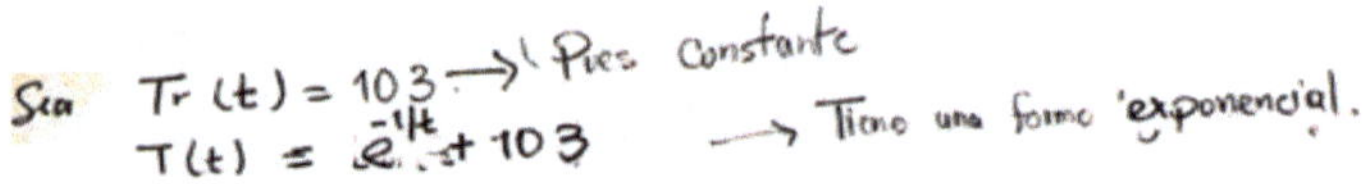

H4 student

Answer	Answer's transcription
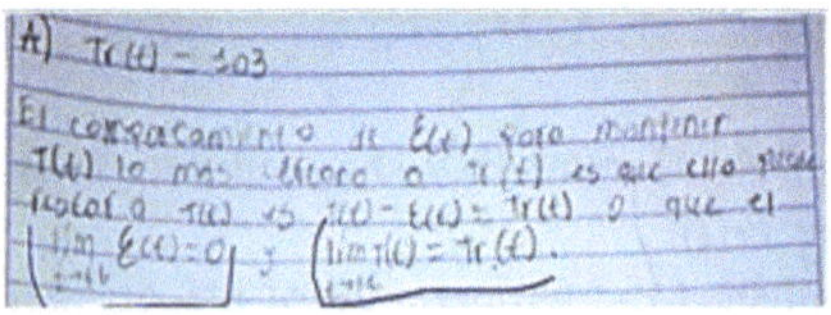	$Tr(t) = 103$ $T(t) = e^{-\frac{1}{t}} + 103$ has an exponential form The behavior $\varepsilon(t)$ to keep $T(t)$ closer to $Tr(t)$ this can subtract to $T(t)$ is $T(t)$-$\varepsilon(t) = Tr(t)$ or that $\lim_{t\to\infty} \varepsilon(t) = 0$ and $\lim_{t\to\infty} T(t) = \lim_{t\to\infty} Tr(t)$.

Source: Students' work

Table 15.8 H1 answer: tendency behavior function of the temperature of the focus description

Answer	Answer's transcription
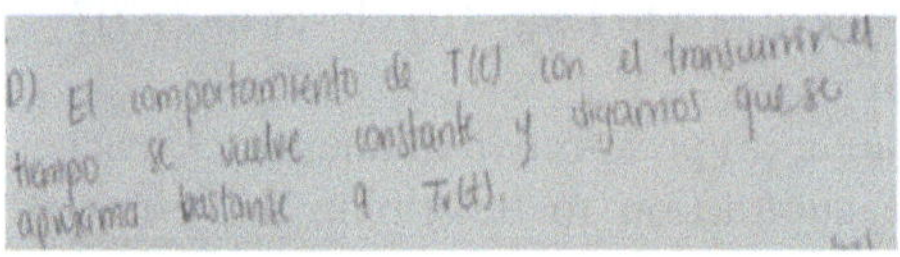	The behavior of $T(t)$ when the time passes by becomes constant and comes closer to $Tr(t)$.

Source: Students' work

tendency with $Tr(t)$, when time tends to the infinite. Table 15.8 shows H1 answer about the tendency behavior function of the temperature of the focus description.

The construction pattern that signified to the asymptote in Se_1 that is re-signified in Se_2 where the tendency of a population dynamics towards a population support is reproduced. The subtraction is also the way of comparison that models the pattern $P(t) = C(t) + S(t)$. $P(t)$ represents the population dynamics $C(t)$ the capacity of population support and S(T) the super saturation. Figure 15.8 shows the capacity of support C(t) and population P(t) relationship.

The expressions $(t) - Tr(t) = \varepsilon(t)$, $P(t) - C(t) = S(t)$ mean a graphic and analytical behavior pattern where the subtraction is part of an asymptote criteria. They will be $f(x)$, $g(x)$ and $h(x)$ functions, $f(x) = g(x) + h(x)$, so $h \rightarrow 0$ when $x \rightarrow \infty$. If $\lim_{x\to\infty} (f(x) - g(x)) = 0$ then $g(x)$ is asymptote of $f(x)$.

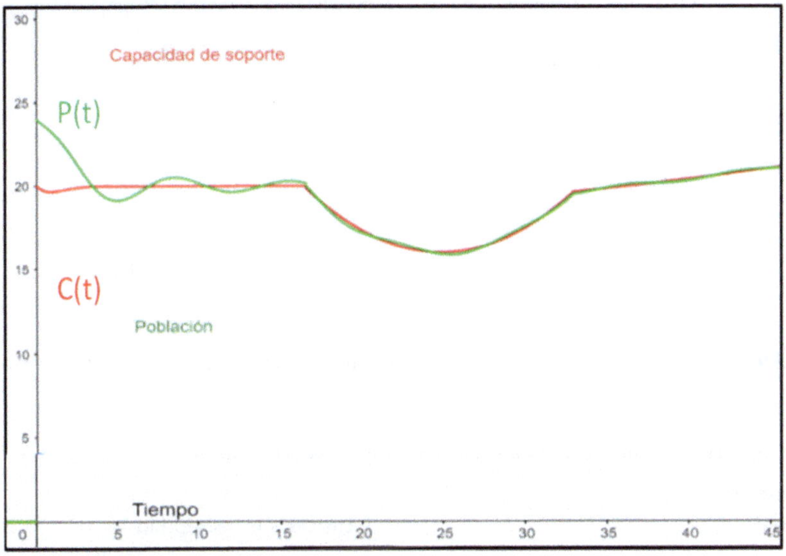

Fig. 15.8 Capacity of support C(t) and population P(t) relationship. (Source: Chávez-Martínez, 2022)

The Se_3 demands a way of comparison by quotient that creates a second asymptote criterion: *They will be f(x), g(x) and h(x) functions of x, if* $\lim_{x\to\infty} \frac{f(x)}{g(x)} = 1$ *then g(x) is asymptote of f(x).* The evidence of this mathematics knowledge that is developed by the prospective teachers is tasks that were prosed in Se_3 and the way they used their knowledge to solve them.

(A) The graphic of three curves was presented to the prospective teachers: one for the infection rate and two for possible recovery rates from a pandemic. Figure 15.9 shows the infection rate and possible recovery rates from a pandemic.
    In the instruction of Se_3 was demanded to choose between two graphics (red or green) the recovery rate that represents the best scenario of a pandemic. The prospective teachers' answers show argumentations from the graphic in terms of the speed of the graphics its closeness to zero or the comparison between its variations. Table 15.9 shows H2, H4, and H5 students' answers to the task (a) of the Se_3.
(B) Given the function $f(t) = e^{-t} + \frac{1}{t+1}$ for the infection rate. Choose between the functions $k_1(t) = e^{-t}$ o $k_2(t) = \frac{1}{t+1}$, the recovery rate that represents the best scenario of the pandemic. Determine what was asked for using the criterion $\lim_{t\to\infty} f(t) - k(t)$.
(C) Does this criterion allow determining the recovery rate that represents the best scenario for the pandemics? Why?

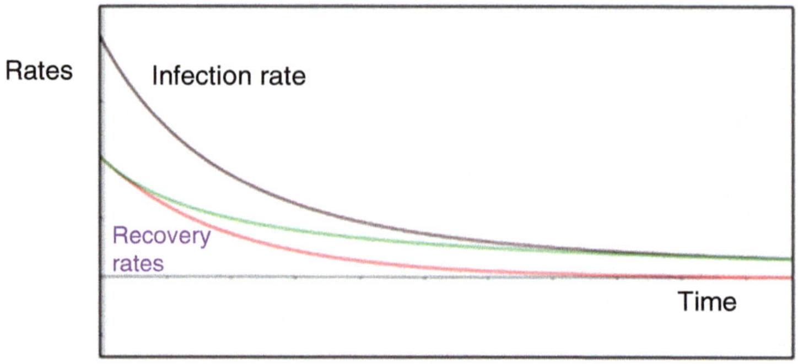

Fig. 15.9 Infection rate and possible recovery rates from a pandemic. (Source: Chávez-Martínez, 2022)

Table 15.9 H2, H4, and H5 students' answers to the task (a) of the Se_3

H2 student	
Answer	**Answer's transcription**
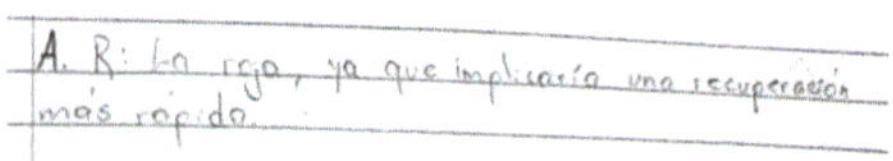 A. R: La roja, ya que implicaría una recuperación más rápida.	The red since it would imply a faster recovery.
H4 student	
Answer	**Answer's transcription**
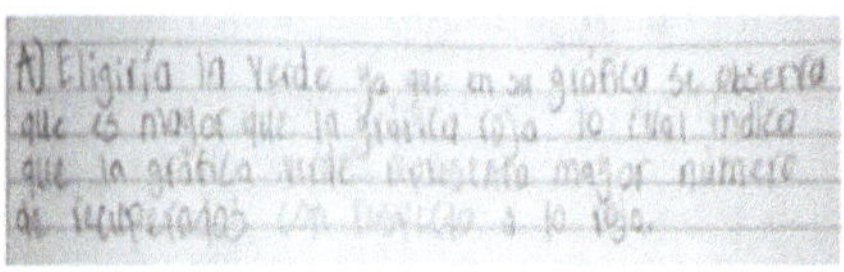 A) Elegiría la verde ya que en su gráfica se observa que es mayor que la gráfica roja lo cual indica que la gráfica verde representa mayor número de recuperados con respecto a la roja.	I would choose the green since in its graphic it is observed that is bigger than the red graphic which indicates the green graphic represents a bigger number of recoveries tan the red one.
H5 student	
Answer	**Answer's transcription**
 A. La verde ya que se mantiene al inicio más lejos de 0.	The green since it remains further from zero at the beginning.

Source: Own construction from the work of the students

In the task (B) previously above, the objective is that the prospective teacher use the first criterion of asymptote in order to create a debate about the way of the use of knowledge in the first two situations. In this sense the participants discuss about the reach of the criterion finding that it does not facilitate to determine the recovery rate that represents the best scenario in a pandemic. However, it is recognized the difference in the speed of the rate increase. Table 15.10 shows H1 student's answer to the task (B) and (C) of the Se_3.

The expression *although* $k_2(t)$ *decreases less* suggests that the speed of the tendency re-significate the uses of the asymptote that refer to its form. In this

Table 15.10 H1 student's answer to the task (B) and (C) of the Se_3

Answer

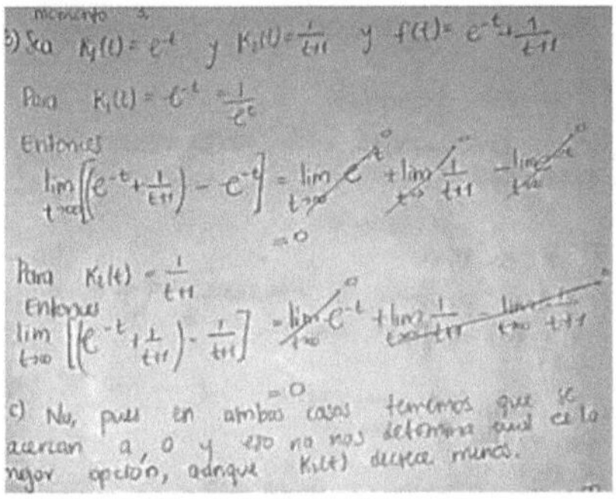

Answer's transcription

(B) This will be $k_1(t) = e^{-t}$ and $k_2(t) = \frac{1}{t+1}$ and $f(t) = e^{-t} + \frac{1}{t+1}$

For $k_1(t) = e^{-t} = \frac{1}{e^t}$

Then

$$\lim_{t\to\infty}\left[\left(e^{-t} + \frac{1}{t+1}\right) - e^{-t}\right] = \lim_{t\to\infty} e^{-t} + \lim_{t\to\infty}\frac{1}{t+1} - \lim_{t\to\infty} e^{-t} = 0$$

For $k_2(t) = \frac{1}{t+1}$

Then

$$\lim_{t\to\infty}\left[\left(e^{-t} + \frac{1}{t+1}\right) - \frac{1}{t+1}\right] = \lim_{t\to\infty} e^{-t} + \lim_{t\to\infty}\frac{1}{t+1} - \lim_{t\to\infty}\frac{1}{t+1} = 0$$

(C) No because in both cases they come closer to 0 and that does not determine which the best option is **although $k_2(t)$ decreases less.**

Source: Own construction from the work of the students

sense, it is affirmed that the asymptote determine the form and the speed of tendency of some curves (Cordero et al., 2010a). In this last task of Se_3 an own epidemiology notion is presented: the basic number of the reproduction the same that allowed expanding the situation offering three possibilities to decide which recovery rate is the best option. Now the task (D) is presented and the development of two future teachers of the study group.

(D) The basic number of reproduction (R_0) represents the number of people that an infected person can infect.

R_0 is obtained through the quotient:

$$R_0 = \frac{\textit{infection rate}}{\textit{recovery rate}}$$

$\lim_{t\to\infty} R_0 > 1$ indicates that there is an increase of the number of the infected cases (epidemic).

$\lim_{t\to\infty} R_0 \leq 1$ indicates that there is a decrease of the number of the infected cases (epidemic balance).

Considering what was stated before, answer the following: calculate $\lim_{t\to\infty} R_0$ to determine which of the functions $k_1(t) = e^{-t}$ o $k_2(t) = \frac{1}{t+1}$ represents the best scenario of the epidemic in termos of the recovery rates

(a) What meaning can you relate to each quotient?
(b) How is the graphic behavior of $f(t)$ related to $k_2(t)$?

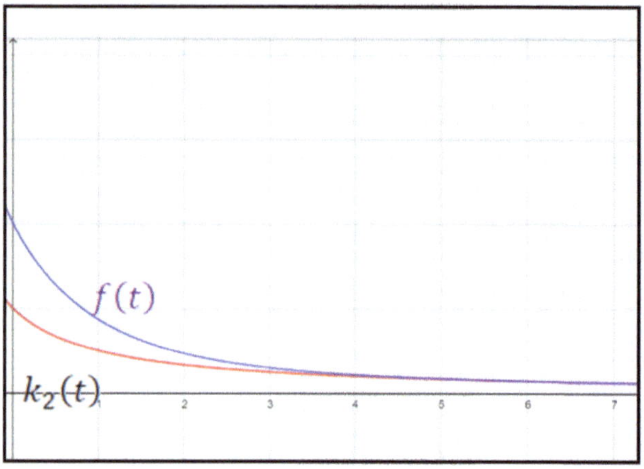

(c) Is there an asymptote relationship between $f(t)$ and $k_2(t)$? Explain.

Proposed tasks in the (D) section have the intention that the future mathematics teachers establish an asymptote relationship between the graphics referring to the tendency behavior. This is comparing the possible recovery rates with the infection rate through the quotient and also to determine the rate that tends faster to the infection rate. Table 15.11 shows the prospective mathematics teacher H5 answers are exposed.

Table 15.11 H5 Student's answer to the task D (a, c and d) of Se_3

Student 5	
Answer	Answer's transcription
	$\lim_{t\to\infty} \frac{e^{-t}+\frac{1}{t+1}}{e^{-t}} = 1 + \frac{e^t}{t+1} = \infty$ $\lim_{t\to\infty} \frac{e^{-t}+\frac{1}{t+1}}{\frac{1}{t+1}} = 1 + \frac{t+1}{e^t} = 1$ (a) It presents an asymptote behavior. (b) Yes, since they are two very similar graphics and one gets closer to the other one.
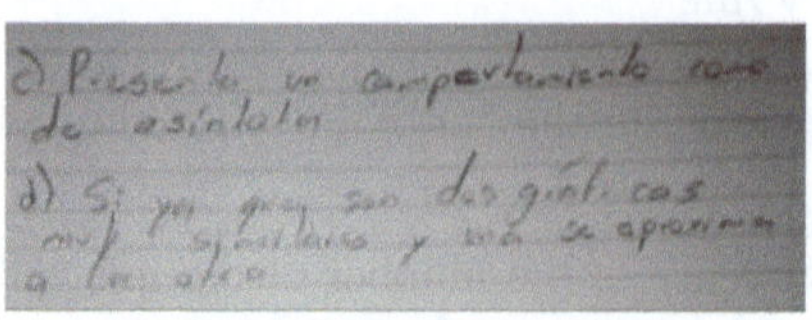	

Source: Students' work

Table 15.12 H3 student's answer to the task D (b) of Se_3

H3 student	
Answer	Answer's transcription
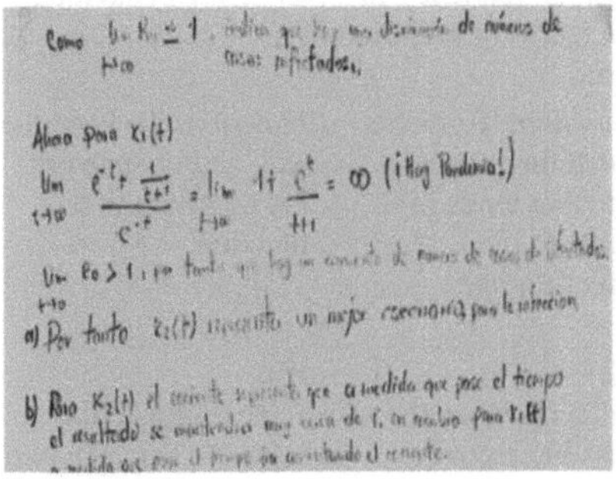	For $k_2(t)$. As $\lim_{t\to\infty} R_0 \leq 1$, indicates that there is a decrease in the number of infected cases. Now for $k_1(t)$. $\lim_{n\to\infty}\left(\frac{e^{-t}+\frac{1}{t+1}}{e^{-t}}\right) = \lim_{n\to\infty}\left(1+\frac{e^{-t}}{t+1}\right) = \infty.$ *There is Pandemic!* $\lim_{t\to\infty} R_0 > 1$, therefore [...] there is an increase of the number of the infected cases. (a) Therefore $k_2(t)$ represents the best scenario for the infection. (b) Para $k_2(t)$ the quotient represents as the time passes by; the result will keep close to 1. On the other hand for $k_1(t)$ when the time passes by the quotient will be increasing.

Source: Students' work

The re-signification of the asymptote uses make us to recognize not only the form but also the speed of tendency between the graphics what gives avoided meanings by the school mathematics discourses. Students' community gives these meanings when interpreting the tendencies between the recovery rates and the infection rate. Table 15.12 shows H3 student's answer to the task D (b) of $Se_{3.}$

15.5.4 Uses of the Asymptote Relationships: Transversality

The re-signification that was exposed within the previous situations resulted from a debate between workings (tasks) and forms (tasks types) (Cordero et al., 2010b) summarized in the following use: *to reproduce tendency behaviors.* In other words, the tasks that promote the construction from the asymptote are closely related with keeping the form and the speed with the tendency of a function to another one.

This is made by the comparison of the functions' behavior *in the infinite* and the determination of its tendencies through the subtraction and/ or the quotient. What stated before is an autonomous argumentation that re-significates the usual teaching of the asymptote since the transversality of knowledge favors the relationships between the uses of the asymptote. Table 15.13 shows the uses of the asymptote relationships since three specific situations.

Table 15.13 Uses of the asymptote relationships since three specific situations

Se_1	Se_2	Se_3	The Asymptote
Use: To reproduce a desired temperature.	*Use:* To reproduce the tendency of a population dynamics.	*Use:* To reproduce a positive tendency in the recovery rate of a virus.	*Use:* To reproduce tendency behaviors.
Function: To keep the temperature of the focus ***T*(*t*)**, as close as possible to the reference temperature ***Tr*(*t*)**.	*Function:* To keep the fluctuation of the population around the capacity of support.	*Function:* To distinguish the recovery rate that tends to faster to the infection rate.	*Function:* To keep the way and/or speed of the tendency of a function.
Way: 1. To compare the temperature of the focus with the reference temperature through the time. 2. To determine an error of temperature of the focus through the function's subtraction. $\lim_{t\to\infty}[T(t) - Tr(t)] = 0$ 3. To reestablish the temperature of a focus	*Way:* 1. To compare the population and the capacity of support through the time. 2. To determine a supersaturation in a population through the subtraction of functions. $\lim_{t\to\infty}(P(t) - C(t)) = 0$ 3. To reestablish the population to its capacity of support.	*Way:* 1. To compare the possible recovery rates with the infection rate through the quotient. 2. To determine the rate that tends faster to the infection rate. $\lim_{t\to\infty}\left(\frac{f(t)}{k(t)}\right) = 1$	*Way:* 1. To compare the behavior between two functions when the independent variable tends to the infinite. 2. To determine its tendencies through the subtraction and/or the quotient. $\lim_{x\to\infty}[f(x) - g(x)] = 0$ $\lim_{x\to\infty}\left(\frac{f(x)}{g(x)}\right) = 1$

Source: Authors own elaboration

15.5.5 *How the Asymptote Increases the Future Mathematics Teacher Knowledge?*

After applying the socialization school situation design, who learn to teacher were asked What is the asymptote? Finding the summary in the tendency behavior of the functions where the graphic organizes behavior that favor to anticipate variation phenomenon in specific domains, scenarios and situations. For example, H1 affirms the following:

> It would be a function since (. . .) there exists another function that takes values in its domain, this functions *tends* to the first function. We can also say that the asymptote *is the tendency that a function has* to evaluate its border for extremely big values.

In this direction, H3 also says that "When there are two functions (. . .) we can say that the asymptote defines the behavior of the other function that means delimiting according to the asymptote function behavior". Then, rethinking the teaching of mathematics demands to value the mathematics knowledge from who learns to teach.

Its impact will be in the re-signification process of the mathematics knowledge which emerge from putting this into practice in situations based on epistemologies where the uses norm the planning, delivering and the evaluation of the teaching. What sated before derives into that the prospective teacher builds a vision about the mathematics knowledge from the transversality of knowledge. In this sense the category of modelling is relevant because it bases the meaning of learn to teach epistemologically and ontologically.

15.6 Conclusions

This chapter calls the attention about the discipline identity factor, theoretical construct that found its ontological and epistemological basis on the re-signification of mathematical knowledge. That means, mathematical knowledge emerges when people face the final objects and opens gaps to the permanent process.

Discipline identity causes that who learns to teach about a vision change about the meaning of what to learn is and what to teach is finding demonstrations of this process in factors such as the epistemological diversity and transversality of knowledge. Two cases have been exposed at the last section.

The relevance of the category of modelling and the discipline identity is to be a common thread about the prospective mathematics teacher knowledge but also in the resistance to the mathematical discourse. This dominant Epistemology causes adhesion to the school knowledge therefore to develop permanent programs that impact the initial teacher training is something urgent.

The impact of the category of modelling and the discipline identity will be defined when to learn to teach is not related to emulate mathematics objects but to value people's knowledge and promote reciprocity and horizontality about mathematics knowledge. Here are our contribution of the education transformation.

References

Blanco, G., & Mercedes, M. (2005). La formación de profesores de matemáticas. Un campo de estudio y preocupación [Mathematics teacher training. A field of study and concern]. *Educación Matemática, 17*(2), 153–166.

Chávez-Martínez, H. (2022). *Un diseño de situación escolar con perspectiva de socialización: una resignificación de la asíntota en estudiantes de docencia de la matemática* [*A school situation design with a socialization perspective: a resignification of the asymptote in mathematics teaching students*]. Tesis de Maestría no publicada, Centro de Investigación y de Estudios Avanzados del Instituto Politécnico Nacional (Cinvestav-IPN), Mexico.

Contreras, I., Rittershaussen, S., Montecinos, C., Solís, M., Núñez, C., & Walker, H. (2010). La escuela como espacio para aprender a enseñar: visiones desde los programas de formación de profesores de educación media [The school as a space for learning to teach: visions from secondary education teacher training programs]. *Estudios Pedagógicos, XXXVI*(I), 85–105.

Cordero, F. (1998). El entendimiento de algunas categorías del conocimiento del cálculo y análisis: el Caso del comportamiento tendencial de las funciones [The understanding of some categories of calculus knowledge and analysis: The case of trend behavior of functions]. *Revista Latinoamericana de Investigación en Matemática Educativa, 1*(1), 56–74.

Cordero, F. (2003). *Reconstrucción de significados del Cálculo Integral. La noción de acumulación como una argumentación [Reconstruction of the meanings of Integral Calculus. The notion of accumulation as an argumentation]*. Grupo Editorial Iberoamérica.

Cordero, F. (2016). La función social del docente de matemáticas: pluralidad, transversalidad y reciprocidad [The social function of the mathematics teacher: plurality, transversality and reciprocity]. In S. Estrella, M. Goizueta, C. Guerrero, A. Mena-Lorca, J. Mena-Lorca, E. Montoya, A. Morales, M. Parraguez, E. Ramos, P. Vásquez, & D. Zakaryan (Eds.), *XX Actas de las Jornadas Nacionales de Educación Matemática* (pp. 23–30). SOCHIEM, Instituto de Matemática de la Pontificia Universidad Católica de Valparaíso. ISSN: 0719-8159. Retrieved from http://ima.ucv.cl/congreso/xxjnem/

Cordero, F. (2022). *La Matemática, sus Usos y Significados. Un Programa Socioepistemológico de la Matemática Educativa* [*Mathematics, its uses and meanings. A socioepistemological program of educational mathematics*]. Manuscrito en preparación.

Cordero, F., Del Valle, T., & Morales, A. (2019). Usos de la optimización de ingenieros en formación: el rol de la ingeniería mecatrónica y de la obra de Lagrange [Uses of optimization for engineers in training: The role of mechatronics engineering and Lagrange work]. *Revista Latinoamericana de Investigación en Matemática Educativa, 22*(2), 185–212.

Cordero, F., & Domínguez, I. (2001). Algunas construcciones de comportamientos asintóticos senoidales en estudiantes de precálculo y cálculo. En G. Beitía (Ed.), Acta latinoamericana dematemática educativa (pp. 325–332). México: Grupo Editorial Iberoamérica.

Cordero, F., Dubinsky, E., & Ferrari, M. P. (2010a). *Arreglos numéricos, transformaciones de figuras geométricas y comportamientos de las funciones como recursos de álgebra abstracta [Numerical arrays, geometric figure transformations and function behaviors as abstract algebra resources].* Díaz de Santos.

Cordero, F., & Flores, R. (2007). El uso de las gráficas en el discurso matemático escolar. Un estudio socioepistemológico en el nivel básico a través de los Libros de texto [The use of graphs in the school mathematics discourse. A socioepistemological study at a basic level through textbooks]. *Revista Latinoamericana de Investigación en Matemática Educativa, 10*(1), 7–38.

Cordero, F., Gómez, K., Silva-Crocci, H., & Soto, D. (2015). *El Discurso Matemático escolar: la Adherencia, la Exclusión y la Opacidad [The school mathematics discourse: Adherence, exclusion and opacity].* Gedisa.

Cordero, F., Mena, J., & Montalto, M. E. (2010b). Il ruolo della giustificazione funzionale in una situazione di risignificazione dell'asintoto [The role of functional justification in a situation of re-signification of the asymptote]. *l'insegnamento della Matemática e delle Scienze Integrate, 33*, 457–488.

Cordero, F., Muñoz, G., & Solís, M. (2002). *La integral y la noción de variación [The integral and the notion of variation]* (Serie: Cuadernos de Didáctica). Grupo Editorial Iberoamérica.

Cordero, F., & Silva-Crocci, H. (2012). Matemática Educativa, Identidad y Latinoamérica: El quehacer y la usanza del conocimiento disciplinar [Educational mathematics, identity and Latin America: The work and the use of disciplinary knowledge]. *Revista Latinoamericana de Investigación en Matemática Educativa, 15*(3), 295–318.

Cornejo, J. (2014). Prácticas profesionales durante la formación inicial docente: análisis y optimización de sus aportes a los que aprenden y a los que enseñan a "enseñar" [Professional practices during initial teacher training: Analysis and optimization of their contribution to those who learn and those who teach to "teach"]. *Estudios Pedagógicos, XV*(I), 239–256.

Domínguez, I. (2003). *La resignificación de lo asintótico en una aproximación Socioepistemológica.* Tesis de Maestría no publicada [*The resignification of the asymptotic in a socioepistemological approach*]. Unpublished master's thesis, Centro de Investigación y de Estudios Avanzados del Instituto Politécnico Nacional (Cinvestav-IPN), Mexico.

Marcia-Rodríguez, S., & Cordero, F. (2021). *Use of the definite integral in mathematics teaching students. A category of accumulation: towards an autonomous and creative learning* (Artículo enviado a revisión).

Marcía-Rodríguez, S. L. (2020). *Resignificación de la Integral en una comunidad de estudiantes de docencia de la matemática. Una categoría de acumulación y la perspectiva identidad disciplinar.* tesis de maestría no publicada [*Resignification of the Integral in a community of mathematics teaching students. A category of accumulation and the prospective disciplinary identity*]. Unpublished master's thesis, Centro de Investigación y de Estudios Avanzados del Instituto Politécnico Nacional (Cinvestav-IPN), Mexico.

Mendoza, J., & Cordero, F. (2018). La modelación en las comunidades de conocimiento matemático. El uso de las matemáticas en ingenieros biónicos. El Caso de la estabilidad [Modelling in the mathematics knowledge communities. The use of mathematics in bionic engineers. The case of stability]. *Revista Latinoamericana de Etnomatemática, 11*(1), 36–61. Retrieved from http://www.revista.etnomatematica.org/index.php/RevLatEm/article/view/501

Mendoza-Higuera, E., Cordero, F., Solís, M., & Gómez, K. (2018). El uso del conocimiento matemático en las comunidades de ingenieros. Del objeto a la funcionalidad matemática [The use of mathematical knowledge in engineering communities. From object to mathematical functionality]. *Bolema—Boletim de Educação matemática, 32*(62), 1219–1243.

Mendoza-Higuera, J. (2020). *El uso de la matemática en la Ingeniería Biónica. De la estabilidad a la reproducción de comportamientos en un sistema de control. Una Categoría de Modelación [The use of mathematics in Bionic Engineering. From stability to reproduction of behaviors in a control system. A Modelling Category*]. Tesis doctoral no publicada, Centro de Investigación y de Estudios Avanzados del Instituto Politécnico Nacional (Cinvestav-IPN), Mexico.

Mercado, R. (2002). *Los saberes docentes como construcción social [Teaching knowledge as a social construction]*. Fondo de Cultura Economica.

Mota, C. (2019). *La Matemática escolar y la Modelación: De la integral a una Categoría de Acumulación* [*School mathematics and modelling: From the integral to a category of accumulation*]. Tesis de maestría no publicada, Centro de Investigación y de Estudios Avanzados del Instituto Politécnico Nacional (Cinvestav-IPN), Mexico.

Opazo-Arellano, C. (2020). *Identidad disciplinar en la formación inicial docente: Una resistencia al discurso Matemático escolar* [*Disciplinary identity in the initial teacher training: A resistance to the school mathematics discourse*]. Tesis doctoral no publicada, Centro de Investigación y de Estudios Avanzados del Instituto Politécnico Nacional (Cinvestav-IPN), Mexico.

Opazo-Arellano, C., & Cordero, F. (2021). Estudiante de docencia en matemáticas y la construcción de la identidad disciplinar [Mathematics teaching student and the construction of the disciplinary identity]. *Estudios Pedagógicos, 47*(1), 109–131. https://doi.org/10.4067/S0718-07052021000100109

Opazo-Arellano, C., Cordero, F., & Silva-Crocci, H. (2018). ¿Por qué estudiar la identidad disciplinar en la formación inicial del docente de matemáticas? [Why study disciplinary identity initial mathematics teacher training?]. *Premisa, 20*(77), 5–20.

Opazo-Arellano, C., Cordero, F., & Silva-Crocci, H. (2021). Valorar los usos y significados del conocimiento matemático del estudiante de pedagogía en matemáticas: Una tarea pendiente [Assessing the uses and meanings of mathematical knowledge of mathematics education student: A pending task]. In F. Cordero (Ed.), *La Matemática Educativa y Latinoamérica. Un diálogo entre grupos de investigación: Chile y México* (pp. 113–134). Gedisa.

Opazo-Arellano, C., Marcía-Rodríguez, S., & Cordero, F. (2020). Adherencia al discurso Matemático escolar: el Caso de la integral definida en la formación inicial docente [Adherence to school mathematical discourse: The case of the definite integral in initial teacher training]. *UCMAULE, 59*, 31–55. https://doi.org/10.29035/ucmaule.59.31

Silva-Crocci, H. (2021). Matemáticos Educativos en formación y programas de investigación en Latinoamérica: Reciprocidad en la Resistencia y organización antes los efectos del fenómeno de adherencia [Educational mathematicians in training and research programs in Latin America: Reciprocity in resistance and organization before the effects of the phenomenon of adhesion]. In F. Cordero (Ed.), *La Matemática Educativa y Latinoamérica. Un diálogo entre grupos de investigación: Chile y México* (pp. 135–147). Gedisa.

Soto, D., & Vilches, K. (2018). Resignificación de algunas nociones matemáticas en escenarios de la Biología [Resignification of some mathematical notions in biology scenarios]. In L. Sema (Ed.), *Acta Latinoamericana de Matemática Educativa* (pp. 1324–1331). Grupo Editrorial Iberoamérica.

Valencia Álvarez, A. B., & Valenzuela González, J. R. (2017). ¿a qué tipo de problemas matemáticos están expuestos los estudiantes de Cálculo? Un análisis de Libros de texto [What kind of mathematical problems are calculus students exposed to? An analysis of textbooks]. *Educación matemática, 29*(3), 51–78. https://doi.org/10.24844/em2903.02

Chapter 16
Contemporary Learning in the Interaction of the Human with Data, Via Technology-Mediated Graphics: The Discourse-Representation Dialogue in Mathematics

Arturo Mena-Lorca, Jaime Mena-Lorca, and Astrid Morales-Soto

16.1 Introduction

We assume that education is a human right (United Nations, 1948). Furthermore, we assume that mathematics education is also a human right (UNESCO-ICSU, 1999), and a "fundamental prerequisite for democracy" (p. 34). Countries and economies recognize these rights and strive to satisfy them.

Our country ratifies it (Gobierno de Chile, 2006). Nevertheless, in the region, data show that goodwill does not suffice to achieve such an important goal. Data also suggest that a country's educational system may not only fail to equalize opportunities for all but also tends to be regressive and to augment inequality.

16.2 A Big Challenge

Chile has had an ever-increasing performance in the measurements of the Program for International Student Assessment, PISA, which it joined in 2003. Such results are consistent with UNESCO's Latin American Laboratory for Evaluation of Education Quality, LLECE, and Trends in International Mathematics and Science Study, TIMSS (OECD, 2017). However, that growth is slow, and, in absolute terms, the performance is not good.

Almost half of the students are below level 2 on the test, either minimum or below the scale. That means that an approximate ratio of 1:2 represents the children who

A. Mena-Lorca (✉) · J. Mena-Lorca · A. Morales-Soto
Pontificia Universidad Católica de Valparaíso, Valparaíso, Chile
e-mail: arturo.mena@pucv.cl; jaime.mena@pucv.cl; astrid.morales@pucv.cl

M. Rosa et al. (eds.), *Mathematical Modelling Programs in Latin America*,
https://doi.org/10.1007/978-3-031-04271-3_16

can handle at most routine problems in which all the information is in sight. The approximate ratio of Chilean children who reach the two highest levels in the test is like the ratio of children below the scale in the countries with the best results.

If we compare the approximate ratio of students below level 2 in the test in different groups of countries, we have that: in Latin America, it is 2:3; in OECD countries, 1:4; in countries (of the sample) with a GDP like Chile, 1:3. There is a gap of 2 years in mathematics schooling with the OECD average and 3 years with high-performing countries such as Taipei (OECD, 2017). We must add significantly lower performances of students with fewer economic resources (OECD, 2017), of women as compared to men, and of students of technical-professional education as compared to those of scientific-humanist education (MINEDUC, 2019).

Now, the World Economic Forum claims that, by 2025, due to a shift in the division of labor between humans and machines, “85 million jobs may be displaced by, while 97 million new roles may emerge that are more adapted to the new division of labor between humans, machines and algorithms” (World Economic Forum, 2020, p. 5). This projection involves a severe social problem (APEC, 2020), of which not all educators seem concerned.

16.3 A Researchers’ Community Task

Researchers involved in mathematics education in the region must take responsibility so that their tasks do not circumscribe to the advancement of the discipline. For instance, if mathematics teachers think that teaching mathematics consists of teaching mathematical procedures (in Chile, which is true for 1:3 teachers) (Ávalos, 2014), or do not tune their student assessment with the facilities provided by computational aids, there is not much room for improvement. The situation vaguely resembles Achilles and the tortoise, but with Achilles already ahead and accelerating and the tortoise doing calisthenics.

Nowadays, general curriculum development tends to stress the importance of independent thinking, collaborative work, use of technology, some integration of scientific disciplines and mathematics, *modelling* to understand phenomena (MINEDUC, 2019). No doubt, problem-solving and *modelling* are mathematics activities; communicating and arguing are also, in this case, mathematics. Nevertheless, the status of representing, which, for many a reason is considered crucial for teaching and learning mathematics, has varied in mathematics history.

Our proposal centers on mathematical modelling with permanent technological support and direct reasoning on graphs of various kinds and involves a rather drastic reduction of routine calculations. However, since the mathematical community took a critical distance from arguments *more geometric* (especially in the nineteenth century), we examine the issue here.

16.4 The Various Societies

Cell phones and computers are essential when considering contemporary culture. Some people are thinking about an ecosystem, the 5.0 or Super-Smart Society (Council for Science, Technology and Innovation, 2018), that would be the culmination, soon, of earlier stages of human development. Society 1.0 would be Hunter-gatherers; 2.0, the Agrarian Society; 3.0, the Industrial Society; and 4.0 would be the current Information Society.

Naturally, educational policies must respond to this type of phenomenon. Suppose we are in transit from one to another of these Societies. Limiting ourselves to insisting on the uses of the preceding one is necessarily a setback since the stage is moving. The dynamics of transition often repeat on a smaller temporal scale, which is related to the evident fact that the knowledge of one generation is not enough to solve the problems that the next generation must face. If education planning does not consider this issue, it will hardly reach its goal, the tortoise walks, modern Achille boarded a plane (Mena-Lorca, 2022).

Now, changing the era has a profoundly democratic quality: a technological advance may initially produce inequality, but when the community already assimilates that advance, that inequality disappears (Turchin, 2016), to give way, in principle, to a better condition. Is it not the goal of all education to move to a better stage than the present, personally, and collectively, and to strive tirelessly to reduce inequality?

Anthropologist Donald states other *eras* or *cultures* of humankind (Donald, 1991) that appear as responses to increasingly complex scenarios. The emergence of language initiates the mythical culture; language depends on the development of symbolic communication: *What things are and what they mean can be elaborated and outsourced.*

Theoretical culture appears with writing; the necessity of recording complex facts and phenomena (trade, astronomy) leads to creating external symbols of which mathematicians were the first; memory externalizes. The written record improves the ability to relate ideas and supports better analytical thinking; the development of science inextricably links to these records; teaching emphasizes computational skills.

The primary function of memory is no longer remembering information but assisting in complex mental processes. What characterizes theoretical culture (Donald, 2007) is its "massive external memory storage" (p. 218), which in turn "becomes by far the most important factor of an individuals' cognition" (p. 212).

Now, it is evident that, at present, the strong development of information technology and the Internet translates into modifications in the acquisition of knowledge. The brain may freely focus on things for which it previously had less time and spend considerably less time on routines and more on creativity (Villani et al., 2018). We add that, with computers, processes that previously only human brains could perform (spell checking, arithmetic, or symbolic calculation), in turn, externalize.

Furthermore, the Internet works as a great calculator with countless controls supported by various groups (scientific and others) relating to tides, earthquakes, climate, economy, and many others. Thus, with digital technology, *processes* that previously only human brains could perform externalize.

16.5 Graphic Thinking

16.5.1 Graphical Representations in Mathematics

Representations play a crucial role in learning mathematics (Duval, 2017). However, the history of mathematics has led it to explicitly depart from graphical representations, at least in some now fundamental school and university curriculum areas. Euclid developed a kind of *geometric algebra* (van der Waerden, 1961; Høyrup, 2017).

Numbers are what we can construct with straightedge and compass: "all of them" are commensurable segments of m, n, $m + n$, mn, m/n units (all of them reducible to the same denominator). Nevertheless, we can also construct $\sqrt{2}$, which cannot be written as m/n (Aristotle, 2000, 1.23.41 to 26). Thus, a segment of 2 units and an arc whose measure is $\sqrt{2}$ do not intersect.

16.5.2 "Modern" Graphics

In 1739, Nicole Oresme, in his *Tractatus de configurationibus qualitatum et motuum*, states a mathematical representation of movement (Clagett, 1968). It includes uniform movement and uniformly and not uniformly accelerated movement. For Oresme, every measurable thing (except numbers) is a continuous *quantity*. He does not refer only to physical bodies' movement, but also to the intensity of linear qualities: the intensity of the light of the sun, the whiteness of a substance, the grace of God.

Oresme represents the "intensity" (Clagett, 1968, p. 169) of a quality by a rectangular line segment orthogonal to a baseline where the *latitude*, i.e., the quantity of any linear quality is continuously represented. Thus, this representation not only preludes the Cartesian plane but is also a tool for modelling quite a wide spectrum of phenomena. Then, Oresme proceeds to think of such diverse linear qualities in terms of their corresponding geometrical representations (Suárez, 2014).

Tartaglia, in his *Nuova Scientia* (1537), takes a necessary step, drawing the trajectory of a bullet as a curve, when even experienced gunners thought, according to their reading of Aristotle, that a cannonball went up in a straight line and then fell vertically (Santbech, 1561). Fermat (1636) and Descartes (1637) changed the sense of graphing in much of mathematics.

For example, an ad hoc (quadrant of a) plane allows drawing curves, condensing loci into formulas, and expressing succinctly many geometric properties that the Greeks studied. For example, equation $\frac{x^2}{a^2} + \frac{y^2}{b^2} = 1$ has all the information needed about an ellipse. It is also an expression of a gardener's construction, although decoding the formula takes some work.

Subsequent advances, particularly by Newton and Leibniz, showed how the 'algebraic' treatment allowed the study of a geometry that cannot be faithfully represented due to its *infinitesimal* smallness.

16.5.3 Two Major Problems

When Infinitesimal Calculus appeared, making a rigorous building in the style of Euclid was not a priority: the vast number of applications was interesting enough. However, later, the need to lay firm foundations became apparent. Two were the most decisive problems; one refers directly to graphic representations; the other closely relates it.

A significant problem was the severe difficulties encountered when handling mathematical series as if it were simply a matter of adding finitely many terms. For example, Euler (1769) registers that $s = 1 - 1 + 1 - 1 \ldots = \frac{1}{2}$, since one would have $1 - s = s$. On the other hand, it became increasingly apparent that working on infinitesimal calculus occasionally resorted to (Euclidean) geometric arguments.

For example, Bolzano (1817), trying to complete one of Gauss's proofs of the Fundamental Theorem of Algebra, notices that this used the following: if a real polynomial is negative for some value a of x, and positive for a value b, then, in some determined value c in the interval $[a, b]$ (or $[b, a]$), the corresponding curve would intersect the axis OX, and therefore would have a root there.

This fact is today a corollary of the intermediate values theorem. However, it requires (similarly to locating $\sqrt{2}$ among the commensurable numbers) to verify that the intersection of the curve with the axis is possible. Mathematics needed to understand continuity/completeness: analysis called for a deeper comprehension of numbers.

The *arithmetization of Analysis*, attributed to Weierstrass, was initiated in Cauchy (1821) and culminated in Peano (1890). It provides a numerical-only foundation for the infinitesimal calculus, whose properties are based only on number systems' axiomatics. It distrusted geometric arguments, and such well-founded skepticism remained.

Additionally, around 1870 began a long-lasting debate related to other foundational problems, and the need to work symbolically and, as far as possible, without inadvertent assumptions, increased (Bochenski, 2018).

All this leaves us, then, a severe dilemma for teaching, which cannot simply betray the desire for mathematical rigor but requires graphic representations.

Nonetheless, the problem solved by the *arithmetization of Analysis* is precise and should not be understood too generally.

Mathematicians, when working, often use pictures, the profusion or scarcity of drawings depending on the specialty, but this will not show in mathematical papers. Regardless, there are phenomena whose representation must be well understood. For example, the convergence of a sequence depends *on a neighborhood of infinity*; what happens before any constant on the domain of the sequence does not affect that convergence; hence, the graph could never represent it.

Also, the graphing of a continuous curve cannot support the claim that a polynomial curve does not have infinite (small) jumps, peaks, or sinuosity. For a considerable part of mathematics, the graphical representation is not understandable without theory's considerations. Hence, using graphical representations for mathematics teaching cannot be justified solely for pedagogical reasons since hiding what we must see is the risk.

16.6 Representations Today

Strictly speaking, the questioning of graphing that we have been doing refers to continuity/completeness and categories within them–differentiability. That leaves aside, of course, the non-continuous case. There are graphical representations in which continuity/completeness might be considered, but it does not hinder their usefulness; for example, (topological) *graphs*, Leibniz-Euler-Venn diagrams. However, these diagrams seek to represent basic notions such as belonging and set inclusion; thus, completeness, which the diagrams might suggest, does not affect those notions.

Now, the idea of representation itself turns into mathematical objects, such as in the representation of groups, of Lie algebras, of categories, and others: *representing* objects allow us to study their properties. A somewhat different approach to imagine how vital representation can be in mathematics is an observation made by the authoritative Saunders Mac Lane (1971).

For example, Mac Lane (1971) states that the notation $f: X \to Y$ for functions quickly replaced the notation $f(X) \subset Y$, and added: "It expressed well a central interest in topology. Thus, a notation (the arrow) led to a concept (category)" (p. 29). Of course, an unavoidable reflection here, as a somehow related example, is the importance of the decimal positional numbering system for the development of mathematics.

16.7 The Case of Differential Equations

The arguments in favor of graphical representations are not valid for the usual topology of Euclidean spaces, questioned by the arithmetization process. We analyze this question here since mathematical modelling tends to resort to differential equations that occupy a prominent place in the arithmetization of Analysis, both in questioning geometric representations and difficulty presented by the convergence of series.

As we will see, mathematicians have used representations to formulate a differential equation, solve it, and advance the theory. Accordingly, the representations are diverse: some very precise, others rather qualitative, freehand drawn; the former seek to understand better the results obtained, the latter develop geometric intuition, helps to reason, stimulates discovery, and even shortens proofs.

At the end of the seventeenth century, Calculus was a way of solving geometric problems: analytical expressions allowed establishing relationships between geometric objects (Tournès, 2009). In differential equations, the main question was solving the inverse tangent problem: constructing a curve whose tangents satisfy a given property. Thus, the solutions obtained had to be interpreted in geometric terms, for which simple curves were drawn and intersected.

Hence the importance of the tractrix and its generalizations, which make it possible to draw transcendent curves. These appeared in that problem but were outside the scope of Cartesian geometry, limited to algebraic curves (Tournès, 2009). Riccati (1752) will prove that general tractrices allow to obtain every curve defined by a differential equation. Differential equations texts frequently used representations (Manfredi, 1707, e. g., collects three dozen illustrations at the end).

In the middle of eighteenth century, there is a marked transition from the geometric to the algebraic: the equation is no longer the means, but the end; the equation is a definition of the curve, and now it is a question of studying functions, even in the absence of geometric representation (Guicciardini, 1994).

Algebraic algorithms now serve to obtain the solutions, finite or not (series, continued fractions). The calculation becomes very technical, for example, with Euler. On the other hand, the study focuses on the complex domain, and the solutions cannot be represented as real curves, and Euler, e. g., does not include any drawing in his writings on the subject (Euler, 1769). This causes an obstacle to visualization and intuition (Israel & Menghini, 1998).

Poincaré (1882a, b) reverts to representations from a different perspective: the qualitative study of the global properties of integral curves. He observes that most differential equations cannot be integrated by known functions and proposes to study locally each function defined by a differential equation, according to tradition, and the global aspect of the set of the integral curves. His diagrams belong in the real domain and are freehand drawn since only the relative positions of the curves matter. He makes sure that the points he defines (singular, limit cycles, and cycles without contact) are represented (Tournès, 2012).

Poincaré (1892) does not abandon the analytical style but makes representations that enhance the intuition of the general shape of the integral curves that must plot in these diagrams; these allow the detection of global properties not yet proven and guide the quantitative analysis in some relevant areas. His geometry is of a new kind, and he names it, following Leibniz, *Analysis situs*. His method allows him to show some errors made, for example, in Hill's careful and relevant calculations, due to a lack of geometric intuition (Tournès, 2012).

Poincaré's works were also the basis for the local and global analysis of nonlinear differential equations, including the stability theory for fixed points and periodic orbits (Kohno, 1994). For his part, Aleksandr Lyapunov studies the behavior of solutions in a neighborhood of an equilibrium position and thus extends the stability of solutions to what today is called Lyapunov stability.

This implies that initially close points continue to be so (which is not as frequent as one might imagine) and defines what today is called the *Lyapunov exponent*, which serves to determine whether elements of the state space of a dynamical system that are arbitrarily close remain close in the subsequent movement, and that serves to anticipate the separation that may occur (Lyapunov, 1966).

However, Poincaré's work did not receive immediate acceptance since it did not depart further from traditional analytical work (Israel & Menghini, 1998). Neither did Lyapunov's, which was published first in Russian and then in French. However, when these works became known and developed by the community, they were suitable for studying Lorenz's strange attractor, one of the first cases of fractals known by the community.

Nowadays, it is unfeasible to imagine the study of dynamical systems and complexity without an abundant presence of representations. In this regard, note that the theory of fractals was already in the works of Gaston Julia (1918) and Pierre Fatou (1919), but that it was necessary to wait years for their study to become popular, which happened, precisely, when it was possible to represent them via electronic resources.

Possibly a phrase by Lebesgue, who read Fatou's thesis with great interest, is the most emphatic way of appreciating the importance of representations in this context: "It seems to me that Fatou supposes gladly enough that everybody knows or sees the same thing as he does and gives perhaps fairly little explanations" (Lebesgue, 1991, p. 138).

16.8 Graphs in the Teaching of Mathematics

Today, mathematics classrooms use algebraic and dynamic geometry software. Nevertheless, the students' work proceeds mainly in the way it used for decades and recurring to software only for the mathematical calculations. This is very regrettable and entails a loss of opportunities. For example, up to the 1980s, in a Calculus course, it was necessary to work rigorously and laboriously to draw a function to use it to solve a problem. Software that directly delivers the curve makes the possibility of using it as a model more immediate.

Here we are interested in pointing out that mathematical modelling offers an opportunity to gain experience mathematics in the same way in which, in general terms, mathematics is elaborated (Borromeo Ferri et al., 2020). Mathematical modelling radically changes the focus of learning, and the profuse use of computers brings in the progressive disappearance of many outdated practices.

16.9 Our Project

16.9.1 Socioepistemology

Socioepistemological Theory of Educational Mathematics, or Socioepistemology, aims for the student to achieve learning that is *functional* to him/her, that is, that allows him/her to build his/her meanings and mobilize knowledge to face situations and solve problems in his/her daily life or in other domains where it is required. It is learning that is incorporated organically and that transforms the student's reality (Cantoral, 2014; Cordero, 2006).

That aim opposes the usual *school mathematical discourse*, dME, in the classroom (Cordero, 2006). The dME is the result of an educational tradition that seeks to ensure that students learn a subject first and then apply it in a situation that, for them, is artificial or scarcely meaningful. Hence, TSME aims to redesign the dME through an epistemological rupture with the usual paradigm of school mathematical knowledge, promoting functional mathematical knowledge. The redesign of the DME requires a *re-signifying* of knowledge, a process that seeks for the elements that favor in the students the construction of mathematical knowledge (Cantoral, 2014; Suárez & Cordero, 2010).

Modelling is a category of mathematical knowledge (Cordero et al., 2022; Morales et al., 2012, 2016; Morales & Cordero, 2014), and an activity that generates mathematics knowledge when facing a mathematical task in which the individual puts his/her knowledge into play. Graphing is a category of modelling. It allows redefining the teaching of mathematics and, with it, redesigning the dME (Mena-Lorca, 2016).

16.9.2 Purpose

Our aim is that learners be part of the construction of mathematical knowledge (Aravena & Morales, 2018). Also, we are always trying to refine our understanding of the direction been taken by the mathematics education's vector. Learning mathematical procedures is part of the experience of doing mathematics. Nevertheless, the purpose of learning for the ordinary citizen never was, and today, less than ever, to learn procedures per se, but to develop tools to solve problems (and acquire the usual competencies related to it).

We privilege mathematical modelling as a particularly suitable means to serve the purposes of mathematics education. We think that, in a way, classroom modelling provides an opportunity to learn mathematics as is done, that is, facing a problem, examining the hypotheses, conjecturing, and verifying, establishing conclusions. On the other hand, there are powerful reasons why some organizations and economies are calling attention to the need of approaching big data in Statistics, computational thinking, and modelling (Araya et al., 2020; González et al., 2020).

We cannot assume that learning to model with this specification is achievable with current curriculum objectives and implementation, at least, not in our region. Notwithstanding, we think that the shadow of a transition to Society 5.0 is an opportunity to get rid of some computations, eventually long, complicated, and/or outdated, and explore paths for resolution and, more generally, to carry out creative work.

Nowadays, individuals must understand complex phenomena, which requires mixing general or specific knowledge; Mathematics helps make connections. Moreover, we have observed how different communities, scientists, engineers, economists, and professionals in general use graphs and various apps to extract information, make decisions, or communicate results, that is, how they use them in their daily lives work (Mena-Lorca et al., 2021). Thus, we must re-design the DME, and, for that, we need to re-signify mathematical knowledge.

So, we elaborated a series of measures, a strategy. We will follow Oresme's idea: to think about phenomena starting from their corresponding representations. For that, we will need to get deeper into contemporary mathematical practice and add digital technology's considerations.

16.9.3 Modelling

Mathematical modelling study starts with simple, baseline models, some of which are well known, each associated with a graph. However, in practice, phenomena become more complex when connected, which usually requires assembling models. A basic example is Newton's laws: a slight alteration in the falling of a body may traduce in it no longer being modellable with a parabola. The parabolic model must couple with another that accounts for the variation.

Additionally, many situations that we need to model consist of a couple of different interacting phenomena. For instance, the logistic population growth model is a baseline model that can be mathematically studied and plotted for ideal situations. Now, growth of population depends on infectious processes (Gao et al., 1995; Mena-Lorca & Hethcote, 1992; Mena-Lorca et al., 1999). Moreover, for a pandemic, e. g., we need to consider the interaction of the dynamics of human populations with known local behaviors of economic growth and birth and death rates, which, in turn, are regularly adjusted.

The latter is related to the complexity of many phenomena that we are interested in understanding and modelling, even in an elementary way, such as disasters of various kinds (earthquakes, hurricanes, volcanic eruptions, avalanches, fires, Covid-19). Approaching complexity is unavoidable, and chaotic behavior may arise even when the basal models are relatively simple. For example, complexity occurs in the problem of three gravitational masses attracting each other.

Also, in the interaction of three species via predation: a specific logistic equation models each one, but the global dynamics of the behavior of the three species together is chaotic (Ramos-Jiliberto et al., 2008). Something similar happens with Lorenz's physical model for climate (Lorenz, 1963), which is somehow simpler since it considers only polynomial expressions up to degree 2.

Nota bene, our modelling category does not restrict to the usual cycle connecting reality and mathematics. Instead, we veer attention to re-signifying of uses, centering on the functionality of the reciprocal relation between mathematics and daily life, in the transition from one situation S to another, S′ (Cordero, 2016; Mena-Lorca, 2016). We will give an example in the following.

16.9.4 A Strategy

To address the complex situation described so far, we developed a strategy. We foster collaborative work, both for students and teachers, and adapt the learning evaluation to that scenario. We resolutely prioritize developing mathematical thinking over retaining and performing calculation procedures whose usefulness for tackling problems is not ensured and can be done faster and correctly by digital technology.

We have observed that when modelling a phenomenon, going firstly to symbolic expressions, whether known or unknown, not only does not help to understand that phenomenon but also tends to seize the students' thinking. Symbols may compress much meaning, but the students hasten to manipulate it in front of a symbolic expression, distracting themselves from the phenomenon. Thus, we use graphic representations to avoid this issue. We encourage students to learn mainly on representations that they can make and manipulate on their own, extract qualitative information from them, and, whenever possible, reason directly with the graphs of the models.

We do not focus directly on mathematical theory but keep in mind that, ultimately, a graphic representation might not faithfully represent a mathematical phenomenon: it is the mathematical considerations that support the analysis and conclusions in the context of the model. Students begin by familiarizing themselves with simple models, as we suggested before.

Then, we invite them to vary the model's parameters, and later, to consider situations that need a model to interact with another one. For instance, active noise control, which eliminates exterior noise by using a sound created by an electronic circuit that superimposes and, adding to it, cancels it, helps to re-signify the addition of sinusoidal functions.

Thus, students must first know of some 'baseline' models—linear, quadratic, exponential, and others, and become familiar with them and explore their parameters' variation according to a given situation. Then, we ask them to observe and analyze graphs usually taught in mathematics courses and realize that some are associated with baseline models of specific disciplines and that they refer to local behaviors only.

Later, we present situations that need modelling a phenomenon from a local view, in variables and time. From there, the students will address problems that are disciplinary (biological, economic, physical, and, naturally, mathematical) and/or real-life (decision-making for investment in health, education or retirement, control of infectious diseases).

Students explore, conjecture, argue graphically. By varying the parameters, they develop graphic categories of functions (affine, quadratic, exponential, trigonometric, e. g., which they name at will) and associated models (linear, parabolic, periodic) that are related to algebraic expressions they may have. They interpret trigonometric functions in a Cartesian plane and a goniometric circle and classify them (sine–cosine, tangent–cotangent, and secant–cosecant).

In so doing, they create additional categories of asymptoticity, periodicity, amplitude, and the like. Superimposing graphs, they identify trigonometric identities, solve equations and inequalities, and even glimpse and later develop the Fourier series (Mena-Lorca et al., 2021). It is particularly interesting for modelling that they experience the variation of parameters with functions such as $f(x) = a \sin(bx + c) + d$; $g(x) = a \exp(bx) + c$ (Mena-Lorca et al., 2021).

Subsequently, that way of studying turns into the analysis of 'real' phenomena. Thus, the student can glimpse and examine, for example, the eventual periodicity of a particular behavior (Mena-Lorca, 2016; Morales & Cordero, 2014; Morales et al., 2012). In this regard, observing the instability that can occur due to slight variations in the initial conditions, an observable phenomenon in our strategy, is very instructive to acquire a contemporary scientific perspective (Mena-Lorca et al., 2006).

16.10 Functioning-Form Dialogue, Re-signifying, and Functionality

If we take, for example, a sinusoidal function, the variations of the parameters a, b, c allow us to understand the behavior of the curve better but also to re-signify it through considering its meaning in different scenarios such as musical and others. This example shows a glimpse of a far-reaching general situation, which we refer to in our earlier comments on differential equations: the lack of a functioning-form dialogue (Cordero et al., 2010a) was an obstacle for some developments, such as finding the solutions to some of those equations.

Poincaré, by considering both the local and global aspects of a differential equation and freehand drawing the relative positions of the curves, clearly advocated such a dialogue, thus enhancing the intuition of the general shape of the corresponding integral curves and detecting global properties which needed to be proven. That dialogue guided his study and allowed him to correct mistakes due to a lack of knowledge of the *form*.

Similarly, Lyapunov, extending local stability to global (Lyapunov) stability determines whether the transformation of the state space of the corresponding dynamical system is continuous to anticipate eventual separations. Moreover, the fact that the work of Julia and Fatou on fractals needed half a century to be comprehended is, clearly, another example of the absence of the functioning-form dialogue provided by the graphs. Similarly, Lorenz's attractor had to wait 10 years for people to understand its *form* and thus comprehend its functioning.

The difference between the two last cases is that Fatou and Julia's arguments were purely mathematical, but Lorenz's analysis comes from modelling.

In our project, re-signifying draws mainly from the specific situation under study. We use diverse data and graphics from the media and some databases (https://www.gapminder.org/tools/, https://www.worldometers.info/, and similar) that represent a variety of models (physical, economic, climatic, and others). This allows the student to handle the information better and recognize functional mathematics in his/her daily life (Mena-Lorca et al., 2021).

As we said, our modelling category focuses on the re-signifying of uses, and concentrates on the functionality of the reciprocal relation between mathematics and daily life, in the transition from one situation S to another, S′ (Cordero, 2016; Mena-Lorca, 2016).

Take, for example, the quadratic function. Let S be the parabolic aspect in the reflection in a mirror, and S′ the graphic-cinematic in the falling of bodies. We do not focus on the quadratic function as the mathematical model, there are reference frames, and, in both cases, the domain is bounded, so what is involved is just a section of the graphic of that function.

Instead, we concentrate first on that the epistemic bases of S have an optical origin, and its variability refers to the position of the mirror's derivative function, whose variation is, in turn, constant, to get the appropriate concavity. Now, for S′, the second derivative is constant by Newton's law. The model refers only to first and second derivatives, and thus, we have Taylor's series development in a bounded interval (Mena-Lorca et al., 2021).

16.10.1 Graphing and Technology

The Internet provides graphs that represent phenomena like those taught in school classrooms at various levels. Nevertheless, it also provides many graphs generated by basal models of physics, economics, biology, and science not approachable in school education. Moreover, the Internet makes available graphs of data collected by sensors that function as the initial values of specific numerical methods to tackle models used to analyze complex behavior (such as tsunamis, storm surges, or economic issues that affect the stock market).

In such cases, it may be challenging to determine the initial conditions or the trend of the phenomenon; the behavior may or may not be stable (something analogous to the Lorenz model behavior may occur, although, in these cases, the phenomena are, nowadays, better known). On the other hand, the use of technological resources helps to specify and manage different registers and representations developed to understand and test feasible conclusions of working groups.

Also, we developed specific apps to raise modelling categories via graphs. These not only represent and reference specific mathematical contents and their properties but also make it possible to reason and argue directly from graphic families with graphic arguments (Mena-Lorca et al., 2021).

All this allows both students and mathematics teachers to talk about global and local aspects and become more aware of the properties of the different categories of models and graphs. For example, trending behaviors such as periodicity, asymptoticity (Cordero et al., 2010b), and limit processes such as Taylor series expansion in the local case, and Fourier series (Morales & Cordero, 2014).

The web offers information about specific and, thus far, local processes that help make statements that are sufficient for decisions of some kind. However, in the long term, the behavior may be different. A parameter-dependent dialogue of a local situation and a global situation is required; eventually, that dialogue may involve two curves that vary differently.

Mathematically, it is well known, for example, the behavior of the sine and quadratic functions, which allows studying the behavior of phenomena modeled with any of these functions, with data provided by the Internet. However, if we consider a phenomenon consisting of two others, reach one modeled with one of these functions, the result is not entirely known (not sinusoidal nor quadratic).

Today's apps provide resources that enable a vision of the phenomenon's behavior for cases such as this. Thus, we must study the use of complex mathematical knowledge, highlighting the work with complex models. More generally, if two phenomena, modeled with known functions f and g, respectively, interact, we cannot assume that the resulting phenomenon is easily understood. For example, in the current global pandemic, the infection should grow exponentially, and the control sought (via vaccination, use of a mask, isolation, and others) points to achieving a Gaussian distribution, but that does not happen (Mena et al., 2021).

The phenomenon under study has a mathematical basis, which allows making graphic representations; however, understanding it starts with the graphic representations. Indeed, the phenomenon studied is frequently multi-dimensional; it depends on several variables. To understand its complexity, we vary one or more parameters to observe the sensitivity of the phenomenon to that change. This effect can be chaotic; Lorenz himself (1963) had a 12-variable climate model; his discovery comes from having removed the last three decimal places from six that one of the variables had.

We use graphic representations to analyze the effect of variation of the parameters. Rotating the axes allows better appreciation of the phenomenon, the quadratic law of the fall of a body is not apprehended in the same way in the vertical fall as in the displacement of a projectile. On the other hand, the graph makes it possible to section according to different hyperplanes and reconstruct the phenomenon using level sets as in working with a scanner, that is, in the way of Oresme's.

Mathematics has tools to integrate the information on the behavior of the phenomenon thus obtained. All of this is makes up the dialogue between functioning and form (Cordero et al., 2010a). We observe that this is how experts from various specialties understand and handle the phenomena that interest them and thus value mathematics. On the other hand, without digital technologies, it might be challenging to understand some phenomena (10 years after its publication, Lorenz's seminal work had been cited in just three meteorology papers).

The case of the pandemic represents a more general and more recurrent situation. The general population may access the web's data and graphs over phenomena that it needs to understand (although it does not know the initial graphs and sources). These phenomena may need a couple (or more) of local modelling processes, both related to phenomena under study, but the result of the interaction is not entirely clear due to the interaction of the models.

To deal with the complexity described, simulations and graphs associated with them are generally required. For example, for phenomena such as species competition, either for resources or as in predator-prey phenomena (González-Olivares et al., 2013; Ramos-Jiliberto et al., 2004, 2008), and more particular ones, such as habitat fragmentation due to highroads (Mena-Lorca et al., 2006), or spread of infectious diseases such as Covid-19 (Mena et al., 2021).

16.10.2 Integrating Resources

Bearing the above in mind, we have integrated tablets and smartphones for the construction of mathematical knowledge. We operate in a freely available platform, integrated to Moodle, which uses dynamic geometry software and a symbolic calculation system. To these, we integrated a series of graphic apps of our creation, manipulable via parameter variation, by using sliders (https://www.matemati.cl/webapp/).

This integration makes it possible for us to leave graphing, arithmetic, and algebraic' computations (factoring, solving equations, calculating finite sums, derivation, integration, and others) to the electronic resources (Mena-Lorca, 2016; Mena-Lorca et al., 2021). When at work, the student has several windows simultaneously opened to support the functioning-form debate and, with it, the re-signifying of mathematical objects (Mena-Lorca, 2016; Mena-Lorca et al., 2021).

16.10.3 The Strategy at Work

We have employed this methodology with our students (either undergraduate or registered in a masters' degree in mathematics education) to analyze phenomena of a very diverse nature: economic, geographic, distribution of goods, availability of services, decomposition of interacting physical particles, and many others. With this, we have brought them closer to understanding the world and valuing mathematics, especially appreciating specific theorems, sometimes previously unknown, that clarify their view on specific phenomena (Mena-Lorca et al., 2021).

The goal is not to learn mathematics per se but to appreciate the role of mathematics in understanding the world. The view focuses on the phenomena under study and how the communities of experts work on them. So, we start with the information found on the web about ad hoc models, in which specialists use graphs to communicate. To understand the phenomenon and its complexity, we encourage to learn to connect the information and data with the underlying mathematics; for that, the graphic representation comes in handy (Mena-Lorca, 2016; Mena-Lorca et al., 2021).

16.10.4 An Opportunity for Teachers

An appropriate platform, directly connected to the Internet, allows the student to self-evaluate in various subjects, answering questions asked by the system; feedback may include algebraic properties and graphical representations. In turn, the teacher can estimate the possible responses of the student and eventual ad hoc returns. Moreover, free interaction with the Internet, so that integrating data and information from other disciplines is more straightforward, thus including teachers from other places and other areas makes it easier, for example, to design STEM activities (Mena-Lorca et al., 2021).

A realistic scenario is that teachers work together and take advantage of that space to plan their lessons, share designed material, and develop learning sequences, all provided with technological resources for computing and graphing. This scenario allows for their professional development, saves them time, and makes it possible to use, adapt and improve material produced by other users (Mena-Lorca et al., 2021). Several in-progress modelling-related graduate theses, both master's and doctorate's, under our guidance, use this approach.

16.11 General Assessments

We have verified that our work focuses on the reciprocal relationship between mathematics and everyday life and produces functional learning and a break in the usual dME. In addition, the student values mathematics and uses graphs to connect models, eventually interdisciplinary (Mena-Lorca et al., 2021).

By working as we describe here, the students develop functional mathematical knowledge, specifically by arguing with the aid of graphs (Morales et al., 2012), and, more precisely, the functioning-form dialogue. Naturally, this is important at all levels; however, it is especially relevant in higher education, where a significant number of students drop out of their careers due to their inability to handle the (calculation procedures of) mathematics.

We have gained understanding from the confluence of collaborative work, information from the web, and integrating existing resources with specially designed apps. Thus, we can claim that every participating student has developed his/her mathematics thinking and has a better understanding of the role of mathematics in the world.

Acknowledgment This chapter was partially funded by the National Research and Development Agency, ANID, CHILE, via Fondecyt Grant #1220565.

References

APEC. (2020). *COVID-19 hastens automation*. APEC Policy Support Unit. Retrieved from https://www.apec.org/Press/News-Releases/2020/0626_Future

Aravena, A., & Morales, A. (2018). El plano cartesiano en estudiantes de quinto básico: su resignificación en una situación específica. *Bolema, 32*(62), 825–846.

Araya, R., Isoda, M., & González, O. (2020). A framework for computational thinking in preparation for transitioning to a super smart society. *Journal of Southeast Asian Education, 1*, 1–16.

Aristotle. (2000). *Prior analytics*. The Internet Classic Archive. Retrieved from http://classics.mit.edu/Aristotle/prior.html

Ávalos, B. (2014). La formación inicial docente en Chile: Tensiones entre políticas de apoyo y control. *Estudios Pedagógicos (Valdivia), 4*(Especial), 11–28.

Bochenski, J. M. (2018). *A history of formal logic*. Forgotten Books.

Bolzano, B. (1817). *Rein analytischer Beweis des Lehrsatzes, dass zwischen je zwey Werthen, die ein entgegengesetzes Resultat gewahren, wenigstens eine reelle Wurzel der Gleichung liege*. Wilhelm Engelmann.

Borromeo Ferri, R., Mena-Lorca, J., & Mena-Lorca, A. (Eds.). (2020). *Fomento de la Educación-STEM y la Modelización Matemática para profesores. Fundamentos, ejemplos y experiencias*. Kassel University Press.

Cantoral, R. (2014). *Teoría Socioepistemológica de la Matemática Educativa. Estudios sobre construcción social del conocimiento* (2nd ed.). Gedisa.

Cauchy, A. L. (1821). *Cours d'Analyse de l'École Royale Polytechnique : I.re Partie. Analyse algébrique*. Imprimérie Royale.

Clagett, M. (1968). *Nicole Oresme and the medieval geometry of qualities and motions*. University of Wisconsin Press.

Cordero, F. (2006). La modellazione e la rappresentazione grafica nell'insegnamento-apprendimento della matemática. *La Matemática e la sua Didattica, 20*(1), 59–79.

Cordero, F. (2016). Modelación, funcionalidad y multidisciplinariedad: el eslabón de la matemática y el cotidiano. In J. Arrieta & L. Díaz (Eds.), *Investigaciones Latinoamericanas en Modelación Matemática Educativa* (pp. 59–88). Gedisa.

Cordero, F., Cen, C., & Suárez-Téllez, L. (2010a). Los funcionamientos y formas de las gráficas en los libros de texto: una práctica institucional en el bachillerato. *Revista Latinoamericana de Investigación en Matemática Educativa, 13*(2), 187–214.

Cordero, F., Mena, J., & Montalto, M. (2010b). Il ruolo Della giustificazione funzionale in una situa-zione di risignificazione dell'asintoto. *L'insegnemento della Matematica e delle Scienze Integrate, 33B*(4), 457–488.

Cordero, F., Mendoza-Higuera, J., Pérez-Oxté, I., Mena-Lorca, J., & Huincahue, J. (2022). A category of modelling: The uses of mathematical knowledge in different scenarios and the learning of mathematics, this volume, Part IV, Chapter 12.

Council for Science, Technology and Innovation. (2018). *Japan Science and Technology Policy: Realizing Society 5.0*. Government of Japan. Retrieved from https://www8.cao.go.jp/cstp/english/society5_0/index.html

Descartes. (1637/2018). *Le Discours de la méthode. Pour bien conduire sa raison, et chercher la vérité dans les sciences*. Createspace Independent Publishing Platform.

Donald, M. (1991). *Origins of the modern mind: Three stages in the evolution of culture and cognition*. Harvard University Press.

Donald, M. (2007). The slow process: A hypothetical cognitive adaptation for distributed cognitive networks. *Journal of Physiology Paris, 101*, 214–222.

Duval, R. (2017). *Understanding the mathematical way of thinking—The registers of semiotic representations*. Springer.

Euler, L. (1769). *Institutiones calculi differentialis cum eius usu in Analysi finitorum ac doctrina seriourum*. Academiæ Imperialis Scientiarum. Retrieved from https://archive.org/details/bub_gb_sYE_AAAAcAAJ/mode/2up

Fatou, P. (1919). Sur les équations fonctionnelles. *Bulletin de la S. M. F., 47*, 161–271.

Fermat, P. (1636/1891). *Oeuvres de Fermat. Tome Premier*. Gauthier-Villars et fils. Retrieved from https://archive.org/details/uvresdefermat00natigoog/page/n8/mode/2up

Gao, L. Q., Mena-Lorca, J., & Hethcote, H. W. (1995). Four SEI endemic models with periodicity and separatrices. *Mathematical Biosciences, 128*(1–2), 157–184.

Gobierno de Chile. (2006). *Ley 20.129. Establece un Sistema Nacional de Aseguramiento de la Calidad de la Educación Superior*. Biblioteca del Congreso Nacional.

González, O., Isoda, M., & Araya, R. (2020). A new framework for statistical thinking in the time of big data and the digital economy. *Journal of Southeast Asian Education, 1*, 61–70.

González-Olivares, E., González-Yáñez, B., Mena-Lorca, J., & Flores, J. D. (2013). Uniqueness of limit cycles and multiple attractors in a Gause-type predator-prey model with nonmonotonic. *Mathematical Biosciences and Engineering, 10*(2), 345–367.

Guicciardini, N. (1994). Three traditions in the calculus: Newton, Leibniz, and Lagrange. In I. Grattan-Guinness (Ed.), *Companion encyclopedia of the history and philosophy of the mathematical sciences* (Vol. 1, pp. 308–317). Routledge.

Høyrup, J. (2017). What is "geometric algebra," and what has it been in historiography? *AIMS Mathematics, 2*(1), 128–160.

Israel, G., & Menghini, M. (1998). The "essential tension" at work in qualitative analysis: A case study of the opposite points of view of Poincaré and Enriques on the relationships between analysis and geometry. *Historia Mathematica, 25*, 379–411.

Julia, G. (1918). Mémoire sur l'itération des fonctions rationnelles. *Journal de Mathématiques Pures et Appliquées, 8*, 47–245.

Kohno, T. (1994). Topological invariants for 3-manifolds using representations of mapping class groups II: Estimating tunnel number of knots. *Contemporary Mathematics - American Mathematical Society, 175*, 193–217.

Lebesgue, H. (1991). Lettres d'Henri Lebesgue à Émile Borel. *Cahiers du séminaire d'histoire des mathématiques 1re série, 12*, 1–506.
Lorenz, E. (1963). Deterministic nonperiodic flow. *Journal of the Atmospheric Sciences, 20*(2), 130–141.
Lyapunov, A. M. (1966). Stability of motion (Trans. from Russian). Acad. Press.
Mac Lane, S. (1971). *Categories for the working mathematician*. Springer Verlag.
Manfredi, G. (1707). *De constructione æquationum differentialium primi gradus*. Constantini Pisarri.
Mena, G. E., Martínez, P. P., Mahmud, A. S., Marquet, P. A., Buckee, C. O., & Santillana, M. (2021). Socioeconomic status determines COVID-19 incidence and related mortality in Santiago, Chile. *Science, 372*(6545), eabg5298.
Mena-Lorca, A. (2022). Sobre la nueva reforma de la educación matemática; invitación a un debate, 1. *Revista Chilena de Educación Matemática—RECHIEM, 14*(1), 4–16.
Mena-Lorca, J. (2016). Modelación Matemática y la construcción del conocimiento matemático. In J. Arrieta & L. Díaz (Eds.), *Investigaciones Latinoamericanas en Modelación Matemática Educativa* (pp. 139–162). Gedisa.
Mena-Lorca, J., & Hethcote, H. W. (1992). Dynamic models of infectious diseases as regulators of population sizes. *Journal of Mathematical Biology, 30*(7), 693–716.
Mena-Lorca, J., Klenner-Jerez, J., & Alarcón-Relmucao, N. (2021). Modelación gráfica en la enseñanza del cálculo para el siglo XXI. In C. Guerrero-Ortiz, E. Ramos, & A. Morales (Eds.), *Aportes desde la didáctica de la matemática para investigar, innovar y mejorar. Modelación matemática* (pp. 283–310). Grao.
Mena-Lorca, J., Velasco-Hernandez, J. X., & Castillo-Chavez, C. (1999). Density-dependent dynamics and superinfection in epidemic model. *Mathematical Medicine and Biology: A Journal of the IMA, 16*(4), 307–317.
Mena-Lorca, J., Velasco-Hernández, J. X., & Marquet, P. A. (2006). Coexistence in metacommunities: A tree-species model. *Mathematical Biosciences, 202*(1), 42–56.
MINEDUC. (2019). *Bases Curriculares 3° y 4° Medio*. Mineduc.
Morales, A., & Cordero, F. (2014). La graficación-modelación y la Serie de Taylor. Una socioepistemología del Cálculo. *Revista Latinoamericana de Investigación en Matemática Educativa, 17*(3), 319–345.
Morales, A., Mena, J., Vera, F., & Rivera, R. (2012). El rol del tiempo en un proceso de modelación utilizando videos de experimentos físicos. *Enseñanza de las Ciencias, 30*(3), 237–256.
Morales, A., Mena-Lorca, J., & González, A. (2016). Modelación Matemática y la Matemática Educativa de un investigador en Ciencias: en pos de la innovación y de la transdisciplina. *Investigación e Innovación en Matemática Educativa, 1*(1), 192–199.
OECD. (2017). *Education in Chile, Review of national policies for education*. OECD Publishing.
Poincaré, A. (1892). *Méthodes nouvelles de la mécanique celeste* (Vol. 1–3). Gauthier-Villars.
Poincaré, H. (1882a). *Les méthodes nouvelles de la mécanique céleste* (Vol. 1–3). Blanchard.
Poincaré, H. (1882b). Mémoire sur les courbes définies par une équation différentielle. *Journal de Mathématiques Pures et Appliquées, 8*(3), 251–296.
Ramos-Jiliberto, R., Hoecker-Escuti, F., & Mena-Lorca, J. (2004). Why the dimension matters in ecological models? *Revista Chilena de Historia Natural, 77*(4), 711–723.
Ramos-Jiliberto, R., Mena-Lorca, J., Flores, J. D., & Morales-Álvarez, W. (2008). Role of inducible defenses in the stability of a tritrophic system. *Ecological Complexity, 5*(2), 183–192.
Riccati, V. (1752). *De usu motus tractorii in constructione Aequationum Differentialium Commentarius*. Laelii.
Santbech, D. (1561). *Problematum astronomicorum et geometricorum*. Henrichum Petri et Petrum Pernam.
Suárez, L. (2014). *Modelación-Graficación para la matemática escolar*. Madrid: Díaz de Santos.
Suárez, L., & Cordero, F. (2010). Modelación-Graficación, una categoría para la matemática escolar. Resultados de un estudio socioepistemológico. *Revista Latinoamericana de Investigación en Matemática Educativa, 13*(4), 319–333.

Tartaglia, N. (1537/1998). *La Nueva Ciencia* (J. R. Martínez & C. Guevara, Trans.). Facultad de Ciencias, UNAM.
Tournès, D. (2009). *La construction tractionnelle des équations différentielles*. Albert Blanchard.
Tournès, D. (2012). Diagrams in the theory of differential equations (eighteenth to nineteenth centuries). *Synthèse, 186*, 257–288.
Turchin, P. (2016). *Ultra-society: How 10,000 years of war made humans the greatest cooperators on Earth*. Beresta Books.
UNESCO-ICSU. (1999). *Declaration on Science and the Use of Scientific Knowledge*. Retrieved from https://worldscienceforum.org/contents/declaration-of-the-1999-unesco-world-conference-on-science-110056
United Nations. (1948). *Universal Declaration of Human Rights*. Retrieved from https://www.un.org/en/about-us/universal-declaration-of-human-rights
van der Waerden, B. L. (1961). *Science awakening* (2nd ed.). Oxford University Press.
Villani, C., Schoenauer, M., Bonnet, Y., Berthet, C., Cornut, A.-C., Levin, F., & Rondepierre, B. (2018). *Donner un sens à l'Intelligence Artificielle : Pour une stratégie nationale et européenne*. Mission Villani sur l'intelligence artificielle. AI for Humanity.
World Economic Forum. (2020). *The future of jobs report 2020*. Retrieved from http://www3.weforum.org/docs/WEF_Future_of_Jobs_2020.pdf

Chapter 17
Modelling of Natural Phenomena as a Source to Re-signify Mathematical Knowledge

Miguel Solís, Francisco Cordero, Eleany Barrios-Borges, and Adriana Atenea De la Cruz-Ramos

17.1 Introduction

Within the COVID-19 pandemic context that started in 2020, official information in different countries was used through mathematical graphs within other ways of communication. This type of report was spread among specialists and was given to all the population (Gobierno de Mexico, Secretaría de Salud, 2020). Terms as *daily cases curve* or *accumulated cases curve* and expressions as *flattening of the curve* become very popular among people. However, not always they were understood properly.

Without having the intention of doing a sociological analysis to answer why the information through mathematical graphs was not given to the citizens including some professionals where the curricular mathematics played a fundamental role in their university training. Therefore, we focused our attention on mathematics education in different school levels in societies as ours in Latino America.

We compared it with the school mathematics, mathematical knowledge categories that emerged in the representations, by multidisciplinary groups, of the pandemic behavior of the world's cities, to know *daily cases curve* or *accumulated cases curve* and expressions as *flattening of the curve* become very popular among people.

M. Solís (✉)
Universidad Autónoma de Chiapas, Tuxtla Gutiérrez, Chiapas, Mexico
e-mail: solise@unach.mx

F. Cordero · E. Barrios-Borges
Centro de Investigación y de Estudios Avanzados del Instituto Politécnico Nacional, Ciudad de México, Mexico
e-mail: fcordero@cinvestav.mx; eleany.barrios@cinvestav.mx

A. A. De la Cruz-Ramos
Secretaria de Educación del Estado de Chiapas, Tuxtla Gutiérrez, Chiapas, Mexico

M. Rosa et al. (eds.), *Mathematical Modelling Programs in Latin America*,
https://doi.org/10.1007/978-3-031-04271-3_17

This comparison will provide us an epistemological vision of mathematics that is in the multidisciplinary relationships and put people's mathematical knowledge into practice but it is not in the school mathematics. This fact will entail to criticize the school mathematics that insists on to establish an only mathematical epistemology over other possibilities.

In this chapter, we show an example within the pandemic development context, a real or natural phenomenon as in its representation which re-signify the mathematical knowledge. On the one hand, this context is an environment where the appearance of the knowledge category denominated function graph tendency behavior in the representation. On the other hand, the basis of a functional principle: the use and its people's mathematics re-signification.

These categories of re-signification are the support for didactic designs within the school mathematics. Through this fact we reflect about the present mathematics teacher status and a proposal for his training and permanent support giving reference frames of the uses of people's mathematical knowledge as school activities context.

17.2 Mathematics Education Problem and Latin America

Different sources offer revealing data of the National Educational System (Instituto Nacional para la Evaluación de la Educación, 2007) that on one way make a reference frame up of the Mexican Educational system and that can be extended to any Latino-American educational system. For example, in some specific numbers in all the SEN there are 32 million of students from different educational levels through this way: within the basic level an 80%, within the medium upper level a 12% and within the upper level a 10%. This means that from the 15 million of boys and girls who attend primary level just 2.4 million attend university.

Moreover, it is know that from 100 boys and girls who attend primary level, five finish a university career. From these five, one is company manager another one is a lawyer, one is economist and a half engineer half a doctor. None of them studies physics, mathematics or biology. Some revealing data indicates that there is a cultural distortion in terms of the role of science among most of the population. Society itself considers that science is far away from daily life (Cordero, 2015).

It calls the attention that at the seminar about divulgation of science and technology that was done in 1999 (Seminario sobre divulgación de la ciencia y Tecnología, 1999) distinguished that the citizen in general, considers that what it is made in science has nothing to do with what happens in the society but this fact in some sense it thrives nowadays (Cordero et al., 2009).

All of this together with some political and economic aspects of the country lack of social cohesion and social inequity. On the one hand, a notorious reduction of the flow of student population from primary level to university and on the other hand, the skimpy selection of scientific careers of the population that achieves to finish their university studies in some way express a loss of knowledge value, an unequal education. This is the educational problem that boys and girls live in the society and develop in their daily lives. It will be said in this way learn mathematics or not.

To compare this lack of social cohesion and this inequity it will be required necessarily to work more and intensively in the *knowledge socialization* where reciprocal dialogue environments between school knowledge and people's reality (Cordero, 2016a; Cordero et al., 2015). Some *permanent reciprocal simultaneous programs* should be created with the development of science and education that contribute directly to the building of a knowledge society (Cordero, 2015). Any educational program will be articulated to this fact. The functional and what is usual are unavoidable within the socialization studies since they express the knowledge and the own environment of the citizen, the dialectics between the academic and the native (from people) knowledge (Cordero, 2016a; Cordero et al., 2015).

Educational research in generic terms where the socialization plays an important role heightens the problem: the student's learning chapters within classrooms will increase people's routine at the institution and society as an educational referent and the teacher's function will keep the environments where reciprocal relationships systems of the school mathematics and the learners' reality at specific knowledge situations.

17.3 The Forgotten Subject and the Theory

Cantoral (2013) provides basics to the Socioepistemological Theory Educational Mathematics through formulating the research program own theory's nature. This one consists about the study of the social construction of the mathematical knowledge and its institutional spreading. A central construct is the social practice which is a complex system of social, didactics, epistemological and cognitive processes which emerge to make the mathematical knowledge problematic considering the scientific, technical and popular knowledge to summarize it within the human knowledge.

The theoretical perspective has been made by a hundred of made researches since more than two decades by a socioepistemological community with different generations. Cordero (2016b) considers that the social practice construct within the evolution and development of the *Socioepistemological Theory* has provided a meaning considering about the problem that there is a forgotten subject and it is fundamental to rescue it. This thesis has different expressions: reality, what is usual, the uses of the knowledge and in more general terms, people.

Through this thesis a Socioepistemological program called Forgotten subject and knowledge mainstreaming is formulated (Cordero et al., 2020). The main objective consists of revealing the uses of mathematical knowledge and its re-significations within the people's mathematical knowledge communities: at school, at work or the profession and its realities. Through these two simultaneous work lines: the re-signification of the mathematical knowledge and its educational impact. Within the first one the categories of the mathematical knowledge that happen in the communities among different domains of knowledge which play an obliged role: the school mathematics discourse, the discipline field and the community's usual situations and within the second one the multi-factors and stages which help the quality mathematics teaching alliance.

The main research questions of the program consider the following questioning: Which is the epistemological status of the mathematical knowledge function with specific situations around people's environments? What are the extensions of the classroom learning chapters when the forgotten subject recovers? What are the fundamental factors that formulate the new permanent quality program for the mathematics teacher's function?

17.4 The Natural Phenomena Modelling

Nowadays school mathematics has lost the link with the reality. However, if we talk about reality we will restrict it to standardize the mathematics education: all educational levels will be considered and the discipline diversity the same as work and the city. The reality will be interpreted in what is usual within all these scenarios where the usual uses of the mathematical knowledge are expressed: the functional. This means that professionals, workers and citizens' routines will be the reference frame to make the school mathematics recovers the link with the reality (Cordero, 2016a; Cordero et al., 2015).

The pandemic's spreading modelling is mainly required to make decisions. We need to estimate the quantity of infested to prepare the health system to face the pandemics and take decisions to decrease the spreading of the same one. This type of modelling to take decisions but in the Educational Mathematics field is retaken by Niss (2015), who calls it prescriptive modelling what is to design, to prescribe, to organize or structure some world aspects with the purpose to adopt measures based on decisions from some type of mathematics considerations.

An important difference is the essential role of the sensitivity within the evaluation of the models with prescriptive purposes. The sensitivity analysis can be associated with the role that determined variable plays within the model: How does the variation of variables affect within the model? How much does the variation of variables within the model? Some news spread by different means of communication in different countries about the spreading of COVID-19 pandemics show three graphs of estimated number of infected people. They refer to a positive, an ideal and another one which is critical compared to the real possibilities of the health system of the country.

Within these news, the phrase to flatten the curve is cause of many controversies in terms of its interpretation, which can associate with a decrease of the quantity of infected people daily. Which is the dynamics under the phenomenon? How is the model made up? What are the parameters of an epidemiological model? How and

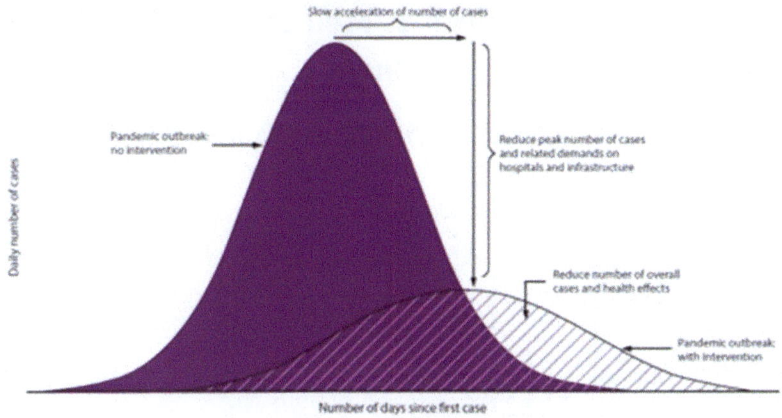

Fig. 17.1 Visualization of the idea to decrease the spreading or to flatten the curve of a pandemic. (Source: Available in https://es.wikipedia.org/wiki/Aplanar_la_curva)

how much do they influence? Were the questions that guided us within the study of the epidemiological models. Figure 17.1 shows the visualization of the idea to decrease the spreading or to flatten the curve of a pandemic.[1]

Vidal et al. (2020) in their search about epidemiological mathematical models draft the mathematical modelling within the study of pandemics as a "model made up by a set of symbols and formal mathematics links which represent an approximation to the existent real links to the study object" (p. 2). These models can be classified into some groups, for example, determinists, stochastics, statics and dynamics. Determinists are those who work with known conditions and data are controlled by the factors that intervene within the study.

Stochastics are associated with the notion of probability; they have a doubtful behavior and the expected result is not known. Statics or dynamics refer to the way time is treated. The static models give a result for all the considered time. The dynamics models give the temporary variable series back considered through the time of the study.

17.4.1 SIR Model

Through time more specific models have been generated for the modelling of the spreading of the illnesses. The Kermack and McKendrick's (1927) Susceptible, Infected and Recovered model (SIR) is still a valid mechanism to take decisions about public policies. Its use is of interest since the low number of variables that

[1]The dark purple curve shows a pandemic that is spreading fast while the other curve shows a slower spreading.

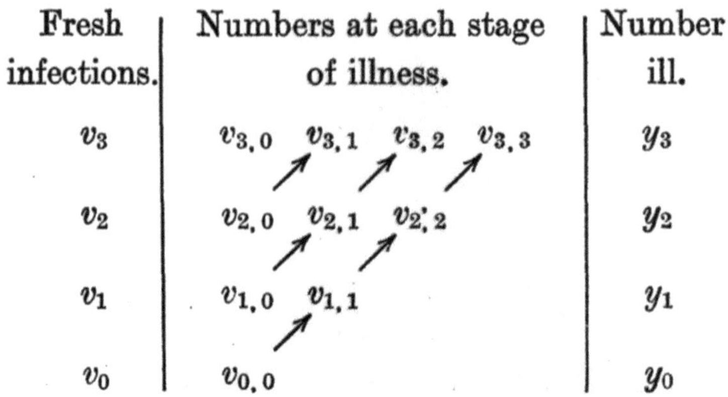

Fig. 17.2 Pandemic process. (Source: Kermack & McKendrick, 1927, p. 703)

describe the model. However, it allows the analysis of the pandemic's behavior in future states and for different values of the variables.

That means how the pandemic can behave for a long period of time in different scenarios of spreading given by the public applied decisions and the population's behavior. Kermack and McKendrick (1927) were more understood about the diverse effects which regulate the spreading of pandemics. The problem can be summarized in the following way: one (or more) infected person in a community of sensitive individuals to the illness. The pandemic spreads from the infected over the sensitive ones. Figure 17.2 shows the pandemic process.

The figure shows the recent infected (v_t), that means those who start the illness moreover, the infected in each stage of the pandemic ($v_{t,\ \theta}$) and the total number of infected people (y_t). The total number of infected is the addition of the infected who are within the different stages of the pandemic.

Kermack and McKendrick (1927) present the spreading model of the pandemic for the case that the elimination rate (l), the adding of the death rate and the recovery rate, and the infection rate (κ) be constant. The probability of an infection is in proportion not only to the number of infected but also to the number of not infected. Through this way they made this system of differential equations to represent the function of susceptible ($x(t)$), infected ($y(t)$) and recovered ($z(t)$) respectively, where N is the total population for all time.

$$\frac{dx}{dt} = -\kappa xy$$

$$\frac{dy}{dt} = -\kappa xy - ly$$

$$\frac{dz}{dt} = ly$$

where

$$x(t) + y(t) + z(t) = N$$

These mathematical models interpret the pandemic behavior. The models are different but what is similar is that the behavior in interpreted by a tendency behavior category. That means, this category is, in one sense, a principle that thrives in the formulation of the models: the tendency of the spreading of pandemic.

17.5 The Mathematical Knowledge Re-significations

The re-significations of the math knowledge happen within people's life. Next, we will present two examples in this sense, on the one hand, where the use of the derivative in school-academic scenarios and engineer and on the other hand the re-signification of the use of the graph within people's daily life.

17.5.1 The Use of the Derivative by Engineer Students and Engineering

Morales-Reyes (2020) recognizes that the uses of the derivative within the Engineering are excluded from school mathematics. Therefore, it considers very important to create a relationship where reality mathematics permeates school mathematics. Moreover it considers both types of knowledge with the same epistemological value. With this purpose a *school situation design with exclusion-inclusion dialectics perspective.* This type of designs base on an epistemology that favors the uses of mathematical knowledge and in a theoretical perspective which guide the focus of the design, counteract the exclusion phenomenon and allows the analysis of the participants'process of re-signification.

The used epistemology comes from the consideration of a modelling category which is an expressed practice such as the argument of a situation made up of significations and re-significations with their respective procedures which are being build according to the operations that participants are able to do with the conditions they are capable to capture and transform with the concepts that have been build up progressively. For the articulation of these elements we turn to the uses of the derivative which emerge in a community of chemical engineers in the graph analysis of the chemical compounds of electric transformers. In this context, some uses of the derivative emerge which are not part of the school calculation.

The results show that the transition among the three stages that made up the design make the participants to face the school mathematical knowledge: they should determine tangent lines but not knowing the algebraic expression of the function and the point of tangency and moreover they should determine graphs behaviors and later states appealing for this the tangent lines. These are aspects that allow to value in what is usual within the discipline of the use of mathematical knowledge however, they do not consist at any didactic main idea not for the school texts or school curriculum.

17.5.2 The Use of the Graph Within People's Usual Habits

According to Zaldívar-Rojas and Cordero (2021) explain that the mathematical knowledge refers to two dialectic elements: the function and form, what is done and how is done, respectively. The implemented activities since the Socioepistemological theory make possible to identify ideal scenarios since the analysis and the discussions about a determined group and knowledge are not limited.

What is above make possible to implement a situation called *primaveras* this situation consisted of observing with attention the movement of a mass connected to a spring to strengthen the discussion about the conditions and elements as the tendencies involved in this situation giving preference to the visualization and work made with specific devices. The design of this situation retakes functional elements of the mathematical knowledge and makes notions of Cartesian graphs. The situation bases on the modelling of movement situations using graphs known as the modelling-graphing category which emphasizes the use of graphs and the tendency behavior.

The situations studied by Zaldívar-Rojas and Cordero (2021), allow to recover expressions associated to the explanation of movements identifies by the students: for example, to analyze the graph function retaking daily life situations particularly those involved in the uses of springs. The use of movement sensors that together with the analysis and arguments of the explanations achieve to identify maintenance movements as well as movement crisis and functions that means, to mention maintenance refers to these representations and movements relationships.

They represent stability, later while the maintenance crisis, the movements are done in a specific way for the students to identify the tendency behavior related to this identification in their daily life therefore they will be able to explain according to the context that surround them.

At this moment, it is needed that the students identify and explains movements that they can make and represent themselves that means; where the graphic start and which the point of reference is and moreover to find and explain the conditions that originate a tendency or variation model since the student can modify those conditions.

To implement situations as what described above make possible that the students when identifying daily life movements with the graphs generated by the movement of themselves retake in their arguments concepts as functions, point of reference and most of the time without being perceived or what has been explained previously.

17.5.3 Epilogue: A Vision of the School Calculation Based on the Re-significations of the Use of Mathematical Knowledge

The calculation status within the didactics is when we talk about of taking knowledge to teaching within the educational system. It is not about the knowledge itself but knowledge with didactic intention. The objective is to puzzle out an epistemology. School calculation means that the calculation (*Calculus*) with an intentional epistemology of being taught and learned since it brings different component among both types of knowledge.

For example, calculation as knowledge has concepts and explicit definitions while calculation as intentional knowledge has implicit categories. About the first one the main components are the mathematical objects such as the function, the border, the derivative and the integral while for the second one is the situational meanings of those mathematical objects such as the prediction, graphing and analyticity (Cordero, 1998, 2001, 2003).

Not appreciating the difference among these types of knowledge can bring ingenuous ideas about the mathematics teaching-learning problem. For instance, writing mathematical texts with a *didactics without students* (Cantoral & Farfán, 2003). Calculation status within educational institutions that still thrives consists of that the *Calculus* in general terms is known as (with the help of textbooks) the *branch of mathematics that deals with differentiation and integration.* In this sense the programs of such content have to do with the concepts of function, border, derivative, integral and convergence.

In this sense the programs of such content have to do with the concepts of function, border, derivative, integral and convergence. They go together with the specific operations: the quotient limit $\lim\limits_{x \to x_0} \frac{f(x)-f(x_0)}{x-x_0} = f'(x_0)$[2] and the addition limit "$\lim\limits_{n \to \infty} \sum\limits_{i=0}^{n} f(x_i)\Delta x_i$, if $n \to \infty$, $\Delta x_i \to 0$",[3] all of this together to the function as the main concept of the Calculation subject. However, in terms of this perspective it is difficult to achieve the main objective of the *Calculus*: the analyticity of the functions $f(x+h) = f(x) + f'(x)h + f''(x)\frac{h^2}{2!} + \ldots$.[4] That objective concerns the calculation epistemology not the school curriculum unfortunately.

---

[2] *Derivative definition*: The function *f defined* in an open interval *I* that contains the *a* point. Then *f* is differentiable in *a* if $\lim\limits_{h \to 0} \frac{f(a+h)-f(a)}{h}$ exists. In this case, the limit appears by $f'(a)$ and it is called the derivative of *f* in *a*.

[3] *Integral definition*: A function *f* enclosed in $[a, b]$ is integrable in $[a, b]$ if $sup\{L(f, P) : P\ partition\ of\ [a, b]\} = inf\ \{U(f, P) : P\ partition\ of\ [a, b]\}$. In this case the common number it is called the integral of *f* in $[a, b]$ and appears by y $\int_a^b f$.

[4] *Analyticity definition*: A function *f* is the adding of its Taylor series in an open interval that contains *a* is analytical in the point *a*.

It is probable that the Taylor series is more advanced mathematics that to be taught and learned requires of the derivation and in one sense of the integration. What meaning does this fact about the topic we are in charge of? This status consists of focusing the attention on the concepts. In this regard, the analyticity is just one more concept (in this case advanced) and does not reflect that the analyticity can be the main idea of the calculation or in other words the calculation knowledge.

The focus on the concepts creates unavoidable sequences dimming the situational meanings where it is debatable the function and form of those concepts (Domínguez, 2003). To make this clearer we will analyze what happens with the concept of the derivative. In general terms and paraphrasing what the calculation texts say (Stewart, 1994) we find that the differential calculation is considered as the study that deals with the question about how a quantity varies in relationship with another one, in this sense it is established that the main concept of the differential calculation is the derivative (Rosado, 2004).

Later, they advise that after learning to calculate derivatives they will be used to solve problems where rates of change intervene. Therefore, they define the slope of the tangent line to a curve $y = f(x)$ in the point $x = a$, as $m = \lim_{h \to 0} \frac{f(a+h)-f(a)}{h}$. Or define the speed of an object with a position function $s = f(t)$ in the instant $t = a$, as $v(a) = \lim_{h \to 0} \frac{f(a+h)-f(a)}{h}$.

Moreover, they advise that the borders of the previous function emerge when a rate of change is calculated within science or engineering such as the speed of a chemical reaction or a marginal cost in economy. Such borders imply the definition of the derivative of a function:

$$f'(a) = \lim_{h \to 0} \frac{f(a+h) - f(a)}{h}$$

Such focus has been studied by diverse ways but agreeing on there are qualitative leaps between the two conceptualization levels which play a role within the sequence: the level process and the object level (Dubinsky & Harel, 1992). Students' difficulties appear when they have to work not with particular functions but with defined functions by any property. It is because of that; the analysis becomes more difficult since it is needed to consider the functions as objects that can be included in more complex processes such as the function types. The cognitive and epistemological dominant focuses are not enough nowadays. It is necessary to integrate approximations to the didactic field that allow us to take into account the role played by the coercion and institutional and cultural aspects within the teaching-learning problems (Artigue, 1998).

However, the focus of the attention is so strong within the border that avoids other situational aspects. It privileges referring procedures to the quotient $m = \frac{y-y_0}{x-x_0}$, through approximation arguments expressed in the quotient border $\lim_{x \to x_0} \frac{f(x)-f(x_0)}{x-x_0} = f'(x_0)$. In this situation it is required a given function f and a specific point $(a, f(a))$ to calculate the derivative or to find the tangent line to the graph of the function at this point.

There are not other procedures as the comparison between two states of a quantity through a subtraction $f(x + h) - f(x)$ with prediction arguments expressed by the analyticity $f(x+h) = f(x) + f'(x)h + f''(x)\frac{h^2}{2!} + \ldots$

For example, to predict the position of a mobile when its initial position is known and its variation at the same time. In this situation the function f is not known just the quantity states $f(x)$ y $f(x + h)$ and the variations $f'(x), f''(x), \ldots$ or are not perceived as the parameters of the f given function variation ($Af(Bx + C) + D$) with arguments of tendency behaviors expressed within the function graph Af.

For example, to determine the value of the quotient A since the function f behaves as the line $Y = ax + f(x_0)$ close to the point $(x_0, f(x_0))$. In this situation the function f is transformed into the function $Af(Bx + C) + D$, where the parameters A, B, C, and D are the suitable for the function y tends to behave as the specific line Y.

As a summary to find the *tangent line, to predict a mobile position* and *to expand a graph* are situations where it is debatable the function and form of the concept of the derivative. The quotient border re-significates[5] through the prediction, graphing and analyticity: the derivative and the tangent line debate against the comparison of the two states and the simultaneous succession of the derivative (González, 1999; Cantoral, 2001; Cantoral et al., 2000; Buendía, 2004; Buendía & Cordero, 2005; Cordero, 2001), at the same time debate against the parameter's variation and the tendency behavior (Domínguez, 2003; Campos, 2003; Rosado, 2004; Hernández, 2004). However, this fact does not make any didactic content not for the calculation texts or the school curriculum.

17.6 An Example: The School Pandemic Curve Modelling

The example that is presented emerges at the beginning without a school intention just with the objective to explain people the technical information about the pandemic, how the human intervention can manipulate the behavior of this one, represented here by the pandemic curve, which is how people's actions can *flatten the curve*. When in this basic explanatory model it is perceived the emergency of aspects such as graph behaviors, simulation, prediction, accumulation among others it is decided to create a school activity based on this basic model with the objective that these present aspects contribute to the construction of school mathematical knowledge (Solís, 2020).

Through the design of this activity, it starts simulating the phenomenon using concrete materials, file cards in this case (these will represent individuals), from this hypothetical data the time is distributed in the file cards and its components are studied from simulations some concepts are studied such as growing, growing speed,

[5]Resignification is not establishing a meaning within a context for later to look for another one and in this way to re-significate what it already has a meaning. But it is the construction of the knowledge within the humans.

maximum values. Then school mathematics conventional representations are made and finally it is discussed about the accumulation and accumulated value. The complete activity (Solis & De la Cruz, 2022) can be read from the book: *La Matemática en la Ingeniería. Modelación y Transversalidad de Saberes. Situaciones de Aprendizaje* (Cordero, Solis & Opazo, 2022), which is presented and described in the next section.

17.6.1 Moments

In this section of the chapter, the moments of the activity are presented and described.

17.6.1.1 Moment 1: Pandemic Modelling That Is Developed Without Contention and Mitigation Actions Which Can Intervene

Time Cases Distribution (Curve Form)

Here the *natural* evolution of the pandemic is presented. That is there are not intentional measures to contain or mitigate it. The way of modelling the Pandemic is distributing over a base line (the time), concrete objects, which represent infected individuals (in this case file cards). The distribution is made from a table of given values.

The practice of counting will allow answering the questions about the peak of the phenomenon such as the number of accumulated cases. The distribution of the file cards shows a characteristic form (Fig. 17.3), in this moment we have not talked about graphs or functions or to establish a conventional system of Cartesian axis in this case time v/s infected.

Towards a Conventional Graphic Representation

To go deeper towards the graphic representation of the phenomenon it is asked to draw the outline of the distributed file cards and some questions are asked about the number of cases in a specific day. When asking to enumerate the files (number of cases) and the columns (pandemic days) a conventional Cartesian frame of reference is being constructed. Figure 17.4 shows the distribution of case outline.

Variation, Accumulation, and Accumulated Value

In this moment, the behavior of the phenomenon is analyzed from the configuration of the file cards distribution. There is not an explicit reference of the graph of a

Fig. 17.3 Cases distribution during the pandemic. (Source: Solís & De la Cruz, 2022)

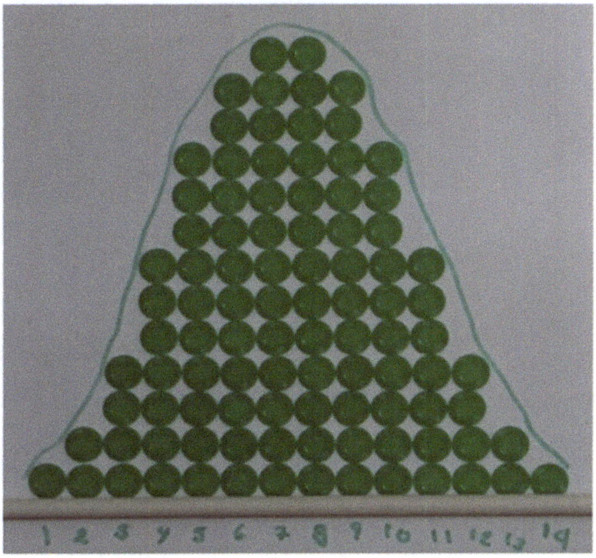

Fig. 17.4 Distribution of cases outline. (Source: Solís & De la Cruz, 2022)

function, we are in what Cordero and Flores (2007) have called *the moment of the symptom of the use of a graph of a function*.

Within a value table (Table 17.1) it is asked about the daily difference in the number of cases and in the other one (Table 17.2) about the accumulated cases every

Table 17.1 Daily difference cases

Day	New cases	Previous day difference
1	1	1
2	2	1
3	4	2
4	7	
5	10	
6	12	
7	13	
8	13	
9	12	
10	10	
11	7	
12	4	
13	2	
14	1	

Source: Solís and De la Cruz (2022)

Table 17.2 Accumulated cases per day

Day	New cases	Accumulated cases per day
1	1	1
2	2	3
3	4	7
4	7	
5	10	
6	12	
7	13	
8	13	
9	12	
10	10	
11	7	
12	4	
13	2	
14	1	

Source: Solís and De la Cruz (2022)

day. The intention is to analyze the evolution within the Pandemic time paying special attention to the daily speed about Pandemic growing.

A questionnaire about the given and found values is focusing the attention on the increase or decrease of the cases, same as the peak of the curve giving answers as what can be formal as the derivative and the slope of the function. At the same time, questions which ask about the moments of more or less speed of growing can have link with the mathematical objects as second derivative, concavity and inflection point.

17.6.1.2 Moment 2: Pandemic Modelling When Some Sanitary Measures Have Been Taken: The Flattening of the Curve

In this moment is a logic that when reducing the movement and encourages the social distance of the community, the speed of the spreading of the Pandemic will be reduced. Some values of the table that propose values according to this situation have been taken. The new situation shows a similar number of total infected at the end of the Pandemic with the situation of the moment 1. What did it change then? The new distribution shows a peak less *high* but the duration of the Pandemic is *longer* now.

New Distribution of Cases Through Time: Flattening the Curve

About the reference of the previous situation that means, using the same files and columns (that will become the time v/s cases in a Cartesian system) we will make the proposed distribution in a given values table. Figure 17.5 shows the distribution and outline at the scenario.

Graphic Comparison of the Two Phenomena

The same as the situation of the scenario 1, the outline of the new distribution is drawn and they are compared. It is now necessary to add more columns what indicates that the pandemic has been longer however, the peak is less high. The number of the total cases represented by the file cards is similar in both scenarios.

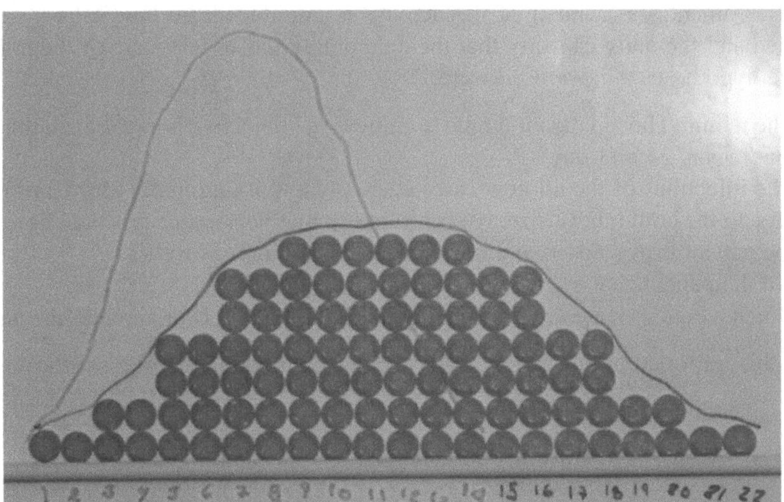

Fig. 17.5 Distribution and outline at the scenario 2. (Source: Solís & De la Cruz, 2022)

The similarities between the two outlines are analyzed that means in both cases the pandemic starts with a growing that is increasing until to achieve the maximum speed represented in the growing by the point of inflection since the speed of growing decreases gradually until becoming zero being represented by the peak of the curve after it is presented a decrease of the cases with a similar behavior to the growing.

With value tables it is asked about the quantitative aspects of the graph of daily cases and accumulated cases. At every moment, the analysis is between the differences and similarities of the graphs. A graphic representation about how the accumulated cases behave has not been worked in this moment.

17.6.1.3 Moment 3: A Flattened Curve

The two graphs represent the way of distributing time (horizontal line) a determined number of file cards (cases). In this context it is about the same population within two different situations (scenarios). In this activity it is shown that flattening the curve does not mean to achieve a smaller peak or less total infected numbers but to distribute all the possible cases looking a smaller peak but within a longer period.

What maintains (almost) the total number of cases (area under the curve) allows perceiving the distribution 1 has been accommodated in the distribution 2 by the effect to flatten it. The activity favors the idea of the curve 1 has been transformed in the 2 by the effect of flattening it.

17.6.1.4 Moment 4: What Is the Best Scenario?

In this moment, the context of the activity is introduced the hospital care of the infected and the daily capacity that the community has to take care of them. Some simple hypotheses that we used were:

(a) The required hospitalization time to achieve the objective is equal in all the cases that means 24 h (1 day).
(b) The infections of the all-new cases occur always at midnight (12:00 a.m.).
(c) Due to the community size, we can suppose that because of practical objectives the patient is diagnosed and hospitalized at midnight as well.
(d) All the patients are discharged by a doctor at midnight.
(e) When a patient leaves the hospital the possibility to accept another one arises.

With this information the students will decide about which the best scenario is.

17.6.1.5 Moment 5: Building the Graphs

Until here concrete material has been used to model the pandemic situation in our hypothetical community. From the file cards the cases were accumulated in columns that presented the days of these cases happened and we obtained a representation of what happened.

In this moment it is pretended to do the abstraction of what happened previously and their conventional mathematics and science representations, that means, the Cartesian plane are used. Since the instructions are between files and columns to Cartesian coordinates.

17.6.1.6 Moment 6: Accumulated Cases Graph: A Growing and Bounded Function

With this new frame of reference, the construction of the accumulated cases graph calculated in the moment 1 and 2 starts. Both graphs are analyzed and compared (daily cases v/s accumulated cases) within the concepts of primitive functions and derivative function are implicit.

Here it is evident the growing and bounded characteristic of this new curse and it is shown that *to flatten the curve* has not to do with the maximum height of the same but its slope.

17.6.1.7 Moment 7: Analyzing Curves of a Real Pandemic

When finishing the activity, the student can draft an accumulated cases curve from a daily cases one with a real phenomenon data measured until the design of the activity. (Dirección General de Epidemiología, 2021).

Intentions and Learning

This activity was designed to be used by different educational level students; the teacher who uses it can do the pertinent changes. This can be used even in not educational situations or in other subjects apart from mathematics. Although the activity does not explicit some concepts the teacher can refer to the activity to introduce formal mathematical objects within the class when the subject requires it.

Some mathematical concepts involved in the activity are among others: function (to distribute the cases through days), increasing and decreasing function, maximum, tangent line, first derivative (growing speed), second derivative, (curve inflection), area under the curve (accumulated value), Cartesian plane (explicit), point in the plane and a graph of a function and the graphic integration (in the sense of the calculus) this present in the last activity.

In this example we show the tendency behavior category in particular presenting the function (pandemic curve) as an instruction (mitigation, contention and even restriction measures) which organizes behaviors (flattening of the curve).

17.7 Mathematics Teacher's Role: A Hope

Mathematics teachers as *homo academicus* (in the sense of Bourdieu (2008)) live within a discipline and social disadvantage. It is not clear who trains them. The training of the mathematics teachers is debatable when it is needed to precise their discipline. There are not agreements, it is debatable if the discipline is the pedagogy or mathematics or even both. This increases in countries as ours where there are mathematics teacher with no teaching training.

With not clear definition a mathematics teacher should be subject of the school mathematical discourse as a consequence the teacher adheres to the school mathematical discourse, is not able to (or does not want) to touch it. But disrupting is condition *sine qua non* to achieve a profound teacher training transformation (Cordero et al., 2015). A hope is to achieve that the mathematics teacher builds a discipline identity whose source of sense can be the social construction of the mathematical knowledge. Permanent programs with these slogans will be the resistance instruments of the school mathematical discourse where *the teacher role will keep the autonomy.*

Then the mathematics teachers' role should *keep the environments* where reciprocal relationships of the school mathematical system and the reality of the learner in specific knowledge situations. However, to keep the maintenance should disrupt the school mathematical knowledge (Cantoral et al., 2015); the uses of people's knowledge (Cordero, 2016a; Opazo-Arellano, 2020; Medina, 2019; Mendoza-Higuera, 2020; Pérez-Oxté, 2021) and in more general terms: mathematics as a human activity, mathematics from contexts and mathematics for all the students (Freudenthal, 1968).

17.8 Research Programs and Recovery Instruments

We do not pretend to propose a new methodology or a new reform of the mathematics teacher training our main objective, on the one hand, is to present a frame of reference from the learner's native knowledge, the person who uses his mathematical knowledge in his profession and the person who uses his mathematical knowledge to live in the society and on the other hand, to present the processes where the horizontal and reciprocal dialogue among the frame of reference, the educational models and the teacher training will happen.

It is important to mention that this frame of reference and this horizontal dialogue do not exist within the educational system so we have to build them. To achieve this task it is required to disrupt the dominant epistemology school mathematics, to be opened to the epistemological pluralism that obliges the inclusion of the *forgotten subject*. This individual uses his mathematical knowledge in diverse functions and forms that school does not imagine nowadays. That is why we say that the construction will derive in a school where the mathematical knowledge will dialogue between the academic discovery and the people's native knowledge revelation horizontally. In this last one, in generic terms, is the subject who learns, works and lives in the city but is out of the school.

Maybe because of that, the childhood and youth mathematical knowledge social representations where women and men admit mathematics is far from reality. The relationships among mathematics as a discipline, school mathematics and daily life mathematics are not clear in the educational programs of societies. The loss of the mathematical knowledge and the educational inequality (just some can learn mathematics) keeps growing. With no doubt we have to do something. Our research program, which we call it from now one *Socioepistemological Program*, consists of three axes: education, research and intervention.

A knowledge society consists of valuing the knowledge and makes it equal, which means, the most important element to achieve that is in our case the mathematical knowledge function. We are required to study and know it and make the frame of reference explicit. Through this we recover the forgotten subject and in consequence school mathematics will be increased. Therefore, we should focus our attention to the knowledge processes of socialization and reformulate the mathematics education programs according to the societies.

Research should be made by constructs whose nature bases the knowledge role. We need to create a source of sense to achieve that. Studies will need to be oriented towards the knowledge mainstreaming to know the re-signification of the mathematics at school, work and city. Every time to progress in the creation of some characteristics of the mathematical function that means to identify the mathematical knowledge in specific situations categories but in generic terms within people's daily lives.

17.9 Conclusions

As a summary to formulate an epistemological pluralism made up by the function, re-signification and mathematical knowledge mainstreaming. Mathematics will have new expressions according to people.

The intervention within the problem will consist of creating recovery instruments that set in a horizontal dialogue the school mathematics and the mathematics in daily life. A natural discipline debate will consist in the articulation of both math. This last one will create a construct of ethnographic nature due to we require it of the function math own by people in their specific field.

Because of the importance of this fact, we will call it *the native's math revel* methodologically. We will analyze "the mathematics" of the school, work and city. The specification in this field will be defines by the permanent uses of people and the maintenance of the field. Fields will be distinguished and usual will be defined with adjectives. These will be the recovery instruments.

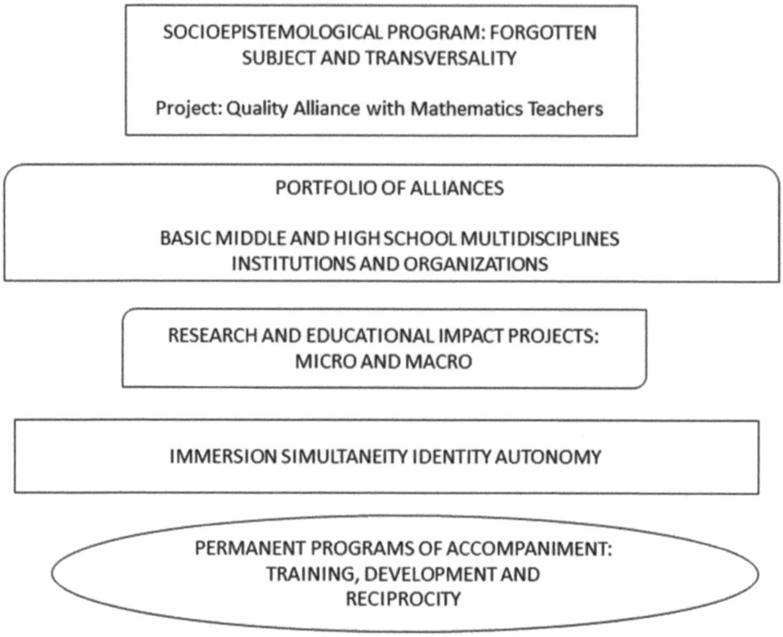

Fig. 17.6 Simultaneous reciprocal permanent program: the mathematics teacher's role. (Source: Cordero, 2016b, p. 29)

The articulation between the institutional and functional will consist of breaking the focus on the object. The new arguments will be about the re-significations of the uses of mathematical knowledge which make the classic orientations of resolutions tense against an innovative orientation, the modelling. The first one worries about the knowledge processes while the second one focuses on knowledge function.

As a summary, the center of the problem about mathematics education consists of the forgotten subject thesis therefore we will recover it. We cannot see the mathematics teacher as someone who does not have knowledge, who needs more training to cover his empty fields. It is necessary to understand it as a mathematical knowledge community that constructs its own mathematical categories of its environment normed by the reciprocal relationships between the school and reality knowledge.

The educational mathematics discipline with its research programs will play a fundamental role to achieve that. This should guide the necessary articulations in three big actions: *alliance, immersion and reciprocity* with the mathematics teacher communities in all the educational levels (Fig. 17.6).

These actions should be around the social and economic moments that the world lives and agree with the communities (where all research groups and teachers participate) strategies that set in the same academic and social status the training of researchers in our discipline field and make the teaching professional with all the educational levels. The actions and strategies will be transformed into concrete ideas quality Alliance with the educational system teachers, which will help to precise if the new development program is suitable and the mathematics educational impact. Figure 17.6 shows simultaneous reciprocal permanent program: the mathematics teacher's role.

We will design Alliance portfolios with each educational levels (Primary, Secondary and Higher Education), institutions and organisms, with research programs with different levels of impact together its multifactorial permanent programs (identity, inclusion, socialization and emancipation) focused on the reciprocal dialogue construct between the classroom and reality in levels: disadvantage, hope, possible and autonomy. *Mathematical knowledge mainstreaming* will be the new epistemological status.

These *programs are systems* that favor the mathematical knowledge function where some diversity, mainstreaming and the other consideration happen. These will be own programs of the knowledge communities where reciprocities between mathematics and reality happen, private knowledge categories of these communities that later become discipline jargons and places that express social, political and cultural movements. The immersion with the mathematical knowledge communities will thrive to the alliance of the quality of mathematics teaching in balance with the development of the education mathematics discipline.

References

Artigue, M. (1998). Enseñanza y aprendizaje del análisis elemental: ¿qué se puede aprender de las investigaciones didácticas y los cambios curriculares? *Revista Latinoamericana de Investigación en Matemática Educativa, 1*(1), 40–55.

Bourdieu, P. (2008). *Homo academicus* (Ariel Dilon, Trans.). Editora Siglo XXI.

Buendía, G. (2004). *Una epistemología del aspecto periódico de las funciones en un marco de prácticas sociales (Un estudio socioepistemológico)*. Unpublished doctoral dissertation, Cinvestav-IPN, Mexico.

Buendía, G., & Cordero, F. (2005). Prediction and the periodical aspect as generators of knowledge in a social practice framework. A socioepistemological study. *Educational Studies in Mathematics, 58*, 299–333.

Campos, C. (2003) *La argumentación gráfica en la transformación de funciones cuadráticas. Una aproximación socioepistemológica*. Unpublished master's thesis, Cinvestav-IPN, Mexico.

Cantoral, R. (2001). *Matemática Educativa. Un estudio de la formación social de la analiticidad*. Grupo Editorial Iberoamérica.

Cantoral, R. (2013). *Teoría Socioepistemológica de la matemática Educativa. Estudios sobre la construcción social del conocimiento*. Gedisa.

Cantoral, R., & Farfán, R. (2003). Matemática Educativa: Una visión de su evolución. *Revista Latinoamericana de Matemática Educativa, 6*(1), 27–40.

Cantoral, R., Farfán, R., Cordero, F., et al. (2000). *Desarrollo del Pensamiento Matemático*. Editorial Trillas.

Cantoral, R., Reyes-Gasperini, D., & y Montiel, G. (2015). Socioepistemología, Matemáticas y Realidad. *Revista Latinoamericana de Etnomatemática, 7*(3), 91–116.

Cordero, F. (1998). El entendimiento de algunas categorías del conocimiento del cálculo y análisis: el caso de comportamiento tendencial de las funciones. *Revista Latinoamericana de Investigación en Matemática Educativa, 1*, 56–74.

Cordero, F. (2001). La distinción entre construcciones del Cálculo. Una epistemología a través de la actividad humana. *Revista Latinoamericana de Investigación en Matemática Educativa, 4*(2), 103–128.

Cordero, F. (2003). *Reconstrucción de significados del Cálculo Integral. La noción de acumulación como una argumentación*. Grupo Editorial Iberoamérica.

Cordero, F. (2015). *La ciencia desde el niñ@. Porque el conocimiento también se siente* (1st ed.). Gedisa.

Cordero, F. (2016a). Modelación, funcionalidad y multidisciplinaridad: el eslabón de la matemática y el cotidiano. In J. Arrieta & L. Díaz (Eds.), *Investigaciones latinoamericanas de modelación de la matemática educativa* (pp. 59–88). Gedisa.

Cordero, F. (2016b). La función social del docente de matemáticas: pluralidad, transversalidad y reciprocidad. In S. Estrella, M. Goizueta, C. Guerrero, A. Mena-Lorca, J. Mena-Lorca, E. Montoya, A. Morales, M. Parraguez, E. Ramos, P. Vásquez, & D. Zakaryan (Eds.), *XX Actas de las Jornadas Nacionales de Educación Matemática* (pp. 23–30). SOCHIEM, Instituto de Matemática de la Pontificia Universidad Católica de Valparaíso. ISSN: 0719-8159. Retrieved from http://ima.ucv.cl/congreso/xxjnem/

Cordero, F., Albores, A., Asomoza, R., Briceño, E., Cabrera, L., Canché, E., Cen, C., Gómez, K., Miguel, M., Silva, H., Simón, G., Soto, D., Viramontes, D., & Zaldívar, D. (2009). Cinvesniñ@s. Una experiencia de difusión del conocimiento científico. *Avance y Perspectiva, 2*(4), 33–47. ISSN: 1870-5499.

Cordero, F., & Flores, R. (2007). El uso de las gráficas en el discurso matemático escolar un estudio socioepistemológico en el nivel básico a través de los libros de texto. *Revista Latinoamericana de Investigación en Matemática Educativa, 10*(1), 7–38.

Cordero, F., Gómez, K., Silva-Crocci, H., & Soto, D. (2015). *El Discurso Matemático Escolar: la Adherencia, la Exclusión y la Opacidad*. Gedisa.

Cordero, F., Henríquez, C., Solís, M., Méndez, C., Opazo, C., & De la Cruz, A. (2020). La modelación en la matemática educativa: sus programas de investigación y la docencia. El rol de la transversalidad de saberes matemáticos. *Acta Latinoamericana de Matemática Educativa, 33*(2), 584–594.

Cordero, F., Solís, M., & Opazo, C. (2022). *La matemática en la ingeniería. Modelación y transversalidad de saberes. Situaciones de aprendizaje* (1st ed.). Gedisa.

Dirección General de Epidemiología. (2021). Covid 19, México: Datos epidemiológicos. Retrieved from https://covid19.sinave.gob.mx/

Domínguez, I. (2003). *La resignificación de lo asintótico en una aproximación socioepistemológica*. Unpublished master's thesis, Cinvestav-IPN, Mexico.

Dubinsky, E., & Harel, G. (1992). The nature of the process conception of function. In E. Dubinsky & G. Harel (Eds.), *The concept of function: Aspects of epistemology and pedagogy* (Vol. 25, pp. 85–106). Mathematical Association of America.

Freudenthal, H. (1968). Why to teach mathematics so as to be useful. *Educational Studies in Mathematics, 1*, 3–8.

Gobierno de México, Secretaría de Salud. (2020). Mensajero de la Salud. Nuevo Coronavirus 2019-nCovid. Retrieved from https://www.gob.mx/cms/uploads/attachment/file/571451/Mensajero_Coronavirus.pdf

González, R. (1999). *La derivada como una organización de las derivadas sucesivas: Estudio de la puesta en escena en funcionamiento de una ingeniería didáctica de resignificación*. Unpublished master's thesis, Cinvestav-IPN, Mexico.

Hernández, D. (2004). *Las argumentaciones gráficas de los estudiantes en las relaciones de f y f' para las funciones x, x^2, y x^3*. Unpublished master's thesis, Cinvestav-IPN, Mexico.

Instituto Nacional para la Evaluación de la Educación. (2007). *Panorama Educativo de México 2007. Indicadores del Sistema Educativo Nacional*. Retrieved April 12, 2008 from http://www.oei.es/pdfs/panorama_educativo_mexico2007.pdf

Kermack, W. O., & McKendrick, A. G. (1927). A contribution to the mathematical theory of epidemics. *Proceedings of the Royal Society of London. Series A, Containing Papers of a Mathematical and Physical Character, 115*(772), 700–721.

Medina, D. (2019). *Transformación educativa del docente de matemáticas. Un episodio: el uso de la compensación como una resignificación de la media aritmética*. Unpublished doctoral dissertation, CINVESTAV-IPN, Mexico.

Mendoza-Higuera, E. J. (2020). *El uso de la matemática en la Ingeniería Biónica. De la estabilidad a la reproducción de comportamientos en un sistema de control. Una Categoría de Modelación*. Unpublished doctoral dissertation, Cinvestav-IPN, Mexico.

Morales-Reyes, J. L. (2020). *Resignificación de los usos de la derivada en un diseño escolar con perspectiva de dialéctica exclusión-inclusión: predicción, comportamiento tendencial y analiticidad*. Unpublished master's thesis, Cinvestav-IPN, Mexico.

Niss, M. (2015). Prescriptive modelling: Challenges and opportunities. In G. A. Stillman, W. Blum, & M. S. Biembengut (Eds.), *Mathematical modelling in education research and practice: Cultural, social and cognitive influences* (pp. 67–79). Springer.

Opazo-Arellano, C. (2020). *Identidad disciplinar en la formación inicial docente: una resistencia al discurso matemático escolar*. Unpublished doctoral dissertation, Cinvestav-IPN, Mexico.

Pérez-Oxté, I. (2021). *Anticipar-periodizar: una socialización de saberes matemáticos entre la Ingeniería y la docencia*. Unpublished doctoral dissertation, CINVESTAV-IPN, Mexico.

Rosado, P. (2004). *Una resignificación de la derivada. El caso de la linealidad del polinomio en la aproximación socioepistemológica*. Unpublished master's thesis, Cinvestav-IPN, Mexico.

Seminario sobre Divulgación de la Ciencia y la Tecnología. (January 1999). *Primer seminario sobre divulgación de la ciencia y la tecnología*. Auditorio Bruno Mascanzoni, Mexico, D.F.: Instituto Mexicano del Petróleo.

Solís, M. (October 10, 2020). Modelo de fichas [Video file]. Retrieved from https://youtu.be/7XxkroqJd1M

Solís, M., & De la Cruz, A. (2022). Modelación, simulación y graficación de fenómenos naturales. El caso de una epidemia. In F. Cordero, M. Solis and C. Opazo (Eds), La matemática en la ingeniería. Modelación y transversalidad de saberes. Situaciones de aprendizaje. (pp 325–345): Gedisa.

Stewart, J. (1994). *Cálculo*. Grupo Editorial Iberoamérica.

Vidal, M., Guinovart, R., Baldoquín, W., Valdivia, N. C., & Morales, W. (2020). Modelos matemáticos para el control epidemiológico. *Educación Médica Superior, 32*(2).

Zaldívar-Rojas, J., & Cordero, F. (2021). An analysis of the use of graphs in people's daily life. *Revista Colombiana de Educación, 1*(83), 1–22. https://doi.org/10.17227/rce.num83-10608

Part V
Conclusion

The re-signification of mathematical knowledge as generated by a selection of mathematical modelling programs in diverse regions in Latin America form both an epistemological and ontological basis to the design and development of teaching and learning of mathematics at all educational levels: basic, intermediate, and higher education. This development is cyclical, continuous, and reflective. The program designs consist of locating moments of transversality of knowledge that favor and value the emergence and re-signification of a diversity of mathematical knowledge. In this context, learning consists of a creation of pedagogical relations between diverse mathematical knowledge, teaching, and didactics that are accompanied by the social function of educators, which consists of maintaining the environments of mathematical knowledge through permanent support and partnership programs. Therefore, the discourse of school mathematics is disrupted and transformed in order to create its redesign in which this fundamental epistemological and ontological basis develops a relation of horizontality-autonomy-cultural knowledge, which is fostered by the principles of mathematical modelling programs. This approach entails an educational change of mathematics, at all educational levels, which considers the development of knowledge of *the others* by including their diverse reasoned and symmetrical decision-making process through modelling as found in diverse regions of Latin America.

Chapter 18
The Mathematical Teaching and Learning Process Through Mathematical Modelling: Educational Change in Latin America

Milton Rosa, Daniel Clark Orey, Francisco Cordero, and Pablo Carranza

18.1 Introduction

Historical evolution enables the development of alternative mathematical knowledge systems that provide explanations of daily problems, and situations, and which lead towards the elaboration of models as representations of our own reality (D'Ambrosio, 2015). After centuries of cultural exchange between Europe, North and South America a unique, sophisticated, and rigorous mathematics and science is emerging in our region.

The modelling process that was introduced a century ago was adapted by and helps members of distinct communities to draw information about their own problems, realities and needs, through the elaboration of representations, which generate mathematical knowledge and incorporates our unique creativity and invention. It is proposed, with empirical evidence, that a category of mathematical modelling has developed here in Latin America that assesses both the horizontal and reciprocal relations between mathematics (school/non-school contexts) and the real-world.

These relations, on the one hand, provide scholars with powerful opportunities that make use of epistemological and ontological changes, where mathematical knowledge of the *others* is recognized, equally, on a horizontal plane. On the

M. Rosa (✉) · D. C. Orey
Universidade Federal de Ouro Preto, Ouro Preto, Minas Gerais, Brazil
e-mail: milton.rosa@ufop.edu.br; oreydc@ufop.edu.br

F. Cordero
Centro de Investigación y de Estudios Avanzados del Instituto Politécnico Nacional, Ciudad de México, Mexico
e-mail: fcordero@cinvestav.mx

P. Carranza
Universidad Nacional de Río Negro, General Roca, Río Negro, Argentina
e-mail: pcarranza@unrn.edu.ar

M. Rosa et al. (eds.), *Mathematical Modelling Programs in Latin America*,
https://doi.org/10.1007/978-3-031-04271-3_18

other hand, they encourage both mathematics teachers and students to understand and investigate how they can build their own mathematical categories that are found in environments governed by the many reciprocal relationships between academic knowledge and functional knowledge, which can be considered as the common ground for the community of knowledge creators.

18.2 Contextualizing Three Latin American Modelling Programs

The dimensions of mathematical modelling described in the chapters of this book show us the relations that can develop a strong frame of reference that can guide the educational change in Latin America. This change agent makes use of autonomous actions compared to the emulations of typical mathematical procedures in our classrooms.

Therefore, this approach recognizes that there may be many more, but begins with three Latin American modelling programs: ethnomodelling, transversality of knowledge, and reasoned decision-making. Each one, with its respective theoretical and methodological foundations uses ethnomathematics and ethnomodelling, interdisciplinarity, and socioepistemology seek to give and attribute meaning to learning and the daily activities performed by members of distinct cultural groups.

The late Professor Ubiratan D'Ambrosio graciously accepted an invitation to write the *Foreword* to open this book. We believe that this foreword is one of the very last texts he wrote before his passing on May 12, 2021, and we are greatly honored and thank the D'Ambrosio family for this honor in including his thoughts here. D'Ambrosio talks here that this book brings a collection of chapters dealing with mathematical modelling programs in Latin America.

D'Ambrosio also emphasized the importance of a collaborative context for social construction of mathematical knowledge for educational change in society that seeks social justice and promotes peace. He also states that this book shows how historical evolution enables the development of alternative mathematical knowledge systems that provide explanations of daily problems, and situations, and which can lead to the elaboration of models as representations of facts present in our own reality.

We would like to take this opportunity to pay tribute to Ubiratan D'Ambrosio for his contributions to mathematics, mathematics education, and other knowledge areas. Internationally, D'Ambrosio contributed to the development of mathematics education, through his international leadership and worldwide dissemination of ideas related to sociocultural issues, peace, and social justice.

D′Ambrosio's development of ethnomathematics and mathematical modelling, and their application in mathematics education were powerful and profound. It is important to state here that D'Ambrosio is one of the most important and influential mathematicians of the twentieth and twenty-first centuries, mainly in relation to his mentorship, support, and encouragement through investigations related to social, political, and cultural aspects of mathematics in the entire world.

For example, D'Ambrosio's approach sought to promote interactions among all social classes and his preoccupation with people's welfare, the preservation of natural and cultural resources can be synthesized as peace in its several dimensions, such as inner peace, social peace, environmental peace, and military peace, which is defined as total peace.

In this regard, D'Ambrosio shared with us his vision as to how fundamental it should be that the teaching and learning of mathematics values the greater sociocultural context of mathematical knowledge and that we must encourage and connect this aspect to diverse learning processes for goals important not just to formal or academic curricula so that we are able to achieve social justice and peace in this time of crisis.

Continuing with the contextualization of three Latin America mathematical modelling programs, the introduction as the first chapter of the first section of this book, entitled: *Modelling in the Life of People: An Alternative Program for Teaching and Learning of Mathematics*, written by Francisco Cordero, Milton Rosa, Daniel Orey, and Pablo Carranza, proposes a category of mathematical modelling that assesses the horizontal and reciprocal relationships between mathematics and the real world. These relations provide epistemological and ontological changes in which mathematical knowledge of the *others* is recognized on a horizontal stage.

On the other hand, these changes oblige mathematics teachers and students to understand as mathematical knowledge producers that through modelling processes that enable us to build our own mathematical categories of their own contexts, which are governed by the reciprocal relations between academic knowledge and functional knowledge that is part of their daily life. The dimensions of these relations guide educational change that encourages forms of mathematical teaching in autonomous actions compared to the emulations of typical mathematical procedures in the classroom.

Chapters of this book show the creation of diverse ways of dealing with the social construction of mathematical knowledge in Latin America. Thus, the authors of the 18 chapters in this book, who represent the diversity of Latin America, are from nine countries: Argentina, Brazil, Chile, Colombia, Costa Rica, Cuba, Ecuador, Honduras, and Mexico. It is important to highlight that authors from Cuba and Honduras are also developing their investigations at Cinvestav in Mexico.

They were invited to share their ideas, perspectives, and discuss investigations that represent a rich sample of three Latin American mathematical modelling programs: ethnomodelling, transversality of knowledge, and reasoned decision-making processes. Based on empirical evidence, each one of these dimensions are also theoretically based on ethnomathematics, socioepistemology, and the attribution of meaning to the learning process.

18.2.1 Ethnomathematics and Ethnomodelling

The six chapters of the second section of this book come from common conceptions of ethnomathematics and ethnomodelling and its empirical investigations, theoretical and methodological approaches, and research questions.

The second chapter entitled: *Conceptualizing Positive Deviance in Ethnomodelling Research: Creatively Insubordinating and Responsably Subverting Mathematics Education*, written by Milton Rosa and Daniel Clark Orey, discussed important dilemmas in mathematics education and how they form a certain bias in relation to the acknowledgment of the importance of local mathematical knowledge orientation in many research paradigms. Thus, a search for innovative pedagogical actions such as ethnomodelling can be useful for recording historical forms of mathematical ideas, procedures, and practices developed in diverse cultural contexts.

Yet, it is important to emphasize how basic assumptions found in ethnomodelling do not present an attempt to replace school and/or academic mathematics. At the same time, it is necessary to acknowledge the existence of important contributions found in local mathematical knowledge in the development of mathematics through history for its inclusion in the school mathematics curriculum.

In this context, a certain insubordination triggered by investigations conducted in ethnomodelling is creative and often evokes disturbance that causes a sense of responsible subversion in the revision of rules and regulations in the context of mathematics curriculum.

This approach enables educators to use acts of positive deviance to develop pedagogical actions. A sense of positive deviance involves an intentional act of rule breaking in order to serve the greater good of students. This process increases the potential for continual growth in the debate about the nature of mathematics we teach as it relates to local needs and the cultural aspects of where to who we serve as educators. It proposes a dialogue between the local and global approaches to the construction of mathematical knowledge through unique dialogical approaches using ethnomodelling.

The third chapter entitled: *Ethnomodelling as an Alternative to Basic Education: Perceptions of Members of a Research Project*, written by Zulma Elizabete de Freitas Madruga, presents the perceptions of members of a ethnomodelling research group at the Universidade Federal do Recôncavo Baiano, in the state of Bahia, Brazil. A group composed of schoolteachers, undergraduate students, graduate students, and mathematics professors participated in the project. Here, the primary goal is to demonstrate what participants in the project found after the conclusion of the first stage of the project.

This stage consists of theoretical studies on ethnomodelling, which members in the group see as a methodological alternative to mathematics teaching in basic education. Interviews were conducted with all 15 members of the project. These interviews were analyzed by means of discursive textual analysis. The results indicate that the project contribution to the initial or continuous education for both teachers and researchers is related to the discussion and exploration this innovative alternative approach to mathematics education.

The fourth chapter entitled: *Ethnomodelling Aspects of Positionality Between Local and Global Knowledge Through Glocalization: A Case of a Farmer Vendor*, written by Diego Pereira de Oliveira Cortes and Daniel Clark Orey, is both theoretically and empirically based on a master's degree research entitled: "*Re-signifying the concepts of function: A mixed-methods study to understand the contributions of the dialogic approach of ethnomodelling*" conducted by Cortes (2017).

Through ethnomodelling, the authors seek to discuss the connection between local, global, and glocal mathematical knowledge as defined by Rosa and Orey (2017). In particular, the authors discuss aspects of a neighborhood farmer's market (local) and the positionality of a farmer vendor as a result of his dialogical contact with aspects of global knowledge studied in academic environments.

In addition, the authors also seek to understand concepts of both familiarity and strangeness that occurred during interactions between the farmer vendor and students in this study, which enabled the development of cultural dynamism that promoted the diffusion of glocalized knowledge during the development of ethnomodelling (Orey & Rosa, 2021).

The fifth chapter entitled: *Ethnomodelling as a Pedagogical Action in Diverse Contexts by Using a Dialogical Knowledge*, written by Ana Paula Santos de Sousa Mesquita, Érika Dagnoni Ruggiero Dutra, Jéssica Rodrigues, and Milton Rosa, shows that the use of ethnomodelling in classrooms promotes the development of strategies that encourages learners to apply mathematical ideas and procedures that help them to elaborate models based on the activities present in their daily lives.

In this chapter, the authors present the results of three studies developed by applying an ethnomodelling perspective that encourages further discussion of dialogical approaches used in different contexts such as coffee culture, peripheral communities, and math trails. The main objective of the first study, which was conducted in the context of coffee culture, was to develop a pedagogical action that help members of distinct cultural groups (coffee and school cultures) to value emic and etic mathematical knowledge in a dialogical manner.

The second study was developed in a peripheral community and aimed to carry-out a sociocritical analysis of the ethnomodelling process as a pedagogical action in the development of mathematical content of the students. The third study sought to analyze and discuss the sociocultural perspective of ethnomodelling that aimed to assisted participants in reading their own reality by directing them to a better understanding of their surroundings through their participation in Math Trails. From the results of these studies, the authors inferred that ethnomodelling provides us with the development of a critical and reflective analysis of the students own communities in a multicultural and interdisciplinary fashion in a holistic and dialogical way.

The sixth chapter entitled: *Ethnomodelling: Weaving Networks Between Academic Mathematical Knowledge and Cultural Knowledge in the Southeastern Region of Tocantins*, written by Alcione Marques Fernandes, Cristiane Castro Pimentel, and Nayane Rodrigues de Deus, presents the research and guidelines developed with the studies and research group in Mathematics and Mathematics Teaching linked to the Professional Masters' Degree Program in the Mathematics Network—PROFMAT, at the Universidade Federal do Tocantins, Arraias Campus, in the research line: Ethnomathematics and training of teachers.

Ethnomodelling is defined as the study of ideas, notions, and procedures used by members of distinct cultural groups used to solve daily problems (Rosa & Orey, 2019), and both facilitated and enabled discussion about the elaboration of stone walls in the cemetery in Arraias as well as the research on the making of handmade jewelry in gold and silver filigree in Natividade, and an ongoing research garden project in a quilombola community.

From the discussions conducted in these investigations, it is possible to highlight the new and growing webs that are slowly being woven among academic and cultural knowledge forms in the southeastern region of Tocantins through ethnomodelling.

In the seventh chapter: *Mathematical Analysis of the Ceramic Designs of the Pre-Columbian Cultures of Ecuador Through Ethnomodelling with a Sociocultural Approach*, written by Juan Ramón Cadena and Ronald Patricio Chasiloa Llumiquinga, shows that ethnomodelling is considered as a conjunction between mathematical modelling and ethnomathematics in its social, cultural, and historical dimension.

Latin America is a promising region where educators can experiment and develop alternative ways in which we can insert methodologies optimizing the learning of mathematics in contexts of culture, history, and diversity and through other ways of understanding this science, especially in the Andean region, where similar realities and scenarios are shared.

This chapter aims to make a theoretical and didactic proposal, based on archaeological data on the designs of ceramics of pre-Columbian cultures of Ecuador, considering the geographical situation close to parallel zero, the astronomical, agricultural, and multi-climatic peculiarities, this study intends to rediscover them under an approach does not exempt from the myths and other forms of sensitivity about the understanding of the world that defines the Andean and Latin American ethnos.

Ethnomodelling will allow making a methodological proposal to understand universal notions such as symmetry and translation present in ceramics. The result of the empirical analysis allows obtaining an etic interpretation (global) of the emic context (local) of these designs. This dialogical systematization helps to treat this knowledge under a sociocultural perspective, with pedagogical elements of didactic concretion that can be applied in the classroom by the teachers.

In addition, through the transdisciplinary interaction between mathematics, archeology, anthropology, and history; significant learning will be accomplished in order to contribute to the achievement of regional identity.

18.2.2 Interdisciplinary Ecosystems

The four chapters of the third section of this book focus on interdisciplinary ecosystems, empirical investigations, theoretical and methodological approaches, and related research questions.

The eighth chapter entitled: *Analyzing the Availability of Renewable Energy Resources in a Project in a Real-World Context: A Framework for Making Sense of Learning*, written by Pablo Carranza and Fabio Miguel presented a proposal developed to facilitate the attribution of meaning to learning: it is the calculation, construction and installation of Savonius windmills. This project may be considered as a solution to the problem of access to water for rural residents with limited economic resources in the Argentinean Patagonia.

This project is a case where mathematics and other disciplines must interact to analyze renewable energy resources and produce a viable solution to the real problems of rural people. This is an example of how the three dimensions or characteristics: temporality, transcendence and functionality of learning can be applied in research in order to facilitate the attribution of meaning. In the case of mathematics, modelling represents an indispensable space for analysis and construction of arguments for the actions to be undertaken.

The ninth chapter entitled: *Descriptive and Prescriptive Modelling in a Math Class Project: Disciplinary Concepts Participating in the Construction of Arguments for Decision Making*, written by Pablo Carranza and Jaime Moreno, analyzes the dynamics of modelling as space for the analysis and construction of arguments in order to have foundations for the important actions to be taken within the framework of a project consisting of calculating, building and installing Savonius windmills.

It is important to state here that these windmills are not imaginary or model nor are they scale models. They can be considered true windmills that work in safe conditions for both students and rural residents.

The rationality of this study does not come from a demand by teachers because it represents a genuine and authentic situation as it comes from the real-world context in which the project is developed. Thus, this chapter analyzes two critical situations: one is related to the fastening system that keeps the mill in its vertical position, allowing it to withstand strong winds of the Argentinean Patagonia while the other deals with the hoisting mechanism of the mill at the time of its installation.

Hence, models that arise in this context are primarily descriptive because they facilitate an understanding of the phenomena and then become prescriptive, because they allow for the construction of reasoning in order to function with solid foundations. This article also highlights how the context determines the use of technological tools and how they favor the approach of different concepts, even if it is the same problem. For example, the case of the guy wires was analyzed by using both Geogebra and Spreadsheet tools.

The tenth chapter entitled: *Designing and Building a Mobile Support for Solar Panels: A Project for 12-Year-Old Students That Requires Concepts, Among Others, of Mathematics*, written by Pablo Carranza and Ailén Morales, develops a fieldwork study with university students and the research team, which was conducted with 12–15 years-old students in a secondary school.

Three dimensions were applied in this study: temporality, transcendence, and functionality of learning, whose characteristics were crossed with the school's curriculum and its possible availability in educational environments. This led to proposing a project in which students had to build a mobile support for a photovoltaic panel in order to provide electricity to a water pump.

This project highlights the importance of considering the integration of disciplines when dealing with real-world situations since it raises questions about the relevance of promoting learning in sealed disciplinary compartments, as traditionally occurs in educational systems, at least in Argentina. Thus, this project can be considered as an example of the interaction between disciplines if researchers want to maintain an argumentative coherence in the development of their projects. They questioned then the traditional monodisciplinary didactic transposition.

The eleventh chapter entitled: *Analogical Modelling and Analytical Modelling: Different Approaches to the Same Context?*, written by Pablo Carranza, Mónica Navarro, and Mariana Letourneau, brings together a set of mathematical modelling projects developed in real-world environments with the objective of proposing a new categorization related to mathematical modelling based on the objects retained in these contexts.

When analyzing the set of modelling processes conducted in these projects, the authors observed that some of them retained relatively observable elements of their context while others were developed from more abstract relationships. The former were called analogical modelling due to their close relationship with objects of reality.

The latter is called analytical modelling due to the higher level of abstraction students developed in these processes. In addition, a review of the development of the modelling projects enabled the identification of a sequence that was characterized by the initial analyzes of the proposed problem through the application of an analogical modelling and then continuing with the use of analytical modelling.

The reasoning produced in an analogical modelling exhausted the possibilities of development, while at the same time, the space where they emerged enabled these processes to move to another type of modelling in order to continue its development. It is important to state here that analytical modelling is more abstract and dynamic, which enriches the analyses of the proposed object of study.

Yet, analogical modelling projects can reach their potentialities, and are fundamental for the development of modelling processes when students build strong arguments in order to understand the abstract relationships necessary for the understanding of the analytical modelling process that are developed later in these projects.

18.2.3 Mathematics and People

The six chapters of the fourth section of this book deal with interactions between of mathematics and people and empirical investigations, theoretical and methodological approaches, and research questions.

The twelfth chapter entitled: *A Category of Modelling: The Uses of Mathematical Knowledge in Different Scenarios and the Learning of Mathematics*, written by Francisco Cordero, E. Johanna Mendoza-Higuera, Irene Carolina Pérez-Oxté, Jaime Mena-Lorca, and Jaime Huincahue, discussed that there is a functionality of mathematical knowledge that is demanded by other domains of knowledge, such as: school/academic, work/profession, and daily life.

The knowledge of any one person is put to use in these scenarios and many times simultaneously. This means that the use of this knowledge is re-signified with characteristics specific to each scenario, and the re-signification emerges because these uses are in a horizontal and reciprocal relationship with each other. However, these resignifications, in general, are not present in school mathematics.

The authors propose a research program to build a frame of reference to legitimize the mathematics in use that takes place in the different scenarios and in the transversality between them. These two aspects define a category of mathematical modelling, which on the one hand, formulates a theoretical variety and, on the other, it will involve a program to permanently change and transform school mathematical knowledge.

Due to the nature of this research, the authors decided to study mathematics in its social construction, through the development of Socioepistemological Theory of Educational Mathematics, which takes as its epistemological and ontological basis human wisdom as a synthesis of wise, technical, and popular knowledge.

On this basis, the authors present powerful empirical evidence on the role of the modelling category. On the one hand, they discuss the resignification of stability, in a control system, through a modelling category, the reproduction of behaviors in a community of bionic engineers.

And, on the other hand, the re-signification of asymptotics and optimization, in a graphical model of chemical elements, through modelling categories prediction, trend behavior and selection to establish the life of a transformer, in a community of chemical engineers. Using an ethnographic and case study methods, immersions were made in these communities to reveal the emergence of modelling categories.

The results show that the aforementioned modelling categories bring into play the use of mathematical knowledge with the realities of the learners. It also offers an environment of meanings to the mathematical objects of school mathematics, as in the case of functions of real variable and linear differential equations are re-signified as instructions that organize and reproduce behaviors. This category is transversal to the different scenarios and could be transversal to the different educational levels (elementary, middle, and high school). In the light of the results, the authors reflect on the need to understand the mathematics teachers as a generator of their own modelling categories and thus pronounce the autonomy of mathematical knowledge as a source of learning.

The thirteenth chapter entitled: *Modelling and Anticipation of Graphical Behaviors in Industrial Chemical Engineering: The Role of Transversality of Knowledge in Learning Mathematics*, written by Irene Pérez-Oxté and Francisco Cordero, presents a peculiar study of the modelling category in the daily professional work of a community of industrial chemical engineers. The peculiarity consisted in constructing a method to be developed from the category of modelling in the profession to the category of modelling in the schools.

The research method is empirical. An immersion was made in a community of industrial chemical engineers who diagnose electrical transformers. With the socioepistemological theory, the authors give evidence of the emergence of the modelling category anticipation of graphic behaviors through the periodization of segmented times. Then, school activities were designed to make this category transversal in engineering students.

The methodological instruments were carried out by using documental analysis technique in order to analyze semi-structured interviews and a focus group. The activities were focused on confronting the modelling category with the usual school mathematics, in order to bring about the valorization of the uses of mathematical knowledge, usually overlooked in school mathematics. The authors highlighted the activities: use of statistical control, use of the relationship graph-failure, and use of the graphical model of diagnosis.

These activities were based on three moments of construction of the modelling category *anticipation of graphical behavior*: prediction of graphical behaviors for diagnosis, trend behaviors for stable behaviors, and selection of ideal behaviors. One aspect that stands out in the results is the confrontation between the uses of anticipation and school mathematics: linear models of approximation with irregular/regular behaviors, definition of the limit of a function with behaviors with a tendency in a period, and second derivative criterion with ideal behaviors and their reproduction.

The fourteenth chapter entitled: *Category of Modelling and Reproduction of Behaviours in Other Disciplines: The Teaching of Mathematics and Engineering*, written by E. Johanna Mendoza-Higuera, Falconery Giacoleti-Castillo, José Luis Morales-Reyes, and Francisco Cordero. The authors outlined how to build a horizontal and reciprocal relationship between school mathematics and engineering.

With the Socioepistemological Theory of Educational Mathematics, they proposed a reference framework that values the functionality of mathematics that engineering demands. On the one hand, they evidenced the emergence of behavior modelling categories in engineering knowledge domains, in specific situations, and on the other hand, they discussed the educational impact of this category on the mathematical training of engineers.

More specifically, the re-signification of linear differential equations forms a trend in models that reproduces behaviors and forms the derivative as a model that means the tangent line as an element that models a trend behavior that predicts a future state is evidenced. As a result of the investigation, the authors concluded that the category of behavior reproduction emerges in situations of electronic engineering and chemical engineering, describing a transversality of uses that re-signify the knowledge of the *Laplace Transform* and the derivative, respectively.

Thus, they concluded that the modelling category is a didactic action that makes visible the transversality of uses of mathematical knowledge and the re-definition of mathematical knowledge. However, the category of modelling does not appear in the usual school mathematics for the training of engineers.

The fifteenth chapter entitled: *The Disciplinary Identity in Initial Mathematics Teacher Training and People's Category of Modelling: A Valorization of the Knowledge of the Learner*, written by Claudio Opazo-Arellano, Sindi Marcía-Rodríguez, Henry Chávez-Martínez, Eleany Barrios-Borges and Francisco Cordero, shows how the problematization of mathematical knowledge, with the category of modelling, causes the emergence of autonomous arguments, in contrast to the emulation of procedures that were previously taught.

The authors justify this fact through the use of *disciplinary identity factor*, a theoretical construct that found its ontological and epistemological foundation in the re-signification of mathematical knowledge. That identity provokes in the one who learns to teach a change of vision about the meaning of what is to learn and what is to teach. It opens its spectrum of school mathematics to epistemological plurality and the transversality of knowledge, of mathematics.

The educational impacts of this category of modelling and disciplinary identity is defined when learning to teach is not associated with emulating procedures of mathematical objects, but with valuing the knowledge of people and promoting the reciprocity and horizontality of mathematical knowledge. The authors show evidence of this through, on the one hand, the re-signification of the accumulation of quantities that change over time to confront the area representation of the integral of a function.

On the other hand, the transversality of the uses of the asymptote in different situations, such as: reproducing a desired temperature, reproducing the trend of a population dynamics, and reproducing a favorable trend in the recovery rate of a virus. These transversalities of the asymptote confront the privileged representation of the symptom as a straight line, in school mathematics.

The sixteenth chapter entitled: *Contemporary Learning in the Interaction of the Human with Data, Via Technology-Mediated Graphics: The Discourse-Representation Dialogue in Mathematics*, written by Arturo Mena-Lorca, Jaime Mena-Lorca, and Astrid Morales-Soto, presents a reflection on the contemporary processes of learning mathematics.

The authors justify the thesis that human beings process information in a dialogue of the discursive and graphic forms in which it is presented. Contemporaneity consists in managing technology to obtain adequate representations of the data and for the symbolic processing of the relationships at stake; in turn, to some extent, this shifts attention away from computational routines and allows focus on contemporary learning requirements: modelling, decision making, computational thinking.

The authors explore correspondences of relationships in the ways of processing information, in history and how the ways of processing information have been redefined with the management of technology. In this sense, mathematical modelling with permanent technological support and direct reasoning on various types of graphs implies a fairly drastic reduction in routine calculations.

The authors also consider previous research to argue that students value mathematics and use graphics to connect models, eventually interdisciplinary. In addition, they develop functional mathematical knowledge, specifically arguing with the help of graphs. Naturally, this is important at all school levels; however, it is especially relevant in higher education, where a significant number of students abandon their careers due to their inability to handle the (calculation) procedures of mathematics.

In the seventeenth chapter: *Modelling of Natural Phenomena as a Source to Re-signify Mathematical Knowledge*, written by Miguel Solís, Francisco Cordero, Eleany Barrios-Borges, and Adriana Atenea de la Cruz-Ramos, shows that in the context of the development of the pandemic, Covid 19, which began in 2020, a real or natural phenomenon, in its modelling mathematical knowledge is re-signified.

On the one hand, this context is found in environments where the emergence of the category of knowledge referred to as trend behavior of the graph of functions in modelling was shown. And, on the other, the basis of a functional principle: the use and re-signification of the mathematics of people. These re-signification categories are the foundation for didactic designs in school mathematics.

With this fact, we reflect on the current status of the mathematics teacher and a proposal for their training and permanent accompaniment, providing reference frameworks for the uses of people's mathematical knowledge as a context for school activities. The authors provided an example, which questions how the actions of the population can *flatten the epidemic curve.*

The questioning is a reality, its modelling is a horizontal and reciprocal relationship between mathematics and that reality. This modelling corresponded to the category of trend behavior, with which the function is re-signified as an instruction that organizes behaviors: epidemic curve, mitigation measures, containment, restriction and flattening of the curve.

The authors also reflect, as a prospective, on the systems that favors the functionality of mathematical knowledge, where pluralities, transversalities, and the consideration of the other occur. They consider that the systems should be typical of the knowledge communities, where reciprocities between mathematics and realities occur; categories of intimate knowledge of those communities that later become disciplinary jargons; and localities that express social, political, and cultural movements.

The fifth section of this book presents a summary of findings, conclusions and reflections of the work presented in this collection. This chapter entitled: *The Mathematical Teaching and Learning Process Through Mathematical Modelling: Educational Change in Latin America*, written by Francisco Cordero, Milton Rosa, Daniel Orey, and Pablo Carranza, discusses the re-signification of mathematical knowledge as generated by a selection of mathematical modelling programs in diverse regions in Latin America form both an epistemological and ontological basis to the design and development of teaching and learning of mathematics at all educational levels: basic, intermediate, and higher education. This development is cyclical, continuous, and reflective.

The designs consist of locating moments of transversality of knowledge to favor and value the emergence and re-signification of a diversity of mathematical knowledge. Learning consists of the creation of new relations between this diverse mathematical knowledge, teaching and didactics accompanied by the social function of educators, which consists of maintaining the environments of mathematical knowledge through permanent support and partnership programs.

Thus, the discourse of school mathematics is disrupted and transformed, to create its redesign whose fundamental epistemological and ontological basis will forms the relationship of horizontality-autonomy-cultural knowledge, fostered by the principles of mathematical modelling programs. This entails the educational change of mathematics, at all educational levels, permanently considering the knowledge of *the others*, and include the reasoned decision, and symmetry and the others as found in diverse parts of Latin America.

18.3 Final Considerations

Undoubtedly, the three mathematical modelling programs presented here provide educational opportunities for reflection, changes and gains, each with its levels of specificity and loyal to its principles. However, in the exercise of putting them together, organized by *axes* has come to define a corpus of mathematical knowledge that envisions educational changes in Latin America.

On the one hand, epistemological and ontological changes, where mathematical knowledge of the *others* is recognized, on a horizontal plane. New empirical relationships between mathematical knowledge and reality occur. Re-signification of mathematical knowledge needs to be dimensioned and valued in the classrooms.

Inclusion of these environments can play a fundamental role, since it includes the mathematical knowledge that emerges in the larger community and the student body and the teaching profession itself. This corpus of knowledge enables mathematics teachers and researchers in Latin America to understand its development and at the same time, as a community, to comprehend mathematical knowledge built in accordance with our own mathematical categories in our own contexts, and which is governed by the reciprocal relations between school knowledge and our own diverse contexts and realities.

This framework guides necessary articulations in autonomous actions in mathematics teaching, hence the importance of generating research on the role of the teachers will lead to the permanence of the environment of reciprocal relationships that happen in mathematical functionality, and the educational changes of mathematics. The spectrum of the corpus is great, yet it is necessary to take advantage of it.

When facing new situations or problems, members of distinct cultural groups come to, indeed, construct their own understanding of these phenomena by applying solutions they developed through history. In the next step, they may use the same procedures to solve similar phenomena previously faced in their own daily lives and contexts by organizing them into methods (D'Ambrosio, 2017).

In this regard, the chapters that compose this book show the necessity to continue the debate of issues regarding mathematical modelling and its diverse conceptions in Latin America. The discussions surrounding these issues show that powerful and valuable research and experience as illustrated here, the importance of the field of mathematical modelling in Latin America, and the possibilities for continued and diverse conceptions emerging here, offers as an instrument towards the improvement of mathematics education that helps to clarify the nature of mathematical knowledge and offers possibilities for educators and learners everywhere.

Finally, taken together, the valuable experience and the accompanying investigations featured in this book provide examples of excellence, as well as new trajectories for further research and education. As well, they offer a glimpse of future direction in the field of mathematical modelling, especially in relation to its role in advancing creative and innovative interrelations between academic science and other worldviews and paradigms.

In closing, this book discussed the contributions of a wide variety of investigations conducted from the perspective of mathematical modelling by researching three Latin American modelling programs with their respective theoretical and methodological foundations such as ethnomodelling, interdisciplinarity, and socioepistemology, which demonstrate the vibrance of innovative and creative development of mathematical modelling in this part of the world.

As many researchers in a large variety of research and educational fields in Latin America assert that, there is, despite our many inequities, frustrations, and difficulties found here, relevant investigations in diverse scientific areas are happening as well. We are proud of our colleagues and their work as shared in this book, and we are also pleased to offer powerful examples of the excellent scientific-mathematics work being undertaken towards the improvement of teaching, learning, and investigating in Latin America.

References

Cortes, D. P. O. (2017). *Resignificando os conceitos de função: um estudo misto para entender as contribuições da abordagem dialógica da etnomodelagem* (Re-signifying function concepts: A mixed methods study to understand the contributions of the dialogical approach of ethnomodelling). Master's thesis, Departamento de Educação Matemática, Universidade Federal de Ouro Preto—UFOP, Ouro Preto, Minas Gerais, Brazil.

D'Ambrosio, U. (2015). Mathematical modelling as a strategy for building-up systems of knowledge in different cultural environments. In G. A. Stillman, W. Blum, & M. S. Biembengut (Eds.), *Mathematical modelling in education research and practice: Cultural, social and cognitive influences* (pp. 35–44). Springer.

D'Ambrosio, U. (2017). Ethnomathematics and the pursuit of peace and social justice. *Educação Temática Digital—ETD, 19*(3), 653–666.

Orey, D. C., & Rosa, M. (2021). Ethnomodelling as a glocalization process of mathematical practices through cultural dynamism. *The Mathematics Enthusiast, 18*(3), 439–468.

Rosa, M., & Orey, D. C. (2017). Ethnomodelling as the mathematization of cultural practices. In G. Stillman, W. Blum, & G. Kaiser (Eds.), *Mathematical modelling and applications: Crossing and researching boundaries in mathematics education* (pp. 153–162). Springer.

Rosa, M., & Orey, D. C. (2019). Ethnomodelling as the art of translating mathematical practices. *For the Learning of Mathematics, 39*(2), 19–24.

Index

M. Rosa et al. (eds.), *Mathematical Modelling Programs in Latin America*, https://doi.org/10.1007/978-3-031-04271-3